GEOBRITANNICA

Geological Landscapes and the British Peoples

Also by Mike Leeder and available from Dunedin:

The Gulf of Corinth Classic Geology in Europe (2009)
ISBN: 9781903544235

Other earth science titles from Dunedin
for the less specialist reader include:

The Abyss of Time (2015)
Paul Lyle
ISBN: 9781780460390

Volcanoes and the Making of Scotland (2015)
Second edition
Brian Upton
ISBN: 9781780460567

For further details of these and other Dunedin
Earth and Environmental Sciences titles see
www.dunedinacademicpress.co.uk

GEOBRITANNICA

Geological Landscapes and the British Peoples

Mike Leeder & Joy Lawlor

DUNEDIN

EDINBURGH ◆ LONDON

Published by
Dunedin Academic Press Ltd
Hudson House , 8 Albany Street
Edinburgh EH1 3QB
Scotland

London Office
352 Cromwell Tower, Barbican
London EC2Y 8NB

www.dunedinacademicpress.co.uk

ISBNs
9781780460604 (Hardback)
9781780465678 (ePub)
9781780465685 (Kindle)

British Library Cataloguing in Publication data
A catalogue record for this book is available from the British Library

Typeset by Makar Publishing Production, Edinburgh
Printed in Poland by Hussar Books

Dedications

To our English, Irish and Scottish families past and present, who in their different ways have nurtured our love for these lands we write about: who have made us what we are today.

Chuig ár teaghlaigh Sasanach, Albanach, agus Éireanneach, anuas agus faoi láthair, a chothaigh ár ngrá i gcomhair na tiortha seo a scríobhaimid faoi, ina mbéalaí éagsúla: a rinne dúinn cad atá againn inniu.

I'n teuluoedd Seisnig, Gwyddelig ac Albanaidd ddoe a heddiw, sydd yn eu gwahanol ffyrdd wedi meithrin ein cariad tuag at y gwledydd hyn rydym yn ysgrifennu amdanynt: sydd wedi ein gwneud yr hyn yr ydym heddiw.

Contents

Acknowledgements   viii

Preface   ix

Foreword   xi

Author Statements   xiii

PART 1   **'...the Little Space of this our Island'**

Peoples and geology: affection for; early progress in; works of the imagination

Chapter 1   **Introduction**   2

Chapter 2   **Affection for Things Geological**   6

Chapter 3   **Early Discoveries**   13

Chapter 4   **Works of the Imagination**   21

PART 2   **Assembling the Geological Jigsaw**

Field geology: the geological map; elucidation of geological history

Chapter 5   **Mapping**   36

Chapter 6   **'Deep Time' and the 'Mobile Earth'**   44

PART 3   **Remembrance of Things Past**

The Island's ancient geography and geological history

Chapter 7   **Past Geography and Geological History**   52

PART 4   **Material GeoBritannica**

The make-up of things: settlement; communications; natural resources; building materials

Chapter 8   **Settlement and Communication**   80

Chapter 9   **Natural Resources – General**   88

Chapter 10   **Building Stone and Aggregates**   92

PART 5   **Mineral GeoBritannica**

Exploitation of minerals: discovery to extraction

Chapter 11   **Metals and Mineral Salts**   107

Chapter 12   **Coal, Peat and Oil**   117

PART 6   **'To show to the world what exists in nature'**
Creative imagination in geological landscapes

Chapter 13	**Architecture and Monuments**	121
Chapter 14	**Sculpture**	133
Chapter 15	**Painting**	143
Chapter 16	**Literature**	155

PART 7   **GeoRegions**
Geo-regions: cameos of landscape, culture and history

Chapter 17	**Introduction to GeoRegions**	165
Chapter 18	**Assynt Foreland and Outer Hebrides**	169
Chapter 19	**North West Highlands and Northern Islands**	174
Chapter 20	**Grampian Highlands and Argyll**	179
Chapter 21	**Midland Valley**	184
Chapter 22	**Formerly Volcanic Islands of the Inner Hebrides**	189
Chapter 23	**Southern Uplands and Galloway**	194
Chapter 24	**Scottish-English Borderlands**	199
Chapter 25	**Lakeland, its Surrounds and the Isle of Man**	204
Chapter 26	**North Pennines**	209
Chapter 27	**South Pennines**	214
Chapter 28	**English Midlands**	219
Chapter 29	**Welsh–English Borderlands**	224
Chapter 30	**Northern Wales**	230
Chapter 31	**Southern Wales**	235
Chapter 32	**South West England**	240
Chapter 33	**Southern England**	246
Chapter 34	**Eastern and Central Scarplands**	252
	Glossary	258
	Bibliography and Further Reading	266
	Index	273

Acknowledgements

To individual artists for giving us their direct permission to reproduce their works *gratis* with warmth and encouragement: Caroline Bailey; Chris Griffin; Lulu Hancock; Katharine Holmes; Adam Kennedy; Keith Salmon.

To other artists for generously letting us include their work *gratis* through the intercession of their galleries: Katrina Palmer via Nicola Celia Wright at Motinternational Gallery, London; George Shaw via Rhian Smith at Wilkinson Gallery, London.

For their photographs: Val Corbett; William Lawlor; Vaughan Melzer. Special thanks to Vaughan for travelling in the Scottish Highlands looking for geologically instructive views.

For helpful comments on the South Uist 'Krokeatis Lithos' fragment we thank Professor Niall Sharples of the University of Cardiff. For information on the Forteviot tomb and the Govan Norse 'Hogsback' tombs thanks to Professor Stephen Driscoll of the University of Glasgow.

For encouragement and for reading through an earlier version of the manuscript with helpful suggestions we are grateful to Professor Julian Andrews, School of Environmental Sciences, UEA Norwich and two anonymous referees.

To institutions and their individual chief contacts for granting and expediting our requests for paid permission to licence images for reproduction: Albright Knox, Boston Museum of Fine Art, Buffalo (Martina Breccari); Bridgeman Images (Rob Lloyd, Holly Taylor); DACS (Ksenya Blokhina); English Heritage (Javis Gurr, Graham Deacon); Fitzwilliam Museum, Cambridge (Emma Darbyshire); Gracefield Art Centre, Dumfries (Dawn Henderby); Henry Moore Foundation (Joanna Hill, Kerry Catling); Higgins Gallery, Bedford (Victoria Kahl); Leeds Museums and Art Galleries/ City Art Gallery (Sheel Douglas); Liverpool Museums (Nathan Pendlebury, Andrew Jackson); National Gallery of Art, Washington DC (Peter Huestis); National Trust (Jenny Liddle); Rugby Museum and Art Gallery (Jessica Litherland); Norwich Castle Museum (David Waterhouse); Orkney Museum (Sheila Garson); Arthur Raistrick Estate (Bill Rea); Scala Archives, Florence (Valentina Bandelloni); Scotland National Galleries (Magda Zabieiska, Manju Nair); Tate Images (Chris Sutherns); University of Bradford, Special Collections Archive (Alison Cullingford); University of Glasgow Forteviot Project (Professor Stephen Driscoll); Victoria and Albert Museum; Wales National Museum (Catherine Sutherland); Webb Aviation (Jonathan Webb); Yale Museum of British Art.

Anthony Kinahan enthusiastically embraced and supported our project for Dunedin Press, with Anne Morton and David McLeod providing careful and expert editing and production skills respectively.

Finally, thanks to Zbiggy Kadysewski for his translation of our dedication into Gàidhlig and to Elgan Davies for rendering it into Cymraeg.

Preface

Interest in Britain's landscapes and their geological foundations has probably never been greater, perhaps even more than during the explosion of late-Georgian/early-Victorian cultural and scientific exploration. In the day-to-day world, the language of geology is familiar: political events signal 'tectonic shifts' measured 'high on the Richter scale of politics'; rumours signify changes in the 'moving tectonic plates of party politics'; 'fault-lines' separate public opinion on major issues like devolution, national independence and EU membership; geological periods are used to mimic movie titles by naming great chunks of much-loved coastline as 'Jurassic Coast'. Yet though our geological understanding has greatly progressed, the historian would rightly point to older records of tectonic, volcanic and meteorological events in classical works (like Tacitus' Annals for example) and by the nature-curious scribe(s) of the later Anglo-Saxon Chronicles who recorded the extraordinary succession of strong earth tremors and storms that affected England in the early twelfth century.

Geology as the foundation to landscape was made clear to the mid-twentieth-century reader in books such as A.E. Trueman's *Geology and Scenery in England and Wales* (1938, 1946); a pioneering classic written for 'the educated layperson' and in print as a popular paperback well into the 1970s. A book that first presented geological history in tandem with archaeology, arts and literature was *A Land* (1951; reprinted 2012) by Jacquetta Hawkes. A bestseller in its time, featuring sculptors and artists like Henry Moore and John Piper, it is now regarded as a classic. Hawkes presents a visionary view, an unashamedly and uniquely personal narrative of Britain's geological history. This included the role that it played in her own psyche and, in her view, that of past and present inhabitants. Amongst more recent narrative efforts, R.A. Fortey's *The Hidden Landscape* (1993; 2010) springs to mind; an affectionate and cultured pen-portrait. Also noteworthy is *Death of an Ocean* by Euan Clarkson and Brian Upton (2010) in which the classic geological landscapes of the Southern Uplands and Scottish Borders are explored in a scholarly but informally written manner using the context of the 'lost' Iapetus ocean.

On TV and in related books over the last twenty years Bill Bryson has explored the general character of the Island's great outdoors and its cultural nexus in intimate and companionable ways. Latterly, geologist Iain Stewart has made fine TV documentaries for general audiences on both the foundations of geology and the pioneering geologists who did important fieldwork in Scotland. Archaeologists Neil Oliver and Frances Pryor have done the same for prehistory, providing rich and warm human tapestries from which to view the late-Pleistocene and Holocene epochs. We also applaud the wide vision of popular presenters in the visual arts, notably Lachlan Goudie's landscape-inspired *The Story of Scottish Art* (BBC4, 2016).

Human influences on landscape history have had fine books written on the subject, beginning with W.G. Hoskins' all-time classic *The Making of the English Landscape* (1955; reprinted 2013). It is clear to any close reader of his book that Hoskins was not a happy man in nuclear-armed, creepingly urbanizing and rapidly motorizing post-Second World War England. He and subsequent authors in the field take the geological foundations of landscape very much for granted, putting the human 'making' of it to the forefront. This is all fine and good, but the approach subsequently led to a rather narcissistic vision of landscape – about the author rather than the landscape itself – an issue explored in the 'New Nature Writing' literary spat of recent years.

The present book attempts to put the geological history, landscapes and materials of Britain (hereafter 'the Island') into historic, societal and artistic contexts. What is it about landscape and geology that is so fascinating? How do field geologists do their science? Why are distinctive physical landscapes and their geological foundations where and how they are? What geological discoveries originated here over the past few hundred years? What is the geological context of the raw materials used by industry and in the architecture of contemporary, historic and vernacular buildings? How have geological landscapes and materials influenced past and present architects, visual artists and writers?

The book is written for those interested in both the arts and sciences and who are naturally curious about such questions. It assumes elementary knowledge of basic geology, history and culture: but nothing more than might have passed through the ears of all who went to secondary school and who have noted developments in these fields subsequently. Help is provided to the general reader in all this by providing a glossary of terms and a bibliography with further reading. The book may also appeal to those who earn a living by making or creating using natural rocks and minerals, or those who have an active interest in exploring and discovering the landscapes around them. You may perhaps be a student of earth or environmental sciences interested in seeing how your subject relates to the Island's wider culture and history; or *vice versa*, an archaeologist, architect, artist or historian interested in the foundations of landscape.

The book's layout is simple: Part 1 sets out the stall; Part 2 develops the geological theme; Part 3 journeys through the geological time; Parts 4 and 5 explore the raw materials provided by geology; Part 6 more closely examines the makers who turned landscape and its materials into buildings, artwork, sculpture, paintings and literature; Part 7 presents cameos of the diverse GeoRegions of the Island. The latter is written for dipping into, perhaps before, during or after a long weekend visit to favourite places. It brings together regional geology, landscape, archaeology history and the imaginative arts.

Chapters often begin with quotations or short extracts from literature selected to illustrate aspects of topics under discussion. The 200 or so, mostly colour, illustrations are meant to be perused in conjunction with the text surrounding them. Many of the images are of unique and precious landscapes and/or of original and pioneering works of art. They are all deemed relevant to the Island story, as is their source, location, nature and copyright. For this reason they are included with informative captions rather than as groups of plates whose captions might be pages away and their details vague.

Finally, it is stressed that the book is neither an encyclopaedia, nor by any means a conventional textbook. It is the author's view of how the foundations of geological landscape have influenced human achievements to make us peoples and our Island the way they are today.

Foreword

GeoBritannica concerns the geological legacy of Britain ('the Island'), an inheritance bequeathed to eleven millennia or so of its post-glacial inhabitants. It deals with the geological foundations of landscape and its raw materials; how both have been made use of by society and individuals in creative acts of the imagination. An objective definition, or even description of landscape, is near-impossible, though all inhabitants live in one or another. Its substrate is always below us: a product of changing circumstances over time and in space, a legacy of successive geological and human ruins.

The physical landscape encourages development of an understanding of the natural processes that have formed the Island. This knowledge enables an approach to an uncertain future, having learnt from past environmental change, encouraging care and consideration for all environments, urban and rural. Cultural and artistic links to landscape are another concern. These run deep: everyone inherits some distinctive landscape in childhood, though most will never own one. But as poet Norman MacCaig asks in his *Man of Assynt* (1965): who, in the end, can 'own landscape'?

Locally or during travel, amazing remnants of past culture may be seen within the landscape. Examples are Neolithic/Bronze Age monuments, often site-designed, like Brodgar in Orkney, Callanish on the Island of Lewis, Castlerigg in the northern Lakes, Pentre Ifan in Dyfed and Stonehenge in Wiltshire. Inscribed white horses and giants enliven the chalky downlands of southern England. Smaller-scale abstract images hammered into northern British rock outcrops tell of unknown rituals or stories. Prehistoric ruins of Iron Age and Pictish forts dominate many rocky crags and tracts of elevated ground, like Dunadd in Argyll, Old Oswestry in the Welsh Borders and Maiden Castle in Dorset. And that is not to mention fortified and ecclesiastical buildings, excavated townscapes and treasure troves from Roman, Viking, Anglo-Saxon, Norman and later times.

To many immigrants, internal and external, such sights are seen from the perspective of previous cultures and memories: indigenous landscapes filtered through remembering eyes. Artists of all kinds return to the history of their landscapes: writing, seeing, depicting and making in many different traditions. The earliest responses to environment and landscape are all sensory: the sky, quality of light, smell, sound (wind, water), texture, wetness, vegetation, weather, colour, outcropping rock formations, the general lie of the land (elevation, shape, continuity). Unique and myriad patterns of landscape, seascape and cityscape combine in shapes, patterns and forms. An infinite combination of these things is what any one of us might recognize as distinctive in particular rural and urban landscape: all absorbed from a very young age.

Landscapes also exist in memory. Sounds, sights, tastes and smells can trigger an intense place-memory (the 'Proust effect'). These are experienced from the perspective of individual cultural and emotional backgrounds; so any one place may engender feelings as various as a sense of homecoming or feelings of fear or dread. Such responses are so numerous they defy classification. In his *Landscape and Memory* (1995) Simon Schama writes: 'it is our shaping perception that makes the difference between raw matter and landscape'. The combination of physical and imaginative landscape can make individuals feel deeply at home and content, providing stability in what is often a shifting world; it can quite literally ground or earth us, since we are part of it. So it may happen in the exploration of quarries, of subterranean landscapes in cave systems and mine tunnels; or in immense coastal landscapes, that essential all-enveloping boundary between land and sea that defines any island categorically.

The book attempts to reclaim Schama's 'raw matter'; to release from landscape what John Keats memorably called 'the intellect, the countenance, of such places …'. Like him, we have avoided exhaustive descriptions or long lists of the attributes of features, just giving the basic arrangement and nature of rocks and deposits upon which we live out our lives and into which our physical remains will eventually pass. Like all foundations, the geology is often hidden, but is always there. So an understanding of landscape origins, history and influence can brightly illuminate the sensory and perceptive memory that is landscape's human essence. Lifetimes are measured over decades; ancestors span the generations; historic time involves a couple of millennia; prehistory and archaeological time are measured over ten or more thousands of years (kyr).

Yet geological time beckons back to *c.*3.6 billion years ago (3.6Ga) and is the most formidable of perspectives. Jacquetta Hawkes approached it in her own inimitable fashion: by lying out in her Hampstead garden on warm summer evenings. There she used her imagination to commune with the creatures now exquisitely fossilized in the London Clay beneath her and that once inhabited the warm seas of the Palaeocene epoch, 60 million years ago (Ma). She found that in order to be in landscape, it is necessary to be within it – in one's mind. One cannot just walk, run, cycle or drive all day through landscape in order to achieve this: it means realizing the whole thing and attempting, however simply, to understand its physical origins and meanings. Personal landscape-realization can be subtle and often painfully slow. How to assimilate and order those familiar contours, slopes, eminences and low places? To begin with, it means crawling, walking, rambling, running, biking, trespassing across them; exploring them thoroughly, perhaps later camping and sleeping out in them, smelling them, secreting oneself within them. As George Eliot writes in her novel *Middlemarch*: 'There is no sense of ease like the ease we felt in those scenes where we were born, where objects became dear to us before we had known the labour of choice…' Then comes the optional phase; attention to explanations of why landscape is like it is. The Pre-Raphaelite and Modernist artists and writers of the last 150 years attempted at times both understanding and creative depiction of landscapes, making their art and writings all the richer for the effort. Regarding gods, religions and landscape, Thomas Browne argued long ago in his *Religio Medici* (1643), a masterpiece of warm and eloquent toleration, that God reveals himself to be read not only in religious books but also in that: '…of his

servant Nature, that universall and publick Manuscript, that lies expans'd unto the eyes of all; those that never saw him in the one, have discovered him in the other…'. Some of us might agree with both Browne and Margaret Drabble who, in *A Writer's Britain* (2009), reflects that 'the perception of God in nature is as close as we get to religious experience…'.

As well as providing the foundations to many contrasting landscapes, the Island's diverse geology is responsible for a great richness of mineral wealth. At one time or another its rocks have sourced significant quantities of metal (gold), metallic ores (copper, iron, tin, lead, silver, zinc, tungsten) non-metallic ores (barite, fluorite, celestite) evaporite salts (potash, gypsum, anhydrite, rock salt) and hydrocarbon (coal, oil, natural gas). Further, ample supplies of different clays and a great array of building stone have served to form utensils and erect dwellings, workplaces, municipal buildings and ceremonial structures down the ages. Such treasures have been extracted from the three main classes of rock: sedimentary, igneous and metamorphic, across the breadth of the land. Such natural assets were revealed and exploited piecemeal as successive generations cleared, farmed, constructed, transported, manufactured, mined and quarried. Not all natural resources are present in every area, and so the basis for trade and inter-communal exchanges and barters were set up from earliest prehistory.

Minerals and rocks were not just practically useful; they inspired visual creativeness during prehistory in the form of monuments, sculpture and rock painting. Later, writing and geometry were introduced as part of the Roman cultural transfer. From then on the oral heritage of singers, poets and story-tellers and the technical skills of engineers, miners and stonemasons could be transmitted far beyond local audiences: written records of the human condition as well as architectural, commercial and artistic efforts. Such activities accelerated as Renaissance and Reformation cultures swept through the lands, eventually given even greater impetus by the Romantic Movement. The latter partly originated in the Island and still controls much of today's imaginative scope and feeling for landscape, environment, and the idea of an inclusive society.

Author statements

MRL The landscape I grew up in was rural south Norfolk. It has a rolling, gentle undulation: minor ups and downs with no steep gradients. There were lazy weed-choked rivers, water meadows and fens in the Waveney's flat-bottomed valley where my father cut hazel sticks for his peas and runner beans. Only a few tens of metres above, on the valley interfluves, were copses in old marl pits and arable fields with deep clayey soils, interspersed every few miles with abandoned Second World War bomber airfields. V-bombers regularly flew through the awesome abundance of sky above, drenching us in engine noise.

By way of contrast, my earliest feelings for rock outcrops, hills, mountains, caves and gorges gave me definite unease. Steep, rocky hillsides along the railway through the Lune Gorge in Westmorland were seen fleetingly, sleepily and with amazement from the compartments of the 'Royal Scot' on our annual summer family migrations to Galloway. Waiting for us near Kirkcudbright were rocky cliffs and outpointing skerries that we clambered over, dived and crabbed from near our grandparents' house at Kirkandrews.

Later, as a fourth-form pupil on a field excursion to the Peak District I watched sixth-formers returning from their geological activities laden with curious and beautiful rock, mineral and fossil specimens. These fascinated me, also the comradely and relaxed aura that hung around the group itself: the young men and masters setting out after our evening discussion to the local pub to drink beer, play darts, smoke and indulge (as I imagined) in further geological debate.

The origins of this book, for me, may be traced back to the year 1975. With three friends we began to plan an ambitious project, a geological history inspired by the integrative teaching and research philosophy of Professor Perce Allen at the University of Reading, a much-loved mentor to us doctoral students there. The book (Anderton *et al.* 1979) broke some new ground with its causal treatment of British geological history analysed from the point of view of the then-newish theories of plate tectonics and basin analysis that were sweeping the geological world. Suddenly it seemed to us (and of course to many others) that hitherto disconnected facets of geological history could be simultaneously explained by the new theories.

After academic careers at English Universities (Leeds, UEA Norwich), often field-working abroad, I am thrilled to be writing once again about the geology of my native island – this time for a more general audience, and in close collaboration with Joy Lawlor, English graduate, art teacher and visual artist.

HJL My knowledge of geology is rudimentary, yet I am primarily interested in what I can touch and see; a grubby child rooting in the soil, preoccupied with the look and touch of things (the visual and tangible) grew into an artist and art teacher. What did develop, though, was a curiosity about what lay beneath, and just as the lineaments of a face can generally betray inner character, the patterns, colours, forms, textures and shapes in the landscape have led me to wonder about the geology. The focus of this interest was sharpened through conversations with Michael Leeder, my geological co-author.

Man's relationship with, and response to, the landscape has found expression through shaping the land itself, through architecture (using local materials or concrete, steel and glass) and innumerable art forms, from landscape painting and temporary installations to graffiti. But a work of art reshapes the world, and there is a constant interaction. I am interested in perception and feel curious about how different artists have been influenced by their environments, and how people's perceptions of landscape have changed over time.

So many things shape us. As a child I lay in bed picturing the wind's journey over fields and hedges, skimming over the land I knew so well before it raced shrieking down the entry way of our 1930s semi in Leicestershire: it was in Church Street, Earl Shilton, along an east–west ridge of Triassic sandstone. To the north, looking towards Leicester, it overlooked a pastoral landscape: trees and hedgerows around smallish fields. Villages like Peckleton, Kirkby and Thurlaston could be located by their church towers and old red-bricked farmhouses dotted throughout the landscape. On clear days I saw the distant pyramids of slagheaps marking the Leicestershire coalfield. Our row of houses was built where cottages once stood. Next doors' family must have been shoemakers, since every spadeful dug from their garden turned up leather offcuts; before the time of factories people worked from home, mostly on stocking frames.

To the south was a different world. Almost to Nuneaton it was built-up: endless rows of semi-detached houses dating from the 1930s and later, with barely a break in the ribbon development. Villages, towns and cities joined together: Earl Shilton, Barwell, Hinckley, Nuneaton then Coventry and Birmingham – seemingly one huge urban landscape. Every Earl Shilton street in the 1950s had its red- or blue-bricked factory. Hundreds of workers were bussed in every day to work them, streets humming to the sound of hosiery or boot-and-shoe machines. In the evening other buses returned miners home with their faces still grubby.

PART 1
'...the Little Space of this our Island'

Here is an overview of the book's main themes. Chapter 1 stresses the variety of geology that underlies the Island's many and varied landscapes. Such physical richness and diversity explains why its inhabitants have prospered and multiplied since the last Ice Age. Chapter 2 argues that geology and landscape are inseparable, and briefly illustrates how the British people's innate fondness for both has expressed itself down the years in contemplative writings, outdoor activities, acts of worship, ritual, artistic creativity, material possessions, constructions, and so on. Chapter 3 shows how early geological discoveries played a key role in the furtherance of the Renaissance ideal that humans should be free to enquire into the nature and origins of the Earth. Finally, Chapter 4 looks broadly at geology and landscape through the enquiring and imaginative eyes, ears, hands and feet of artists, sculptors, writers, architects and stonemasons.

Chapter 1

Introduction

... all the different species of earths, of rocks or stone, which have yet appeared, are to be found in the Little Space of this our Island ...

James Hutton, from his introduction to *Theory of the Earth* (1788)

The State We're In

The Island comprises the former nation-states of (as of writing) England, Scotland and Wales. Its location along the north-eastern boundary of the Atlantic Ocean determines that the lands are temperate, despite sharing similar latitudes to the frigid wastes of Labrador and Russia. The maritime climate is produced in a mid-latitude boundary layer between polar, maritime and continental air masses – an interplay controlled largely by the polar jet stream's vorticity. Yet, though changeable, few catastrophic weather events like monsoon, tornado or hurricane disturb it. With watery dawns and dusks firing their imaginations, its inhabitants have developed an outgoing and adventurous temperament, ideal for wider exploration. Island tectonics are also benign, the crust under gentle overall compression, but distant from Eurasia's destructive plate margins. Daily life is uninterrupted by any thoughts of *in situ* destructive earthquakes or volcanic eruptions (but see further below).

All geological intervals with fossil remains are represented in the Island's bedrock. This variability reflects its locations along the margins of major tectonic plates and oceans over more than a million millennia (>1000Myr or 1Gyr). Before that its primal land was derived from some of the oldest continental crust known in the world, formed some four million millennia ago (4Ga) in what is now Greenland. The roll of honour for global subdivision of geological time into thirteen periods (Fig. 3.2) includes four named by nineteenth-century geologists from Britain. The Devonian is after the English county; another is from the Latin for Wales (Cambrian); two more are named after Iron Age tribes from that country, the Silures (Silurian) and Ordovices (Ordovician). Each of the four periods is richly represented in the Island's rock record, their names each defining a type locality, used and recognized internationally. For example, a Tibetan geologist working in the Kun Lun Mountains would have no hesitation in recognizing the fossils that characterize them and naming her own discoveries appropriately. So geology is universal; it opens up vistas of contemplation in 'deep time' (see Chapter 3).

All current inhabitants of the Island are descended from immigrants – many through distant ancestors known only from DNA analysis, others quite recently. It is all too easy to forget the 'foreignness' of any landscape to more recent arrivals. Though its substrate is physical, the human sense and development of the landscape is incorrigibly historical. It owes as much to its past inhabitants as to current Cymraeg-, English-, Gàidhlig- and Scots-speakers and to those whose first language is none of these. It can instill what the Cymry call 'hiraeth', or 'a sense of belonging' – signifying physical and emotional fondness for a particular region – a combination of the secular landscape, something tangible, with an inner landscape that comes out of the self. The objective physical landscape is transformed by imagination, fantasy and worship: Simon Schama's previously quoted 'shaping perception'. Such landscape is existential, part of everyone. There are 65 million or so living landscapes in the Island, each imagined and experienced differently by each and every inhabitant. Such subjective landscapes are the basis for much that is recognized as art and literature. Though an objective description of landscape is impossible, geology defines its foundations absolutely.

Humans of the Mesolithic culture recolonized the Island around eleven millennia ago (11ka) as climate ameliorated after the last great spasms of the Ice Age. The land was then merely one of several peninsulas of western Europe, along with Scandinavia, Brittany and Iberia. Those Mesolithic colonizers can only have been in awe at the sight of their new homeland, like images taken during modern space exploration: such wonders, possibilities, challenges! Glaciers and permafrost disappeared as ever more temperate vegetation arrived. Together with other incoming life-forms the Mesolithics migrated northwards, many ascending hillsides, and some the mountains.

Within a millennium these ancestors had built round-houses beside elevated coastal clifflines of the preceding interglacial stage some 120 millennia (120ka) before. Here and elsewhere kinship groups could congregate, feed, love, raise families and hunt seasonally migrating herds of the temperate plains visible on distant horizons. They established lake-margin settlements, in and around which they fished, hunted and gathered edible vegetation, fruits and nuts, using distinctive flint weapons and tools. They became intimate with their landscapes, expressed themselves in ceremonial attire and created rock art in upland caves. At this time such people could walk easily from what today is the Eurasian continent to the future Island's western coast. One imagines them staying as guests with kinship groups in roundhouses *en route*. Some lost hunting spears on the way, the points embedded and preserved in coastal peats. These were only thinly covered by later sediment, so some became snagged by modern trawl nets on the North Sea bottom.

Eventually the ever-rising postglacial sea level forced the inhabitants of the southern North Sea basin to gradually retreat to higher ground. Land passages such as the deep valley-crossing over the Dover–Boulogne isthmus were inundated. Global sea level eventually stabilized at about present levels some six and a half millennia ago (6.5ka). Until the twentieth century subsequent immigrants, invaders and socio-cultural fashions from all points of the compass had to arrive by sea. First to arrive, around 6ka, were the Neolithic cultivator/pastoral peoples and their culture. From about 4ka, Bronze-Age and then around 2.8ka Iron-Age ('Celtic') immigrants arrived. Then, joining them down the ages, refugees and migrants by boat, ship and plane, a roll-call of kith and kin: Roman, Anglo-Saxon, Scandinavian, Norman, Huegenot, Fleming, Walloon, Sephardim, Lascar, Chinese, Ashkenazi, European, Commonwealth, European Union, Asian, North African, Middle Eastern.

The descendants of these immigrants today live in rural hamlets and villages, market towns, medieval cathedral cities, great conurbations, coastal ports and resorts, spas, old mining centres, rolling arable downlands, intensely farmed former fenlands, and sparsely inhabited mountains and moorland. The location of many of these was originally geologically controlled by the influence of rocks on topography, water supply, agriculture and

mineral wealth. The latter is especially significant and lasting: witness the population growth of the type localities of the Industrial Revolution – the coal-rich dales and valleys of central and northern England, southern and northern Wales and central Scotland (Fig. 1.1). And, the greatest attractor of all, chief of our cities and gateway to the world, coal-less, iron-less, steel-less metropolitan London.

The Island's varied rocks have been naturally sculptured by the action of ice, rivers, tides and waves to produce contrasting physical landscapes that range from rugged mountains with sub-alpine plateaux, scarplands with ridges and vales, and perfectly flat reclaimed former coastal wetlands. Hidden below the many limestone districts (Mendip, White Peak, Yorkshire and Lancashire Dales, Furness) lie cave systems, subterranean realms formed by watery dissolution: also the now-flooded warrens of countless former mine workings with their abandoned adits, drives, tunnels and stopes. And of course, an immensely long coastline of great beauty and exquisite complexity that even nuclear power stations, rusting piers, jetties and dockyards cannot locally subdue – essential places of trade, tourism and embarkation.

Climatic Shifts and Perturbations

Over the past few hundreds of millennia, the Island's climate has fluctuated between warm temperate and polar, with sea-level 'yo-yoing' by scores of metres. *Homo sapiens* as a species has spent its entire existence gripped in the vice of extreme climate change, most particularly over the past 25kyr. The landscape has responded with startling adaptions in both vegetation and landforms as glaciers advanced and retreated. The periodic rise and fall of surrounding seas exposed expanses of continental shelf as low hills and extensive coastal plains.

During the past seven or so millennia deglaciation has brought some instability and regional contrasts in its wake. Thus, despite stable global sea level (though now increasing millimetrically each year), regional adjustments of the crust to removal of a kilometre or so of ice over its central and northern parts are still causing these areas to buoyantly rise back upwards: vice versa for the south. The effect in both areas is an apparent relative change of local sea level, downwards and upwards respectively. The downward relative movement has led to the formation of raised beaches (Fig. 1.2), the upward to estuarine drowning. The former led poet Hugh MacDiarmid to entitle a long and influential poem, *On a Raised Beach* (1932), to address his moral, political,

thousand years of warmer climate in the Bølling-Allerød

Figure 1.1 Maps to show **A** Areas of highest population density **B** Outcrop of coal-bearing Carboniferous strata. Data generalized from the British Geological Survey (2007a, b) and *The Times Atlas of the World* (1987).

national, artistic and philosophical concerns. He wrote:

> Deep conviction or preference can seldom
> Find direct terms in which to express itself.
> Today on this shingle shelf
> I understand this pensive reluctance so well,
> …These stones with their resolve that Creation
>   shall not be
> Injured by iconoclasts and quacks…

The Island's climate is now largely stable, though like everywhere else there are noticeable changes driven by global warming. In the past, other climatic 'tweaks' have provided some far-reaching reactions, especially over higher terrains. Most emphatic, after a couple of

phase, was the return of ice in the latest Pleistocene, 12.9–11.7ka. This interval is called the 'Younger Dryas', named after a pretty alpine/arctic flowering shrub, *Dryas octopetala*, the mountain avens. Glaciers re-grew in the mountains of Argyll and Cairngorm and the earliest human migrants to the Island were forced to retreat southwards. Subsequent cold snaps of hundreds of years' duration occurred in the overall warmer Holocene epoch around 8.2 and 6.5ka. A more continental but moister climate recurred in the late Bronze Age/early Iron Age around 3ka; winters were cooler and summers warmer. Greater runoff led to expansion of wetlands, growth of raised mires and the abandonment

Figure 1.2 Gruinard Bay, Ross and Cromarty (57.893, −5.492), looking north from Achgarve. Photo: Wikimedia Commons©Synchronium. Tucked between rugged headlands of Lewisian gneiss, the oldest rocks in Europe, the bay is bounded landwards by a c.7ky raised beach above a low cliff line. The Torridonian mountains of Assynt (Suilven in the centre) form the distant skyline.

of pastureland over 250 metres elevation. After the warm early medieval period, the fourteenth century ushered in a particularly stormy and damp climate with Europe-wide famine shortly before the arrival of the terrible 'Black Death' epidemic during the 'Hundred Years War'. The seventeenth to nineteenth centuries were the height of the 'Little Ice Age' (Fagan 2000), an interval of very cold winters in northern Europe (ice-skating and fairs on frozen lakes and rivers).

There have also been some shorter meteorological events that were suddenly severe, related to atmospheric triggers, notably the short growing seasons between 3159 and 3141ka, the appalling acidic atmospheric pollution and related effects of 1783–1784 and the 'lost summer' of 1816. These were all due to the effects of distant volcanic eruptions: 1816 to the explosive destruction of Tambora volcano in the Indonesian archipelago; the others to eruptions of Icelandic volcanoes Hekla and Laki respectively. The sulphurous poisoning that drifted through north-west Europe due to Laki was especially shocking in its effects on humans, livestock and crops. Sir John Cullum reported to Sir Joseph Banks of the Royal Society from near Barton, Suffolk just after midsummer's day in 1783:'

> about six o'clock, that morning, I observed the air very much condensed in my chamber-window; and, upon getting up, was informed by a tenant that finding himself cold in bed, about three o'clock in the morning, he looked out at his window, and to his great surprise saw the ground covered with a white frost…' An aerosol-like acidic condensate had caused the frost-like deposit so that in the vegetable garden the crops looked 'exactly as if a fire had been lighted near them, that had shrivelled and discoloured their leaves.'

Such events make it clear that, though the Island is distant from the immediate hazards of geological processes at work along active plate boundaries, it has no immunity from tele-atmospheric phenomena. This was briefly driven home in 2010 by the chaotic effects on Europe-wide air traffic of another Icelandic volcano, Eyjafjallajökull.

Chapter 2

Affection for Things Geological

All is lithogenesis – or lochia,
Carpolite fruit of the forbidden tree,
Stones blacker than any in the Caaba,
Cream-coloured caen-stone, chatoyant pieces,
Celadon and corbeau, bistre and beige ...

Hugh MacDiarmid, *On a Raised Beach* (1932)

Getting Out and About

The British people's love of landscape manifests itself in many ways: the popularity of outdoor pursuits (Fig. 2.1); field courses and excursions organized by schools, colleges, universities and regional scientific societies; flourishing university courses in earth and environmental sciences; the determined upkeep of public footpaths and access to the countryside by hard-pressed local planning authorities. Also to organizations like the Council

Figure 2.1 A welcome sign for the weary walker, climber or field worker. Buck Inn's annexe, Malham, North Yorkshire (54.060904, –2.153094).

for the Preservation of Rural England, the National Trust and the Ramblers Association. Our network of national parks grew as successive twentieth-century governmental initiatives following powerful lobbying by conservationists. Government-funded organizations; English, Scottish and Welsh Heritage, have a wide remit for landscape protection and for preserving and representing landscape-related issues. Sites of Special Scientific Interest (SSSIs) protect key sites like natural outcrops, abandoned quarries, railway cuttings and the like from despoilation and exploitation. They are now managed by national bodies in each country: Natural England, Scottish Natural Heritage and Natural Resources Wales. Such precious places have a special importance to the Island's geological (and biological) history, illustrating its varied assemblages of rocks, minerals and fossil floras and faunas. It is all a far cry from the often wanton destruction previously unleashed on the Island's landscapes, echoed by A.E. Trueman (1949) as the post-war Island struggled into peacetime:

> In recent years there has been a great quickening of interest in town and country planning, in land utilization, in the scenery of our country and its preservation. The proposals for the creation of national parks and nature reserves are but one aspect of this developing interest.

Since 1835 the public-servant geologists of the British Geological Survey and its predecessors have produced informative guides, maps and essential geological data for both professionals and the public at large. The BGS website contains a huge free database; their 'Geology of Britain Viewer', for example, enables rapid inspection of the entire Island's geology from national down to local (1:50,000) scales. Their hard copy geological maps are works of great beauty as well as being informative, all also available to view online.

Last, but not least, the Geological Society, based in London but with specialist groups far and wide, has discovered a role as promulgator of impartial geological advice and background for both government, public and the press – this despite the overwhelming majority of its gifted patronage coming from the extractive industries. Current 'hot' topics include societal issues like climate change, resource exploitation and nuclide waste disposal.

Given all this interest, almost self-generated in the sense that it seems to be part of any individual's birthright, it is not surprising that many authors, scientists, naturalists, novelists and poets have written about the human links with rocks and landscape in both factual and fictional works. Prehistorical archaeologists increasingly attempt to link their discoveries to the physical landscapes that nurtured them. Countless artists have looked at landscape and rock formations through creative eyes. They give us their own unique takes on the visual impact of geological goings-on, often signifying deeper concerns within themselves. In very many ways they enable the rest of us to see the Island from perspectives completely different to the strictly scientific in the here-and-now. For example, visual artists possess a remarkable gift for rock description and characterization. Here is artist John Piper writing about the colour of rocks in the Snowdonian mountains, which he loved to paint outdoors in the 1950s (Jenkins and Munro 2012):

> The rocks can only look grey in a leaden light, and then do not, commonly. Against mountain grass or scree, against peaty patches near tarns, on convex slopes, in dark cwms, the same kind of rock can look utterly different, and change equally violently in colour according to the light and time of year. The rocks are often mirrors for the sky, sometimes antagonistic to the sky's colour. They can react to a shower of rain like dried pebbles with the wash of a rising tide on them, or having grown half waterproof coats of lichen, change very little in the wet weather.

Incorrigibly Plural

The great variety in the mineral and rock kingdoms can be intimidating to some, intriguing to others. To poet Hugh MacDiarmid, himself something of a linguistic show-off, rocks were beyond explanation, purely visual or tactile. In the chapter header quotation above he raises a loving paean to the mason-crafted shapes and forms of stone and mineral that he can see and touch. Louis MacNeice memorably asserted this variousness in the ending to his *Snow* (1935): 'World is crazier and more of it than we think ... Incorrigibly plural.'

In her short story, *The Stone Woman* (2005) A.S. Byatt writes memorably of stone as a metaphor for grief, ageing, stiffening and of the endless inventiveness of words in the buried meanings of stone-names:

> The minds of stone lovers had colonised stones as lichens cling to them with golden or grey-green florid stains. The human world of stones is caught in organic metaphors like flies in amber. Words come from flesh and hair and plants. Reniform, mammilated, botryoidal, dendrite, haematite. Carnelian is from carnal, from flesh. Serpentine and lizardite are strong reptiles; phyllite is leafy-green.

The material legacy from minerals and rocks has resulted in a cornucopia of natural resources. Many exhibit a richness of texture and colour that enhance the architectural forms of field boundary walls, monuments, sculptures, churches, gravemarkers, palaces, homes and workplaces (Fig. 2.2). Field walls march across many a landscape, ranging in age from Bronze Age low-walling (reaves) of granite 'moorstone' around intake fields on Dartmoor to the high, confident eighteenth- to nineteenth-century enclosure walls of the Southern Uplands, Pennines, Cotswolds and elsewhere.

The reason behind mineralogical variation is that the metallic chemical elements rarely occur in their 'native' form. Gold and meteoritic iron respectively are rare and exceptional examples. The majority of metallic elements join together in various combinations with oxygen, silicon, carbon and sulphur to form mineral silicates (the rock-builders), sulphides, carbonates, oxides and others. These minerals are, by definition, naturally occurring inorganic chemical compounds: rocks are simply aggregates of such minerals. The igneous ('fire-formed') rock tribe originated partly from slow-cooled molten magma below surface in vertical or horizontal sheets (dykes and sills) or as great blister-like upwellings from the upper mantle or lower crust (plutons/batholiths). Or they were erupted as fast-cooling lava flows. Metamorphic ('changed-form') rocks were once something else, changing because of slow chemical diffusion adjacent to igneous intrusions or to the elevated temperatures and pressures experienced in the deep roots of ancient mountain ranges. Sedimentary rocks are hardened deposits (like beach, desert or river sands, now sandstones) derived from the erosion of other rocks, or are chemical or biological precipitates, like many limestones.

The ability to determine the true nature of a rock is a developed skill, like bird-watching or plant-identification;

Figure 2.2 A–D Stone constructions in their natural landscapes. **A** Panorama looking east from above Malham Beck, North Yorkshire (54.067573, –2.159111). Malham Cove is on the left skyline. Medieval and later enclosure walls divide up a limestone landscape dotted with hay barns and remains of Iron Age strip fields. **B** View south from the Great Scar Limestone karstic pavement (with clints and grikes) at the top of Malham Cove (54.073035, –2.157625) Photo: Shutterstock56981887©Richard Bowden. **C** Hay barn with local ashlar quoin blocks and flagstones, Wensleydale, North Yorkshire. **D** Mine engine houses built of masoned local granite on craggy cliffs, Botallack Mine, Land's End, West Cornwall. Photo: Shutterstock 261992150©Robyn Mackenzie.

in fact it can be fun to practise all three on country walks. It requires some practice to get going (like any such endeavour) but once learnt it helps the understanding of landscape evolution. It also contributes to detailed acts of detective-work carried out on materials and implements found on archaeological sites, in ancient buildings or even, remotely, on other planetary surfaces (Moon, Mars, asteroids). Geologists can make excessive use of technical vocabulary, that is true, but as Mac-Diarmid and Byatt illustrate, the names have a certain nobility and pedigree. Anyway, a doctor will call some bodily lump a 'nodule' or 'polyp'; an architect will refer to the vertical struts in a window as 'mullions', and so on. The geologist too must be allowed precise words to differentiate the various types of natural raw materials.

The great experimental physicist Ernest Rutherford probably had the naming-mania of geologists, perhaps also biologists, in mind when he said: 'All science is either physics or stamp collecting.' In a way he was right, for there is not much actual science involved in naming, just pertinent and detailed observation. But, as physicists might do, he rather missed the point; it is what you do with the named object that matters. Even in today's highly physical and chemical subject, geology subsumed into earth or environmental sciences, it is vexing to come across the incorrect naming of the commonest types of rock. It often seems that 'granite' is the fallback term of ignorance, e.g. the yellow-brown sparkling setts of squared-off sandstone that pave streets in many Pennine towns and cities are routinely referred to as 'granite' in articles, novels and poems; no two rocks could be so different.

In this book only the commonest types of rocks and minerals are named (aided by the Glossary) in order not to assail the reader with more than is absolutely necessary to follow the story of GeoBritannica. We feel strong sympathy for the plight of poor Michel de Montaigne in his essay *On the Vanity of Words* (transl. Screech 2003):

> I cannot tell if others feel as I do, but when I hear our architects inflating their importance with big words…I cannot stop my thoughts from suddenly dwelling on magic palaces…yet their deeds concern the wretched parts of my kitchen-door!

Cultural Erratics

Certain rocks and minerals gained precedence for our prehistoric ancestors. Practical considerations determined choice of tough quern stones and hammer heads: sharp, hard flint and glassy volcanic rock cutting tools; mineral ore; coal, and slab-like building stone. Yet other materials appear in grave gifts, anointing powders, ancestral memorials, communal monuments, talismen and votive offerings. Some are 'cultural erratics', analogous to materials brought far from their native outcrops by Ice-Age glaciers. They were traded along or gifted as desirable objects, some perhaps to members of far-away kinship groups: ceremonial axe heads, jet and shale buttons, fine Beaker-style pottery and jewellery. Exquisite gold pieces like the Bronze Age 'Mold Cape' from North Wales have no equal outside of the grave goods from Greek Mycenae.

While most materials gathered from archaeological sites are local or regional in origin, others were *de luxe* imports. Such are the 140 or so jadeite hand axes found at Neolithic sites here and in Ireland (Fig. 2.3). The rare

Figure 2.3 The magnificent early Neolithic Newton Peveril jadeite hand axe; 19.5cm long, 8cm wide and 1.2cm deep. Photo: Wikimedia Commons©Pasicles. Though highly polished and probably ceremonial, the edge shows some wear. Now a permanent resident of Dorset County Museum.

and subtly attractive jadeite has been traced to outcrops in the Italian Alps, where it occurs in a metamorphic host originating from a rock such as serpentinite (one of A.S. Byatt's 'metaphoric' rocks). The hard, greenish mineral that colours it is an uncommon member of the pyroxene mineral family; a prosaic relative is black augite, a major component of basic (dark-coloured) igneous rocks such as basalt, dolerite and gabbro. Jadeite is workable since it occurs in its host as aggregates of microcrystal. The immensely fine polish of the hand axes is achieved because, although hard, the mineral is slightly less hard than common quartz sand. So, by a great deal of effort, probably hundreds of hours, a rough-tooled core can be gradually smoothed and polished to its final shape using different grades of quartzose abrasives, pastes and grinding stones.

A centre for the final manufacture of the Alpine jadeite, imported as rough cores, is south Brittany where a great many finds are located. The stylized, elegant shapes of the finished axes and the time and effort put into fabrication make it certain that the axes were ceremonial in some way. We make use of such ceremonial hardware today: think of the three nations' parliamentary maces. Many axes have been found in an Early Neolithic context. One in particular seems to have been deliberately deposited in the Somerset Levels wetlands adjacent to the 'Sweet Track', a long, early-Neolithic timber causeway dated precisely by tree ring studies to 3806-3807BC.

A native greenish rock used for ceremonial axe production (a discount-alternative to exotic imported jadeite, perhaps?) is the Great Langdale tuff: indurated volcanic ash originally deposited in water from eruptions of the long-extinct Lakeland volcanic arc (Chapter 7). The ash was subsequently altered and hardened by mild metamorphism and the passage of heated waters. These caused precipitation of new mineral forms, the greenish colour coming from epidote, a hydrous silicate. Such 'greenstone' axes are widely distributed across England and southern Scotland; around 500 have been discovered, a quarter of the total hand-axe total. The arduous extractive and finishing effort needed for production of these objects has been called the Island's first 'industry', though the flint miners and knappers of Neolithic Sussex and Norfolk (Chapter 10) might have argued the toss over this claim.

Neolithic and Bronze Age people achieved their closest links with native stones in the thousands of ceremonial and funereal stone-cored monuments that they laboriously erected. These are in special places,

sometimes in view of striking mountain scenery and/or of the sea, notably in Pembrokeshire, Gower, Anglesey, Wiltshire, Cumbria, Lewis, Caithness and Orkney. The highly visible and sometimes geometric monuments have survived millennia out in the open, indicating that the stone constructs were meant as lasting focii for ritual and/or celebration; many can be shown to record annual astronomical events and who-knows-what supernatural ones. Suitable stone was usually quarried locally, from outcrops within a dozen miles or so. Some, like the igneous 'bluestones' (spotted dolerites) from the Preseli Hills, Dyfed, were dragged and floated across hundreds of kilometres to reach the site of inner Stonehenge within the 'sacred landscape' of ceremonial and burial sites of the chalk downlands of Wiltshire. Also preserved in upland areas are inscribed carvings on countless rock outcrops, most obviously on hard but chippable sandstones.

The natural whiteness of minerals like common quartz and calcite and of rocks like chalk and other limestones has long attracted attention. It is particularly striking in the cist burial of an Early Bronze Age (c.1900BC) nobleman or king at the ancient royal burial site of Forteviot, Perthshire (Fig. 2.4). Here,

Figure 2.4 Cist grave at Forteviot, Perthshire (56.340601, −3.535610) in August 2009 after removal of the 4 tonne Old Red Sandstone capstone that lay on the flagstone walls. Photo: courtesy of Professor Stephen Driscoll ©University of Glasgow, SERF Forteviot project.

surrounded by white quartz clasts (laid or cast in by mourners?) the body was placed in a birch-bark coffin. Excavation revealed that the skeleton itself had been completely dissolved by the acidic local groundwater. Two bronze daggers and a substantial amount of meadowsweet flower remnants survived; according to Stephen Driscoll, head of the Glasgow University team who excavated it, the latter are the earliest known floral tribute in Britain.

The white quartz found in many post-Roman/pre-Conquest Christian graves in South Wales bear testament to the longevity of this particular habit for honouring the dead. Or perhaps by this time the offerings were more conscious echoes of Revelations 2:17. In the words of the King James Bible: 'To him that overcometh will I give to eat of the hidden manna, and will give him a white stone, and in the stone a new name written, which no man knoweth saving he that receiveth it.' A late-Roman Christian burial practice, particularly common around the *colonia* of York (with more than fifty recorded interments), was the covering of shrouded corpses placed in stone sarcophagi with slaked burnt-gypsum; our Plaster-of-Paris (gypsum is common in the Triassic strata of the Vale of York). These 'White Burials' (see Mattingly 2007, Chapter 11) preserved the remains as detailed white casts, perhaps seeming to make them most suitable for early resurrection.

John Dryden echoes this in *Astraea Redux* (1660), his fawning pacan to Charles II's return from exile:

> And welcome now (*Great Monarch*) to your own;
> Behold th'approaching cliffs of *Albion*;
> It is no longer Motion cheats your view,
> As you meet it, the Land approacheth you.
> The Land returns, and in the white it wears
> The marks of penitence and sorrow bears.

Orkney poet George Mackay Brown well knew the importance of stone in the Neolithic culture of his native lands. In his posthumously published *The Solstice Stone* (2001) he addresses the topic in a sort of creation myth that foretells the coming of Christianity:

> I was the block rejected by mason, carver,
> Shaper of querns.
> A star unlocked the stone.
> The stone was a white rose.

The outskirts of Roman towns featured memorial stones (usually limestone or sandstone) along the sides of major roads. Until more recent discoveries of writing tablets (Chapter 33) their memorial inscriptions are the chief source of knowledge of these, our Island's first

literate inhabitants. In the Christian era, churches and their graveyard headstones and crosses yield a more formulaic record of both past parishioners and the workings of the national and international stone trade. Occasional examples of 'grave humour' slipped past the often strict ecclesiastical rules concerning gravestone etiquette. Many were collected in a hilarious (but also moving) book of the same title by Fritz Spiegl (1971).

Fossils were widely used down the millennia as charms, grave goods and talismans. Examples are the fossil ammonites found beside pre-Ice Age Mesolithic skeletons from South Wales. Twenty-five millennia later, carved ammonites and other fossils feature in the window stonework of Wreay parish church, Cumbria (Fig. 2.5). Pre-medieval pagan burials in western East Anglia and elsewhere included collections of fossil sea

Figure 2.5 Fossil forms (ammonites, crinoid, coral), a Nautilus and pine cones carved on Romanesque-style window jambs and arch, west end of St Mary's Church (completed 1842), Wreay, Cumbria (54.831943, –2.880781). The stone is a local Permian sandstone; its parallel internal laminations are seen below the sill. The architect, Sarah Losh (1786–1853), long foreshadowed the Arts and Crafts Movement in her work. Rossetti described the church as '…a beautiful thing…'.

urchins from local Mesozoic strata (Fig. 2.6), irregular echinoids in this case. They have been suggested to have had portentous symbolism but, more likely, were simply regarded as beautiful forms.

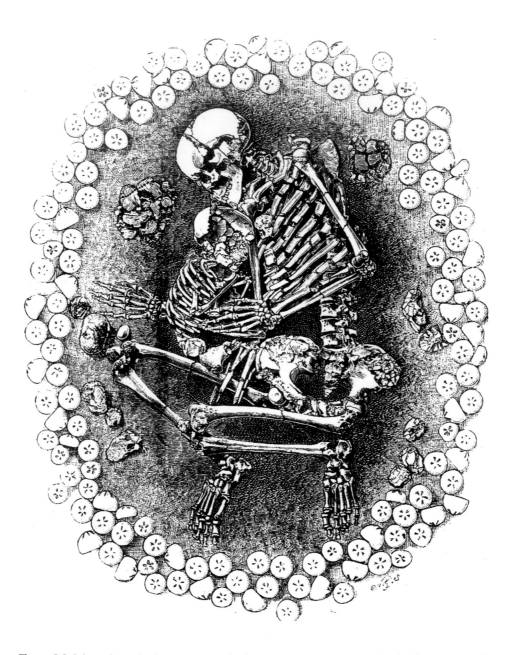

Figure 2.6 Ink and wash drawing by W.G. Smith of a reconstructed burial from the chalk scarplands of Dunstable Downs, Chilterns. The skeleton is a late-Neolithic/early-Bronze Age young woman holding the remains of a young child. Smith ringed the remains with fossil sea-urchins excavated from the grave; a pebble of white quartz was also found.

Chapter 3

Early Discoveries

I have omitted all those things which I have not myself seen, or have not read or heard of from persons upon whom I can rely. That which I have neither seen, nor carefully considered after reading or hearing of, I have not written about.

Georgius Agricola in his preface to *De Re Metallica* (1556)

To cull from books what authors have reported is exceedingly dangerous; true knowledge of things themselves is out of things themselves.

Quoted by Sir Thomas Browne (from Julius Scaliger) in his preface to *Pseudodoxia Epidemica* (1646)

Earliest attempts

The Renaissance pursuit of sceptical enquiry and the individual freedom to do so was furthered in the geological sciences first by Georg Baeur of Saxony in his *De Re Metallica* (1556: the year Archbishop Cramner was burnt at the stake in Oxford). Baeur is better known under the Latinized version of his name, Agricola; both names mean peasant or farmer. He has a right to be regarded as the first geologist through his studies in many branches of the subject. In a more general sense we in Britain also have the wisdom of aforementioned Sir Thomas Browne to consider. Not only did he conduct a pioneering archaeo-cultural investigation of Anglo-Saxon urn-burials in *Hydrotaphia* (1658; though he thought they were Roman); he also insisted upon the primacy of natural observation and the supremacy of independent freedom of opinion in his *Pseudodoxia Epidemica* (1646).

Mineralogy was always at the forefront of pre-modern natural philosophy and in the efforts of humans to make use of the possibilities offered by such natural materials. Red-coloured ochre produced by pulverizing the iron oxide, haematite, was used as pigment by the Mesolithic peoples (Chapters 4, 15). Bronze Age immigrants sought out and dug up mineral ores of green malachite to smelt into copper. In the Iron Age, brownish iron oxides and hydroxides of siderite and goethite as well as haematite

were the chief ores. Renaissance mineralogists and Enlightenment natural philosophers progressed from alchemy to chemical discoveries, eventually paving the way for the industrial manufacture of synthetic pigments. The nobility made many contributions to mineralogy in Enlightenment Britain: notably the Cavendish family. Georgiana, Duchess of Devonshire, for example, was a keen mineralogist in the late eighteenth century, a time of extreme crystal-mania. Natural crystals eventually proved the key to determining atomic structure in the years immediately preceding the First World War: the Braggs, son and father, X-rayed them and calculated their lattice structures.

Practical knowledge of strata (layered sedimentary rocks) and their cross-cutting mineral veins was clearly necessary for efficient and economic exploitation in mines (Chapter 11), but there was little progress in elucidating the nature of the deeper crust. Philosopher and polymath René Descartes, a contemporary of Thomas Browne, was the first to consider the evolution of a layered, spherical Earth in 1644, but other fanciful notions abounded, since direct observation was impossible. Such strictures did not dissuade early members of the Royal Society like Robert Hooke and Robert Boyle from speculating widely on the nature of earthquakes, volcanoes, rocks and minerals. Indeed, Browne began his scientific observations of the Walsingham urn-burials with a paragraph that might have delighted the antiquarian (as archaeologists were then known) but would have dismayed the speculative Stuart geologist:

> In the deep discovery of the Subterranean world, a shallow part would satisfie some enquirers; who, if two or three yards were open about the surface, would not care to rake the bowels of Potosi, and the regions towards the Centre. Nature hath furnished one part of the earth, and man another. The treasures of time lie high, in Urnes, Coynes, and Monuments, scarce below the roots of some vegetables.

(Potosi is a mountain in the Peruvian Andes: at the time of Browne's writing, fabulously rich in silver mined by Inca slaves.)

But things were to change. The skilled Danish anatomist and later cleric, Nicolai Stensen (usually referred to by his Latin name, Nicolaus Steno), published the second ever geological book in 1669. This was his *De Solido* in which he established the principles for recognizing stratal succession (young strata above old; angular relationships between older strata below and younger above), the organic origin of fossil forms, and the hardening of sediment and shell into rock. Though the book was rapidly translated into English (in 1671) Steno's ideas (specifically those concerning fossil petrifaction) proved unpopular until championed by the fossil collector John Woodward in the 1690s.

Practical Ways and Means

A tradition of deep coal mining in many areas of the Island led to several key contributions to development of the subject. As early as 1672 the Scottish academic mathematician and sometime practising mining engineer, George Sinclair, knew much about stratal patterns in central Scotland. He was familiar with local terms like 'cropping-out' (to form an outcrop), 'dip' (strata inclined at an angle from horizontal) and 'strike' (orientation of the stratal plane taken normal to dip). In his *A Short History of Coal* he deduced that coal seams in the Midlothian coalfield were arranged saucer-like, gradually deepening to the centre of a stratal depression. Today such an arrangement implies that the rock strata have been bent by tectonic forces into the downfolds we call synclines. Sinclair also knew from face-working miners that when a coal seam suddenly stopped and was replaced by an altogether different rock it would most probably be rediscovered downwards on the side that the sloping junction with the adjacent rock inclined. Such offsets were known colloquially to Lothian miners as gaes, dykes, ridges or trouble. Today we call them faults: planar or gently curved fractures across which once-continuous rock has been displaced up, down or sideways by tectonically induced stresses.

The first depiction of the likely nature and extent of underground rock, now known as a geological section, was by John Strachey in 1717, of the Somerset coalfield (Fig. 3.1). Whether Strachey was cognizant of the works of Steno and Sinclair we do not know, but in his remarkably prescient section he shows sets of strata cropping out at the surface in valleys. One group of strata contained

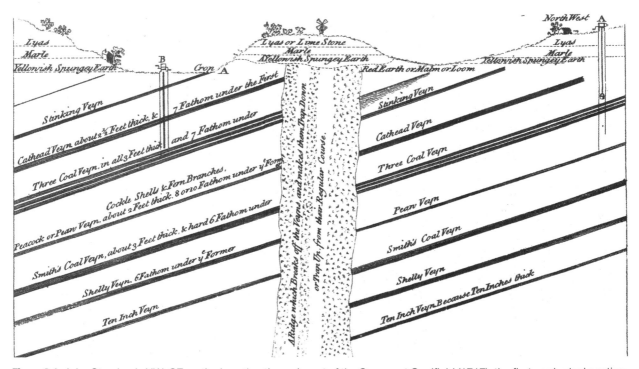

Figure 3.1 John Strachey's NW–SE vertical section through part of the Somerset Coalfield (1717): the first geological section. The tilted Coal Measures (with named seams) are overlain by horizontal Liassic strata (Lower Jurassic) at the top of the section: the first illustration of an unconformity, though not named as such by Strachey.

coal seams (*veyns*) tilted downwards (dipping) to the south-east. Another group drawn above the coal-bearing strata are only gently tilted and so discordant upon them: an unconformable relationship documented and explained by Hutton 70 years later (Chapter 6). Such stratal arrangements attracted practical geometrical solutions to calculate depths of coal seams or other features below the surface (as would certainly have been done long before by Euclidian-trained Roman engineers). The stippled zone in the centre of Strachey's section is a fault that displaces the coal-bearing strata downwards to the right (north-west: compare the *Three Coal Veyn* on either side of it). Strachey describes this as: 'A ridge which breaks off the veyns and makes them trap down or trap up from their regular course.' Such stratal dispositions, breaks and ructions had previously been pondered upon by Steno. To those who saw the world as a biblical creation, they provided hard evidence for creation, deluge, judgement and retribution. To the ordered and rational individuals of the early Enlightenment (like Strachey)

such precise observations suggested only natural (i.e. non-divine) causes.

The eighteenth-century rise of scientific agriculture in the Island (George III, 'Farmer George', was a great influence) was spurred on by an initiative from the Board of Agriculture in 1794 to publish county maps showing soil types and exposed rocks. These must have taken the fancy of mineral, drainage and canal surveyors like William Smith (see below). Such practical men had begun to make slow explorations, starting in their native districts, plotting the distribution of strata on increasingly accurate topographic maps.

Fossil remains and artefacts signified mysterious petrifactions or supernatural happenings until Steno's work became more widely known. Their systematic collection from successive strata in the late eighteenth century by individuals such as Georges Cuvier in the Paris basin and Smith in south-west England eventually proved that a sequence of life-forms could be discerned. Geological time was eventually divided up (Fig. 3.2) into distinct

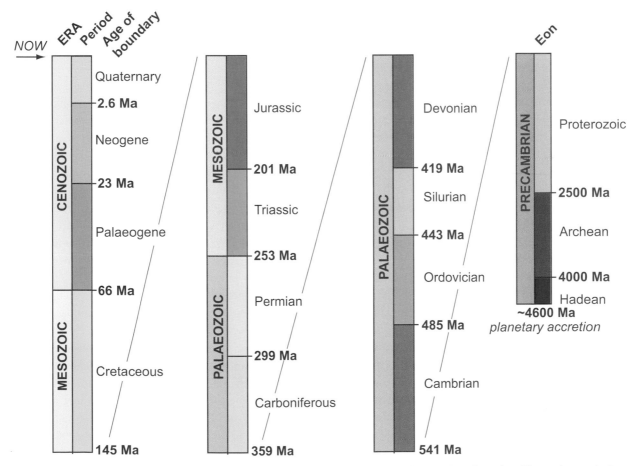

Figure 3.2 The geological time-scale with its eras, periodic subdivisions and their boundary dates in millions of years before present (Ma). Data redrawn from the International Commission on Stratigraphy (2012).

eras on the basis of fossil content – old life (Palaeozoic), middle life (Mesozoic) and young life (Cenozoic) – the terms formally proposed by Smith's nephew John Phillips in the early 1840s. Geological periods are now-adays defined as successions of strata with characteristic fossil remains, their boundary age with a previous period determined by radiometric dating (see Chapter 6). In the mostly unfossiliferous Precambrian the geological record is divided up into eons.

Yet despite these early developments in palaeontology, only two decades later the arguments employed by Charles Darwin in his *Origin of Species by Natural Selection* (1859) placed no reliance whatsoever on the fossil record. It was only gradually realized over the next hundred years, after much systematic study, that evolution, ecological adaptation and mass extinctions were all responsible for the changing distribution of fossil forms through geological time and over the space of the planet's surface.

Three 'Scotch Feelosophers' and an English Mineral Surveyor

It was from the ferment of the eighteenth century philosophical and scientific Enlightenment as practised in the capital of Scotland that three men emerged to give cogent explanations for the chief geological discoveries built up over previous generations: the aforementioned James Hutton (Fig. 3.3); his friend the mathematician and natural philosopher, John Playfair; and the classics undergraduate and reluctant lawyer Charles Lyell. The label 'Scotch Feelosopher' was a sarcastic epithet by journalist and contrarian William Cobbett, aimed chiefly at Adam Smith of Edinburgh.

James Hutton was the original genius: erstwhile farmer, agriculturalist and chemical entrepreneur who had been making geological observations across the breadth of the Island for perhaps thirty or more years. In 1785 he used his many observations to put together a fundamental explanation for the workings of the Earth, reading an original and far-reaching paper, *Theory of the Earth* (published in 1788), to the newly founded Royal Society of Edinburgh. In response to critics, he wrote a later, longer sequel subtitled '…with Proofs and Illustrations' (1795). This was a thoroughly logical work, though its precise logical style (almost Euclidean) was not to some tastes. However, in conversation he was renowned as a fluent, entertaining and stimulating speaker (and listener). His letters to his

Figure 3.3 *James Hutton* (1726–1797) by Henry Raeburn (*c*.1776). Oil on canvas. 125x105cm. Acc. No. PG2686 ©National Museum of Scotland. The sitter has an informal and relaxed pose, left arm hooked around the chair arm, waistcoat unbuttoned. Manuscripts (possibly for his 1785 *Theory of the Earth*), a quill and rock specimens are on the table, some of the latter used later to illustrate his great treatise.

friends (including James Watt) can be hilariously blunt and indiscreet, even by the standards of the day.

Hutton proposed an energetic Earth with a hot interior that periodically renewed its mountainous strata in response to ever-acting physical and chemical erosion. He championed the molten origins of rocks like granite and dolerite (the latter his 'whinstone'): they were igneous, deduced to have solidified by slow crystallization below the terrestrial surface after intrusion in a molten state. Such rocks formed by this mechanism were named 'plutonic' by Irishman Richard Kirwan in 1796 (but who, perversely, never accepted their molten origin); the name a nod to the mineral-rich Greek god of the underworld, Πλούτων, (Ploutōn).

In his various treatises and letters Hutton revealed himself to the world as a truly original and masterful natural philosopher: we can think of him as the first modern geological scientist, notwithstanding Agricola and Steno. From the first principles of deductive reasoning he predicted the necessary replenishment of an ever-eroding Earth by periodic heat-induced uplift and subsidence. The traces of these revolutions were left visible in rock strata which he analysed by widespread and perceptive field observations followed by elegant and universally applicable inductional logic. It is quite wrong of some modern-day science writers to label him as 'only' a theoretician; he regularly got his boots muddy in the field, walking and riding many hundreds of miles on horseback to look at rocks. His 1788 work was thus underpinned by almost 30 years of fieldwork and deep analysis of the wider significance of his particular observations.

After Hutton's death, John Playfair published his own *Illustrations of the Huttonian Theory of the Earth* in 1802. The book became widely read, Playfair illustrating his friend's ideas in plain, passionate and engaging English. It still reads today as a *tour de force* that sets out to demolish all opposition by pure reason. Here is part of his introduction, setting out Hutton's stall:

> ... the earth has been the theatre of many great revolutions, and nothing on its surface has been exempted from their effects.

> To trace the series of these revolutions, to explain their causes, and thus to connect together all the indications of change that are found in the mineral kingdom, is the proper object of a THEORY OF THE EARTH.

All this was in opposition to the influential German mining engineer A.G. Werner of Freiburg, who proposed that all rocks (bar those from modern volcanoes, curiously) were deposited in the oceans as sediment. By this theory, igneous rocks were not fire-formed, they were crystal precipitates from seawater, hence the label 'Neptunists' given to the masses of Wernerian followers worldwide (including the aforementioned Richard Kirwan of Dublin).

Hutton was able to champion the cause of the Plutonists far more effectively than Werner could the Neptunists. Key features in clinching his arguments were clearly visible in the diverse range of Scottish rocks at his doorstep: in the Highlands, Midland Valley, Southern Uplands and Borders. He saw that molten rock had obviously intruded into stratified rocks along sharp contacts, cross-cutting the stratal planes: 116 years earlier only Steno would have known that they were therefore younger. Upon cooling they developed pervasive cracks and mineral veins whilst at the same time the molten fluids heated the 'country rock', notably coking adjacent coal seams in the Midland Valley coalfields. His analysis of unconformities, angular breaks between strata of disparate age (as depicted by Strachey in Fig. 3.1), also

followed an observational route (see Chapter 6 for more on these). Yet again his essential breakthrough was universally applicable; mountain ranges were born from the sedimentary deposits of former oceans for entirely natural reasons arising from the Earth's internal heat. Great lengths of time were obviously needed to encompass repeated cycles of ocean and mountain birth and destruction. Some sense of this Huttonian dynamic is caught in John Keats's sonnet, *Ailsa Craig*, of 1819:

> Hearken, thou craggy ocean pyramid,
> Give answer by thy voice, the Sea fowls' screams!
> When were thy shoulders mantled in huge
>   Streams!
> When from the Sun was thy broad forehead hid?
> How long ist since the mighty Power bid
> Thee heave to airy sleep from fathom dreams...

One senses the young poet had some knowledge of Huttonian geology (see also Chapter 16) and it is possible that in his youth he had perhaps attended, or at least heard of, William Brand's well-balanced lectures at the Royal Institution on the tenets of Hutton's theories versus those of Werner.

By the early years of the nineteenth century the pace of geological discovery quickened. The Geological Society of London was formed by informal geological diners at a pub in Covent Garden in 1807. The society motto, 'Whatever is Under the Earth', was suitable for such solid, doggedly observational men (much as their diners today). Geological sections and accounts of myriad national and international geological localities began their systematic publication in the earnestly fact-gathering 'Transactions' of the young Society. These were distributed far and wide to its entirely male Fellows; the first women were elected as late as 1919.

No British individual approached the sheer scale of compilation achieved by William Smith, self-styled 'Mineral Surveyor', who published his huge Island-wide coloured geological map (though omitting the Scottish Highlands) in 1815. It was an astonishingly accurate map, though understandably generalized in the Lower Palaeozoic rocks of the north and west. These, termed 'Primary Schistus' since Hutton's day, would have to wait nearly fifty years for their basic stratigraphy and fossil content to be elucidated and agreed upon. The map has been headlined in our own day as *The Map that Changed the World* (Winchester 2001). Yet an official map of the geology of the whole of France surveyed by one man, J.E. Guettard, was published thirty-five years earlier in the days of the *ancien régime*, written and 'entrepris par ordre du Roi'. What was original was the novel precision with which Smith mapped strata round topography, fixing their boundaries according to the interaction of topographic slope with the dip and strike of the beds. He also determined, as did Cuvier, that most Mesozoic strata contained characteristic fossil forms that are unique to them. In this way the division of strata into periods (as Smith arranged them in his own museum; Fig. 3.4) was later facilitated when the rock types themselves were seen to change along geological transects. For example, identical ammonite fossils may be found in sandstones, mudrocks and limestones. Smith has also been trumpeted as the 'Father of Stratigraphy'; perhaps his correct title should be the 'Father of British Stratigraphy' since there were many European geologists who were contemporary topdogs in their own countries, not least Cuvier in France.

Playfair's rewriting of Hutton and the Oxford lectures of the eccentric fossil enthusiast William Buckland kindled the geological interest of the last of our trio of pioneer Scots geologists. This was Charles Lyell, a young lawyer settled in London. His father, also Charles, was an accomplished amateur botanist. Young Charles's energy, ambition, networking, wide reading in several languages, and relentless fieldwork over fifteen years led directly to his writing *Principles of Geology* by his early thirties. The first volume of three came out of London publisher John Murray's Albemarle Street offices (still there today) in 1830. Lyell held that field observations of rocks and their interrelations were to be interpreted from presently observed processes and causes – events like erosion, sedimentation, volcanic eruptions and earthquakes. Experiments and measurements could also help, on processes such as sand transport, ripple formation, rock melting and stratal folding (he would undoubtedly have been delighted to welcome modern plate tectonics into this list).

The kind of geology that Lyell practised became known as uniformitarianism. This was a curse of a term (ugly, long and wrong) coined by William Whewell in

Figure 3.4 *View of the Cast Iron Bridge and Museum. Scarborough.* (1828). Coloured lithograph by J. Stubbs. William Smith designed and stocked the Rotunda Museum (54.278717, −0.401817) with specimens. It is built with Jurassic Hackness Sandstone. Given his interest in geology and archaeology, poet and soldier Wilfred Owen may have visited the museum while stationed here in early 1918. In the background to the north is Scarborough Castle, built on a headland formed from a downfaulted outlier of resistant Upper Jurassic sandstones.

a review of Lyell's book. Lyell came to be thought of solely as an espouser of fieldcraft, inductive reasoning, and this so-called uniformitarianism. It was indeed a matter of pure faith to assume that today's Earth was unchanged from the more distant past. But Whewell and many others over the past two centuries have misread Lyell's intellectual honesty and his quest for general scientific explanation (Fig. 3.5; see Chapter 6 for such a contribution to ancient climate).

Darwin the Geologist

Lyell's book took the English-speaking world by storm. In Europe he was fêted and admired for his wide reading, with its free translations from ancient Greek and Latin authors and frequent passages from French savants and German natural philosophers. In the year following publication, Charles Darwin took the just-published volume on his five-year voyage of exploration in Fitzroy's Beagle (the ship's library had 403 other volumes!). It inspired

Figure 3.5 The frontispiece to Lyell's *Principles of Geology* (1830–33) featured a striking engraving of the three surviving columns of a Roman temple at Pozzuoli, beside the Bay of Naples. It is seen here cast on the Geological Society's Lyell medal. About one-third of the way up each column is a sharp boundary between rough and smooth marble. The roughness is due to attack by littoral molluscs during the columns' partial submergence in post-Roman times. The temple was subsequently uplifted to its present position as magma has accumulated deep in the crust beneath.

him to do much original work in Lyellian mode; for example, analysing the effects of great earthquakes in uplifting South American coastlines and causing tsunamis. It helped him develop an ingenious and satisfyingly elegant theory for the origins by volcanic subsidence of coral reefs and atolls. Both the scope provided by more-plentiful past time and the transience of oceanic islands would feed into the threads of logic that he subsequently wove to make the fabric of The Origin of Species. The geological confidence he had gained thirty years earlier made the enunciation of his ideas rapidly acceptable to many after publication in 1859.

So, by the 1830s, geology had become thoroughly fashionable and influential. The Island's glens, dales, cwms and corries echoed to the chip-chip of thousands of geological hammers (Fig. 3.6), an experience dismally described by Wordsworth in The Excursion (1814; extract in Chapter 4). Everyone was at it: men, women, persons of the cloth, gentry, dukes, duchesses; even the enthusiastic and driven Prince Consort had the ear of the leading figures in the Geological Society. As briefly noted in Chapter 1, the Geological Survey

of Great Britain was founded in 1835 on this wave of enthusiasm; it was to foster the practical advantages gained from accurate geological maps. Some landowners knew at first hand of the fortunes to be made from the mining of coal and other minerals in the rocks below their country estates. So it was that the superintendent of the Ordnance Trigonometrical Survey, one Captain T.F. Colby, appointed H.T. De la Beche, at the time Secretary of the Geological Society, to put 'geology' (i.e. the chief upper and lower stratal boundaries of stratigraphic units) onto the Survey maps of SW England. These were the first steps towards establishment of a national geological survey. It was to take over 100 years of intensive fieldwork to produce the first official map of the whole of the Island based upon continuous mapping at the old standard six inches to one mile scale. First published on two arm-length colour sheets at a scale of 1:625,000 in 1948, the latest (5th) Edition appeared in 2007. Simplified and summary versions using data from these great maps are presented at several points in the present volume.

Figure 3.6 An early-Victorian geologist (H.T. De la Beche?), watched by a curious youth, chips away at the unconformity between vertical Carboniferous (below) and horizontal Triassic (above) rocks in South Wales (Bailey 1952). Today we know little of the field safety afforded by top-hats, yet, like Wordsworth, we frown on wanton despoliation of outcrops by hammering.

Chapter 4

Works of the Imagination

The landscape…is not a totality that you or anyone else can look at, it is rather the world in which we stand, taking up a point of view on our surroundings, and it is within the context of attentive involvement in the landscape that the human imagination gets to work in fashioning ideas about it. For the landscape…is not so much the object as the 'homeland' for our thoughts.

Roger Ebbatson, *Landscape and Literature 1830–1914* (2014)

Imaginative Links and Changing Responses

Ever since humans arrived back in the British peninsula after the 'Younger Dryas' cold period, they have been responding to landscape, choosing where to live, building communities and burying their dead. How people lived and died was shaped by knowledge, skills and beliefs passed on by successive generations. But humans possess imaginations, enquiring minds and the need to create (and destroy); their relationships with the material and spiritual worlds must constantly change. An imaginative response to a landscape is inevitably a response of its time and culture. But time passes and, as geology informs us, so does everything else; there is always a state of flux. So even the hardest rocks steadily weather, sometimes shifting, rupturing or deforming by the motion of tectonic plates. Ironically, rocks are often associated with permanence: 'Rock of ages cleft for me', the 'Kaaba' in Mecca, 'Ed's Stone' (now lost) etc.

To illustrate this, contemporary artist/sculptor Ilana Halperin, working in Glasgow, uses many different approaches to explore and express her intentions, including fieldwork and research. Her work addresses the geological timescale underpinned by her perception of the Earth's rocks forever forming and re-eroding and her curiosity regarding our relationship with rocks. One of her studies juxtaposed ancient Lewisian rocks with some of the newest rocks forming in Iceland. Like many contemporary artists she uses mixed media and addresses global issues.

An exhibition at Tate Britain in 2014, *Ruin Lust*, showed how different artists have been 'drawn to that which we most fear' and have tried to capture a moment or address the reality of such constant change. At the heart of the exhibition was the paradox that, in the act of creation (painting, sculpture, building, literature) the artist/author's own world is reshaped, just as our physical world is changing moment by moment. So there is constant interplay between the changing world and changing humans. Against this endless flux, humans have interacted imaginatively with the environment through shaping, making marks and erecting structures.

The following two paintings express very different responses to the British landscape. The early-nineteenth-century fascination with nature could be underpinned by the understanding of it through geology. Ruskin and the Pre-Raphaelites expressed this in their painting (Fig. 4.1): the beginnings of modernism. Ruskin was passionate about the natural environment, a Fellow of the Geological Society, and hugely influential in the arts world of the later nineteenth century. His ideas were fundamentally anti-industrial, emphasizing the connection and mutuality between humans and the environment. 'Truth to Nature' inspired not only the Pre-Raphaelites but also the later Arts and Crafts movement, some regarding nature as being evidence of God's munificence.

Almost a hundred years later in *Pillar and Moon* (Fig. 4.2) Paul Nash depicts Nature in an entirely different way. Whilst Ruskin's paintings are full of light and optimism and nature is benign, here the light is ghostly and the natural world appears brooding and ominous. It is a picture of sadness, of mourning. The austere sweep of bleak winter trees, nocturnal shadows and field furrows lead the eye inevitably to connect the two pale old spheres. Everything is illuminated by moonlight, especially the stone. Nash wants us to contemplate similarities. His influences were many – from his own experiences and the pervading modernist culture of the time. He embraced Surrealism and Symbolism, though

Figure 4.1 *In the Pass of the Killiecrankie* (1857) by John Ruskin (1819–1900). Watercolour and bodycolour on board. 28.7x24.7cm. Acc. No. 1589 ©The Fitzwilliam Museum, Cambridge. Ruskin painted this outcrop of Dalradian schist with great clarity and attention to the fine detail of its foliation. Note the boundary between wet and dry rock and the ghostly mountain top through a gap in the pines: the artist is hinting 'This is what mountains are made of!'

like John Piper and Graham Sutherland he is regarded as a Neo-Romantic artist, responding to landscapes with feeling (Chapter 15). Profoundly affected as an artist by his experiences in the First and Second World Wars , Nash's work repeatedly explores distances, time and space. Like Ruskin, he too longed for a pre-Industrial Revolution connection with the land. This was not a sentimental longing because he had no illusions: neither nature nor people were regarded as benevolent. And while he gazes at nature, it looks back with a cold

Figure 4.2 *Pillar and Moon* (1932–42) by Paul Nash (1889–1946). Oil paint on canvas. 50.8x76.2cm. Acc. No. 5392 ©Tate, London 2015.

and indifferent eye. Many of his other paintings are of austere, ancient megaliths (Fig. 15.10). He visited several ongoing excavations at Avebury and Maiden Castle in the 1930s, and his paintings convey a deep sense of melancholy; stone and trees were generally used symbolically, and though many show recognizable features of his much-loved South Downs landscape, it was never his intention to simply record what was there.

Monuments, Installations and Place

Works of the imagination can also be monumental: an expression of time and culture but also of place. Stonehenge was built and rebuilt for a millennium from around 5ka. No one knows why 80 or so bluestones were moved over 200km from the Preseli Hills in Pembrokeshire to Salisbury plain around 4.6ka (see Chapters 10, 13). Interestingly, carbon dating of postholes at the site suggests the existence of an earlier (perhaps from 5.1ka) large-scale wooden structure within an earthwork enclosure: Stonehenge 1. The alignment of the later sarsen circle to the summer solstice,

the embanked chalk-cut Avenue linking it to a timber circle two miles away at Durrington Walls, the discovery of human remains and other features have led to much debate and speculation about the culture and beliefs of its builders. Whatever these were, it was clearly an important and symbolic place: a spiritual and ceremonial complex, in the words of Bowden *et al.* (2015).

The Romans (an inclusive lot in terms of religion) sometimes sited their temples close to pre-existing Iron Age sites, sacred springs or rivers. Pagan centres were often chosen as church sites by the earliest Christians. Future archaeologists may ponder the significance of John Frankland's *Boulder* (2008) (Fig. 4.3). It is around 4m high and weighs 100 tonnes or so: a 'YouTube' clip shows it and a sister stone being blasted from their granite home in a Cornish quarry, gently lifted onto low-loaders to begin their overnight drives to the flat spaces of London in Shoreditch Park and Mabley Green. The artist, also a rock climber, wants people to touch them, climb them (there are 33 routes up the Shoreditch stone!), own them; the sculptures have become landmarks and a focus. Like Stonehenge they

Figure 4.3 *Boulder* (2008) by John Frankland. Cornish granite >4m high. Shoreditch Park (51.534178, −0.086658). Photo: ©William Lawlor. Commissioned by PEER and the Shoreditch Trust.

engender a sense of awe – reassuring that within a rock-less metropolis one can confront the massive physical presence of what is essentially an artificial outcrop. It makes the Stonehenge story all the more incredible.

Earliest Artists

Stonehenge and Boulder are both of their time. To create is a way of integrating thought, feeling and personality. It is a response not just to landscape and the environment but also the culture of a particular time. Medieval landscapes were created by eyes which envisaged celestial landscapes. Being human, we inevitably reflect on our mortality, and so the nineteenth-century Romantics' obsession with wild landscapes and ruins created a form of satisfaction out of an experience of change and loss. It is possible that their enjoyment of a sublime view (a fine but dangerous prospect) also expressed a sense of ownership, a new detached and materialistic view of nature shaped by the Enlightenment and the Industrial Revolution. Contemporary art often explores ecological and environmental concerns from a fear for the future and a need to do something about it.

The earliest signs of artistic expression here in the Island are to be found amongst the headless remains of 'Paviland Man' from around 30 millennia ago. The burial, in a shallow grave in the limestone Paviland Cave on the Gower peninsula, South Wales, was before the severest onset of the last Ice Age but at a time of general climatic deterioration (a dozen or so millennia previously the cave had hosted Neanderthal burials). The corpse had been embalmed with powdered red ochre, weathered haematite mineral (Chapter 7). Native Australians also use this to paint their deceased – a symbolic coating of blood. The grave contained baton-like rods and shaped ivory from mammoth tusk; a mammoth skull lay nearby, also periwinkle shells, each pierced to be strung as a necklace.

Early post-glacial cave art is best known from Creswell Crags on the Yorkshire/Derbyshire border. In Creswell's caves, Pleistocene sedimentary deposits yield stone tools that record successive occupations: Neanderthal (40–60ka), pre-glacial hunter gatherers (28ka, roughly contemporary with Paviland Man) and early post-glacial Upper Palaeolithic hunters of the wider European Magdalenian culture (15–11ka). These latter humans engraved cave walls and ceilings with images of deer, bison, horse and various birds, now preserved below a layer of flowstone. Also found from about the same time is the Island's sole example of Palaeolithic figurative art, an astonishing etching of a horse's head done on mammalian rib bone (Fig. 4.4). A recent find of a reindeer image on the Gower is also reported to be Magdalenian in age.

Many questions arise. Why did our ancestors make art? Why draw in such dark, inaccessible sites lit only by tallow lamps or viewed with burning torches? According to archaeologists Paul Bahn and Paul Pettit (2009), co-discoverers of the Creswell artwork, a significant amount of Palaeolithic cave art worldwide is located in hard-to-reach places. The earliest post-glacial landscapes had their moist, exposed rock surfaces quickly covered by encrusting flora such as mosses, lichens and algae – hardly conducive to either execution or preservation of surface rock art. Our Magdalenian ancestors wanted their art to endure, like native North American pictoglyph artists in the semi-arid South West. They had to draw on smooth, dry, dark interior walls.

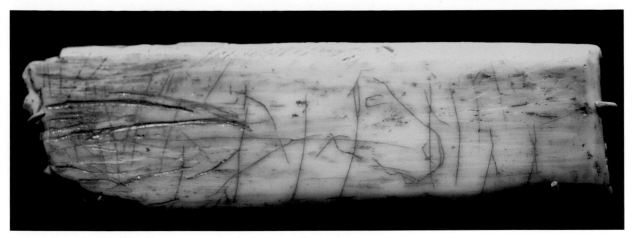

Figure 4.4 A rib fragment engraved with a horse's head, neck and mane, Robin Hood Cave, Creswell Crags, Derbyshire (53.263219, −1.193448). Photo: Wikimedia Commons©DaveKav. Dated as Upper Palaeolithic, *c*.12–12.5ka, the specimen is now in the British Museum.

Preservation and Place

Only some of the Island's ancient buildings, monuments, sculptures, inscriptions and cultural treasures have survived, these often as ruins. Much was lost to quarrying after the Roman withdrawal, during the Reformation, by accidental urban fire down the ages, and by indiscriminate bombing in the Second World War. All over Britain the survivors are lovingly tended in museums, galleries, archaeological and heritage sites and a huge stock of preserved architecture. Most pre-Conquest buildings were of wood and so are lost, but the many earlier stone sculptures and rock carvings speak most eloquently of a people's spirituality and regard for the land. For example, the New Testament carvings on the 5.2m tall *c*.AD 700 Anglo-Saxon High Cross at Ruthwell, Dumfriesshire (Fig. 23.4) may have served as a teaching aid, as well as a symbol of the new faith for both locals and travellers. It probably stood outside the original village kirk; the Angles who first populated the area after the legions left were originally pagans. The cross has both Latin and Runic text carved around its margins, the latter the oldest surviving example of English poetry. Though carved from robust sandstone and made to last, its cruciform top was smashed in 1642. In the nineteenth century, a more respectful age, it was reassembled and placed in a specially deepened enclosure inside the kirk.

Nothing will survive of Richard Long's contemporary sculptures, since his aim is to leave as little imprint on the landscape as possible. One of his earliest was *A Line made by Walking* (1967), a photograph of a strip of grass flattened by his tread as he walked through a meadow. He and many contemporary Land Artists show concern

for the environment and do not want to clutter up the planetary surface with even more sculptures that will last for centuries. Nevertheless it is still possible to travel the length of the Island and observe how the colour and character of ancient monuments and buildings have been dictated by the availability of locally sourced materials. They are intimately connected to landscape; crosses like Ruthwell and local building stone generally were dug and levered out from outcrops up to a few tens of kilometres or so away. These realizations, combined with some understanding of the geology beneath, give us an enhanced sense of natural harmony. The Georgian streets of Bath, for example, were built and faced with locally mined Middle Jurassic oolitic limestone. The Georgian terraces, crescents and squares spread over the city's seven hills are unified in low sunshine by the gold-brown hue of the weathered limestone. By way of contrast, but by no means inferior, Leicester City is predominantly built of red brick manufactured from Triassic claystone. Much of late-nineteenth-century Glasgow comprises brown-red Permian sandstone (once desert sand) imported by rail from Mauchline, Ayrshire.

In *A Land* (1951) Jacquetta Hawkes wrote: 'We are all creatures of the land substantively produced by the terrain on which we live.' Nowadays we know this statement is literally true, for our teeth and bones preserve the chemical characters of trace isotopes of strontium present in the waters of the place that nourished us at an early age. Thus, of the two early Bronze Age male skeletons and their ceremonial accoutrements and jewellery buried near Stonehenge, one (the 'Amesbury Archer') was born in the German Alps, the other (his servant?) drank water from English chalk downlands as his teeth grew in childhood. Neolithic Europe and the Middle East were substantially smaller than we might otherwise think.

Artists Embedded in Landscape

Travel the length of Britain and one is amazed by the variety of landscape. A factual landscape character assessment of the Cotswolds includes the high wolds dip-slope, its ironstone hills and valleys, escarpments and low limestone plateaux. A character assessment of any other region in Britain would list completely different features. Yet always the weather, the time of day and the seasons conspire to alter the mood with different lighting and costume changes. And the sea restlessly paws at the island's coasts, depositing and eroding: a moat to protect us but also a highway. Our shores yield an endless variety of landforms laid bare in sandy or rocky bays, cliffs and islands.

George Shaw's *Tile Hill* paintings (Fig. 4.5) describe quite another landscape – not just the estate in Coventry where he grew up, but also an unflinching eye for detail, showing his deep love for the place. The paintings, sombre and non-pastoral, sometimes show a sinister and dilapidated Britain. Yet each brick is painted intimately and individually. These paintings, he says, show 'eerie scenes of the ruinous edge of modern British life…I feel like I am trying to gather something up… like a sleuth.' Often they show dereliction. His work collectively rings true of (any) urban Midland childhood. Someone pointed out to him that the sun rarely

Figure 4.5 *Scenes from The Passion: Ten Shilling Wood* (2002) by George Shaw (b. 1966). Humbrol enamel on board. 44x53cm. Courtesy of the Wilkinson Gallery, London ©George Shaw.

shone in his paintings. The sun may also rarely shine in many Midland memories, but these are linked with other sensory feelings. What generally doesn't change is the actual lie of the land. The Midlands are not all flat; the distant uplands in Figure 4.5 seem to offer some way out, but the Tile Hill paintings generally have a flatness that resonates: a flatness of light, land, sky; even the spoken Midlands accent. But, travel a short distance, say to Charnwood's uplands, then the Midlands show a different face.

Henry Moore of Castleford, a pioneer British twentieth-century stone carver, was heavily influenced by childhood memories of the often surreal, grotesque and skylining rock outcrops and scarps of the nearby Yorkshire dales, especially Wharfedale (see Fig. 5.1). He became a stone-carving fanatic as a young sculptor, utilizing many different sorts of stone, including the green-tinged Hornton ironstone (Fig. 14.8) from Northamptonshire's Jurassic scarplands. His mature works include large reclining female figures in cast bronze whose curves and nuanced forms seem part of almost any upland landscape they now find themselves in. They seem less at home in lowland England away from scarp, scar, outcrop and crag.

Many artists express a sense of an elemental confrontation with nature. J.M.W. Turner famously liked to experience his landscapes at first hand, for example painting *Snowstorm, Steamship off a Harbour's Mouth* (1842) having had himself lashed to the ship's mast in the middle of a storm. Since then many artists have painted *en plein air* in order to fully experience place. A modern example is Lulu Hancock's embracement of the enormous human-construct that is Mountsorrel Quarry, Leicestershire (Fig. 4.6).

Figure 4.6 *Dusk, Mountsorrel Quarry, Leicestershire* (1999) by Lulu Hancock (b. 1966). Oil on Board. 20.7x30cm. Leicestershire County Council Museums Service. Acc. No. X.F2.2000 © Lulu Hancock. The artist's confrontation at sunset with dusty igneous rock along the arcs of ever-descending working benches in Mountsorrel's vast granodiorite aggregate quarry (52.731304, −1.167673).

Figure 4.9 Artist Katharine Holmes featured the Great Scar Limestone in the gorge of Gordale Scar in many of her studies in the late 1990s. This example is from the catalogue to her 1999 exhibition at the University Gallery, Leeds and reproduced with her permission and © Katharine Holmes. The limestone beds in the gorge are picked out by an ethereal grey-blueness, the rock seeming to reflect the bright sky. See Figure 5.4 for a photograph taken close to the artist's viewpoint.

Figure 4.10 *The Loss Adjusters* (2015) by Katrina Palmer (b. 1967). A site-specific audio walk, Portland, Dorset. Commissioned by Artangel. Photo: Brendan Buesnel © Katrina Palmer. Courtesy of MOT International, London & Brussels. This view shows quarry waste and the working face for Portland Freestone, the white massive limestone, c.9m thick, above the quarry floor: mechanized cutters nowadays access the stone through horizontal adits.

Figure 4.11 *The Deep* (1999–2003) by Terry Farrell (b. 1938) at the confluence of the Rivers Hull and Humber (53.738856, −0.330375). Photo: Shutterstock 369538721©Phil MacD Photography.

ocean, the natural world, the port of Hull itself, and the adjacent Humber estuary. Half-building, half-sculpture, it exploits the estuary-margin site. For example, its glass mirrors the sea and aluminium plating suggests fissured slaty rock plates. The computer-aided design is combined with a profound understanding of materials and a deeply imaginative response to function and place.

By way of contrast, the buildings created by Zaha Hadid generally seem light, playful and characterized by movement. Her zinc-clad ripples on the roof of the Glasgow Riverside transport museum, on the banks of the Clyde, define the confident flow of the whole building. What we create expresses who we are; and how appropriate that a transport museum should itself embody movement.

Consciously or unconsciously, directly or indirectly, the work of artists is embedded in the landscapes they inhabit or have inhabited. How this is expressed is usually as unique as handwriting – sculpted bare rock; conceptual installations exploring the wonder of individual rocks and the ever-changing structure of our earth; paintings that show the elemental majesty

of mountains; exposed quarried rock like flayed flesh; conceptual art exploring the nature of empty spaces once quarried; the tenderly painted detail of a Midlands housing estate; the haunting emptiness of an abandoned building; the more lasting presence of new buildings. All are created out of a profound imaginative response to the British landscape.

A Yearning to be Close to Nature: The Good Life

Rejecting the material trappings of this world and living a simple life close to nature has been a recurring refrain throughout history. At intervals people have reacted against industrialization (notably the Pre-Raphaelites and artists associated with the Arts and Crafts Movement) and yearned for the simplicity of pre-industrial country life and its raw connection with the environment. Before the Industrial Revolution and the mass movement of people to towns, ordinary people used chiefly what was readily to hand for ornaments, building materials and tools. There were exceptions, of course – luxury items and family heirlooms – and it is unwise to romanticize pre-Industrial Revolution rural life in any way; it was short and often brutally hard for most: 'You knew where you were born, you knew where lay your grave.' An intimate knowledge of the land, weather, plants and animals might make the difference between life and death.

William Blake saw the alienation engendered by urbanization and the effects of the Industrial Revolution as evil, urban landscapes of the apocalypse, and he dreamt of a New Jerusalem: 'In England's green and pleasant land'. This probably finds expression in the British love of pastoral images of landscape: it is rare to find a dwelling without at least one picture or print on the theme; Constable is perennially popular.

We live in a period dominated by technology where the relationship between material, function and the environment is sometimes lacking. Contemporary writer/ curator Nicholas Bourraid has explored the long-term implications of this. He asks if technology is separating us from the material world and making us less human.

Perspectives

Landscapes can seem familiar and comforting, expressing a deep sense of belonging. They can also appear hostile and alienating. While acknowledging W. Meinig (1979), quoted in Sullivan (1998): 'Landscape is, first of all, the unity we see, the impressions of our senses rather than the logic of the sciences.' it is also important to be aware that our perception is shaped by many personal and cultural influences. Our responses to landscapes can be shaped by numerous factors, but certainly how we look is augmented by an awareness of other perspectives.

The violent events of war have a special symbolism in imagery and music. The British landscape is littered with hundreds of known battle sites (many more unknown) and military installations old and new. The Second World War art of John Piper, Paul Nash and Graham Sutherland engage in an emotive dialogue with devastated landscapes and hardware; they use colour, composition, shape, texture and line expressively. P.J. Harvey's passionate music lyrics explore both a personal and a savage national past: her 'White chalk hills where I was born' are rooted in the prehistoric southern downlands. Her music in the collection, *Let England Shake,* has been described by Andy Gill as 'a portrait of her homeland as a country built on bloodshed and battle'.

Yet the downs have also been an imaginary place of comfortable retreat, of '…tinkling sheep bells, country pubs and cricket on the village green, an England just out of reach which everyone longed to recapture but which they knew never existed in reality' (this from *A Landscape Character Assessment for Mid-Sussex,* 2005). For many, such stereotypes still represent the ideal Britain, the particular cosiness of the 'Home Counties'. They are reminded of the early-twentieth-century artists' colonies that grew up on the chalk downlands or gathered there to work, hike and socialize, like the Bloomsbury Group at Charleston and the poets Edward Thomas and his friend Robert Frost before the First World War.

Or, on the other hand, rural landscapes became doom-ridden rural sinks in which the likes of Thomas Hardy, D.H. Lawrence and lesser writers imprisoned their luckless and feckless characters. The more romanticized, idealized and often pessimistically narcissistic views of life on and near chalk downland are memorably satirized by Stella Gibbons' *Cold Comfort Farm* (1932).

In *The Englishness of English Art* (1956) Nikolaus Pevsner stated that English Art was determined chiefly by climate and geography. He was (perhaps unwisely and with Germanic thoroughness) trying to uncover some kind of national mind-essence of Englishness. Perception of landscape can also, of course, be political, but when connected to a sense of national identity this can be divisive. National heritage is basically a late-eighteenth/nineteenth-century concept. It promotes a comfortable and pastoral ideal of enduring values (riffing on 'sceptred isle', etc.). One wonders how a recently arrived immigrant views such stuff?

In truth, landscape may be interpreted differently by different ethnic groups. In the 1980s, artist and photographer Ingrid Pollard used Romantic landscape imagery to question how people associate immigrants with cities. Her *Pastoral Interludes* challenges the social constructs of Britishness in a series of photographs of herself set in various Lake District settings. Two highly respected Guyanese artists, Aubrey Williams and Frank Bowling, both arrived in England in their mid-teens. Their work must be considered in the context of Guyana and of the wider Caribbean diaspora within British society. Williams's work is semi-abstract and captures the spirit of the South American landscape; the works of both artists are explosive with colour. Bowling's abstract and expressive paintings evoke places and moods. He says: 'Guyana…became the place where I lived in my imagination.'

Finally, an awareness of different perspectives can help us understand the intentions of artists. Lachlan

Goudie in *The Story of Scottish Art* (2015) described Sir Edwin Landseer's painting *Monarch of the Glen* (1855), a magnificent stag framed against Highland mountains, as 'symbolic of a culture that values rich mens' sport over poor mens' land', an example of 'cultural colonialism'. He went on to point out that most Romantic paintings of Highland Scotland are empty of people for a good reason: they had all been cleared out.

Writers in Landscapes

The Romantic poet William Wordsworth saw in his *Prelude* (1850) the extreme energy of landscape and Nature in the raw. However, he seemed little interested in the actual geology of his mountain paradise, despite the efforts of Adam Sedgwick (he who named the Cambrian geological period, see Chapter 5) to engage him. In his earlier poem, *The Excursion*, he had explained that both field geologists and botanists were unworthy of a place in sublime landscape (see also Chapter 16):

> He who with pocket-hammer smites the
> edge of luckless rock or prominent stone…
> …detaching by the stroke
> A chip or splinter – to resolve his doubts;
> And, with that ready answer satisfied,
> The substance classes by some barbarous
> name,
> And hurries on…

A contemporary of Wordsworth, the poet John Clare of Northamptonshire, was also fixed in landscape. But his home landscape was unfixed due to enclosure and agricultural ruin. You could say, from a reading of his work, that the effect was as that of a modern environmental catastrophe: it served to drive him to despair and deep melancholic depression.

Mid-nineteenth-century novelists, notably Charles Kingsley and George Eliot, reflected on the geological revolution let loose by geologists like Lyell. They engaged their characters in dialogues of self and soul with the history of the earth and its often brutal processes, like the great flood in Eliot's *Mill on the Floss* (1860). Emily Bronte made implicit use of Pennine scenery and moorland crag in her *Wuthering Heights* to perhaps reinforce inescapable and timeless truths of love, passion and loss as she saw them. The physical rawness of Wessex's natural landscapes engendered an extreme *hiraeth* in both the lyric poetry and to the luckless characters that dominate the novels of Thomas Hardy. The opening of *Return of the Native* (1878), for example, describes a heathland landscape (fictional 'Egdon Heath') of unparalleled bleakness and foreboding to some of the main characters – not, though, to the young purveyor of red ochre (powdered haematite) for sheep 'redding'.

Twentieth-century poets have used the idiom and technical rhythms of geology to dig more easily into the hard earth of lyric poetry. Basil Bunting, W.H. Auden, Norman MacCaig and Norman Nicholson, for example, use familiar geology interwoven into their personal journeys through life in North Pennine, Highland and Lakeland landscapes respectively (see Chapter 15). Raw upland moorscapes were brilliantly captured by photographer Fay Godwin in *Remains of Elmet* (1979), her book with poet Ted Hughes. In *Hill-Stone Was Content* the poet confronts the descent of moorland stone into factories and mills in deep-bounding industrial valleys like Calderdale (Fig. 4.12):

> To be cut, to be carted
> And fixed in its new place.
> It let itself be conscripted
> Into mills. And it stayed in position
> Defending this slavery against all.
> It forgot its wild roots
> Its earth-song
> In cement and the drum-song of looms…

Figure 4.12 Sowerby Bridge, West Yorkshire (53.708877, −1.909585). To the left is a sandstone-built early-19th-century mill (now flats). Opposite, through the ginnel and across the River Calder, is a mid-Victorian mill complex with brick detailing.

PART 2
Assembling the Geological Jigsaw

Geology is (usually) an inductionist science: it proceeds from particular evidence to general conclusions. Geological observations made in the landscape lead to the production of geological maps. These enable collection of accurately located rock samples, such as those containing uranium-bearing minerals that led to Arthur Holmes's pioneering efforts to discover the age of Earth's rocky substrates. Chapter 5 relates the primacy of field observations in the production of national geological maps. Chapter 6 outlines how the threads of geological history were unravelled over 200 years of effort, beginning with James Hutton's perceptive field observations. The quest for a rational account of geological history led eventually to the 1960s 'Plate Revolution', when geophysics provided mechanical explanations for events deduced from the rock record.

Chapter 5

Mapping

In the morning, mist lay heavy on the Chilterns … Kick these hills and they bleed white. The mist is like an exhalation of the chalk.

Graham Greene, 24 hours in Metroland, *New Statesman* (1938)

Natural sculptures

Outcrops protrude into the atmosphere from the solid ground underfoot. What you often see (Fig. 5.1) is a weathered mass, often moss-, lichen- or algal-covered, the surface scoured by linear and curved cracks. There are crevices, overhangs, buttresses, vertical columns, detached fallen blocks and scree. Such features, all too familiar to the rock climber, break up the continuity of the rock mass. They hint at an interior made up of infinitely varied shapes and forms: natural sculptures of the kind that so influenced modernist artists and sculptors (Chapters 4, 14).

According to geologist Francis Pettijohn, there was 'nothing so sobering as an outcrop'. He meant the confrontational shock at the sharp-end of field geology: the raw surface of the solid planet exposed. The field geologist has to assess, identify and interpret the rocks on view; the outcrop may be the size of the Grand Canyon or a rut in a path. But whatever their size they must be *in situ*: natural processes and gravity cause sizeable rock

Figure 5.1 The 'Cow and Calf' rocks in Wharfedale above the town of Ilkley, West Yorkshire (53.917011, −1.802856). Photo: Wikimedia Commons©T.J. Blackwell. The prominent Millstone Grit (sandstone) edge sits high on the valleyside. The 'Cow' on the left is outcropping bedrock, the Addingham Edge Grit, a thick and resistant coarse sandstone. The 'Calf' to the right is a fallen block of this. In this way all scarps retreat and valleys widen.

masses to appear lower than their true position, as Figure 5.1 dramatically shows.

Natural outcrops exist because local bedrock has been exposed by erosion. They are uncovered from surficial deposits: soil, peat and deposits of loose sediment laid down by wind, water or ice. Such coverings are less likely in hilly or mountainous country. Here, where bedrock is hard and resistant to erosion, higher surface gradients cause greater erosion by gravity and flowing water. Yet rock masses may also, by reason of high porosity and/or permeability, be able to soak up water and let it pass it through rather than run off causing erosion: chalk downland and harder limestone outcrops are cases in point.

Outcrops themselves are commonest in western areas where high runoff, energetic streams and former glaciers were efficacious in removing (rather than depositing) surficial sediment and soil cover. They are harder to find in eastern, cropped and wooded lowland landscapes. Here, artificial exposures in quarries and cuttings made by road, canal and railway and in temporary excavations are a godsend to the geologist. Yet being an island, there are plentiful coastal outcrops resistant to wave and tidal erosion: visiting geologists from the Mid-Western USA and Central Europe marvel at the scope and splendour of such exposures.

Age Shall Weary Them

Weathered rock reveals the highly active interface between bedrock and atmosphere. It was James Hutton (again) who first realized the key role that rock weathering and breakdown play in the fertility of soils and in the provision of sedimentary detritus for the eventual renewal elsewhere of eroding mountains. Despite their ugly apparent outward chaos, weathered rock and their superincumbent soils are natural ordered products of chemical and physical interaction with atmosphere. During weathering and soil formation, minerals from rock are chemically attacked by natural (and pollutant) acids in the rain (carbonic acid) and by humic acids generated from breakdown of organic matter in rotting vegetation. Breakdown products like clays (Fig. 5.2) and a residue of the more resistant minerals are left in a matrix

Figure 5.2 Deeply weathered granite bedrock, Two Bridges, Dartmoor, Devon (50.558663, −3.964551). The granite has broken down by the chemical weathering of its feldspar crystals to a pasty, granular mixture of kaolinite clay ('china clay'), mica and resistant quartz. Remnants of less-altered granite are also present.

of plant-derived organic humus. Quartz is generally quite stable during this process but other silicate minerals that make up rock, especially feldspars and micas, are less so.

Soil, with all its microbes, plant roots and creatures, is the landscape's valve. It takes in air and moisture through its pores, capturing atmospheric carbonic acid and expelling carbon dioxide gas and ions like potassium and calcium in solution from the weathered rock immediately below it. Atmospheric oxygen (itself produced during plant photosynthesis) allows further alteration (by oxidation) of minerals containing ferrous iron. This produces the insoluble hydroxide residues commonly seen as ochreish stains on weathered rock surfaces. The aqueous solute produced is taken up in the sap of root, bark and trunk and is essential for organic growth. Dead organic material breaks down in turn to release more weathering ions and carbon dioxide and so on, indefinitely. This defines the geochemical cycling of the outer earth: geosphere, atmosphere, hydrosphere and biosphere all taking part. Water is the common thread: without it chemical weathering could not occur and life could never have begun. Further, trapped in the cracks and pores of subducting tectonic plates it eases melting and deformation, enabling the plates to form and move and allowing volcanic-arc eruptions to occur. Water has thus enabled the continental crust to grow by such magmatic additions over geological time.

Geology on maps

Geological maps use outcrops to trace the extent of different rock types. They summarize information on the rocks and their geological arrangement: stratal layering, tilting, succession, folding, faulting; what is called their 'structure' or 'structural architecture'. Such knowledge enables understanding of what might go on below the surface, clearly an interesting and practical exercise.

How is geological mapping done? Every person who has seriously walked the outdoors will be familiar with topographic maps whose contours represent surface relief. It takes a little time and experience to recognize peaks, ridges, drainage divides and valleys from such a map. Once gained, the skill enables routes to be safely taken and engenders a strong sense of the 'grain' of a landscape, its twists and turns, declivities and symmetry. It is onto such topographic surfaces that geological observations are made: ground is covered each day over the course of a field season, excursion or expedition. All accessible outcrops are visited, recorded and traced around hill and mountain using tell-tale features indicative of abrupt relief changes (ridges, terraces, bluffs, hollows) that infer continuity or otherwise between outcrops (Figs 5.3, 5.4). Gradually, contrasts between adjacent localities are noticed; a stratal succession emerges from oldest to youngest. The presence of folding and faulting may become apparent from the evidence

Figure 5.3 Early winter view from above Thwaite, north to Kisdon Hill in Upper Swaledale, North Yorkshire (499 m; 54.392722, −2.155805). The stepped form of the hillslope in the formerly glaciated U-shaped valley is due to resistant limestone and sandstone edges separated by gentler slopes along weaker mudrocks of the Yoredale and Millstone Grit Series (Carboniferous). The neat drystone walls, barns and pasturelands of the valley floor and lower slopes peter out upwards to rough-grazed tops.

Figure 5.4 Gordale Scar, North Yorkshire (54.072347, −2.132063). Craggy limestone outcrops and steep, scree-covered lower slopes like this typify much of upland Britain founded on Carboniferous Limestone. Note the thick limestone beds with their more-or-less horizontal stratification and the more obvious pervasive vertical fractures (joints) caused by proximity to the Craven Fault system.

of crumpled, broken or displaced strata deformed by tectonic forces. Igneous or metamorphic rocks and signs of mineralization may also be noted.

A geological map begins to form as lines are tentatively drawn (or nowadays charted by GPS) to enclose rocks of a certain kind and to distinguish them from adjacent ones. The different types of rock are colour-coded onto the map; symbols added to denote features like stratal tilt (dip). Once completed to the geologist's satisfaction, a local geological map emerges. The geological structure and succession is placed within topographic sections: such geological sections allow appreciation of the shallow structural architecture of the near-surface, like John Strachey's earliest example considered previously (Fig. 3.1). Over time and by the efforts of very many geologists down the ages, whole regions, countries

(Figs 5.5, 5.6) and continents have been mapped and sectioned in this way.

Geological maps that show the extent of bedrock alone are termed 'solid', for they ignore what are termed 'superficial' deposits, chiefly soil and Quaternary glacial detritus and other weathered sediment deposited over the last 500kyr or so. By way of contrast, superficial geological maps show bedrock only where it actually outcrops, emphasizing the soils and younger Quaternary deposits. These were traditionally termed 'drift' maps by the Geological Survey. Here the type of local bedrock must be mapped more tentatively, by extrapolation from hillside outcrop into river beds, landslips, quarries, mines and drill cores. For many practical applications (e.g. for house builders, agronomists and lowland civil engineers) it may not matter what the bedrock is like

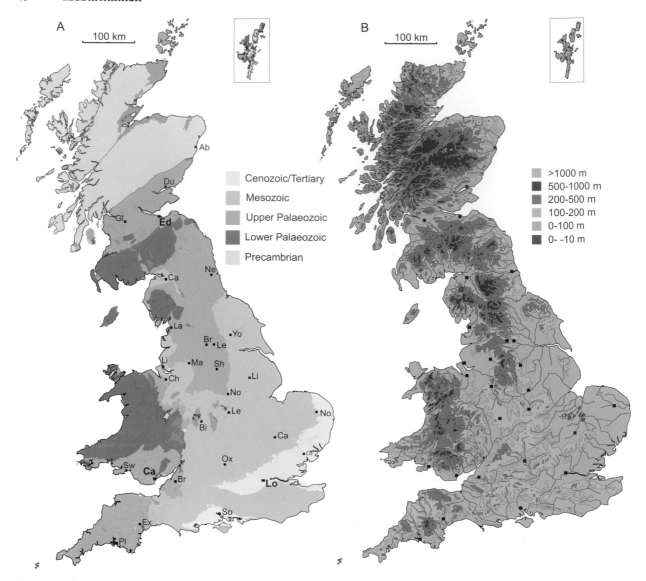

Figure 5.5 Generalized geological and topographic maps at approximately the same scales. **A** 'Solid' geological map generalized to show geological eras. **B** Topography. Data generalized from British Geological Survey (2007a, b) and *The Times Atlas of the World* (1987) sources.

at all; it is more important to know the nature of local soil or near-surface 'drift' including glacial and former permafrost-weathered materials.

Perhaps the best (and completely free!) way to appreciate the different kinds of geological maps available, and, importantly, their relation to topography and landscape, is via the aforementioned (Chapter 2) BGS website and its splendid 'Geology of Britain Viewer'. Having selected an area (it's interesting to choose your home or birthplace) the geology and topography can be viewed separately or in combination at a 1:50,000 scale (2cm per km). By choosing 'bedrock' or 'superficial' viewing options the geology appears correspondingly.

Geology and Topography

Taking the broadest possible view of the Island's topography and geology (Figs. 5.5, 5.6) it is obvious that the western parts are generally higher and older than those to the east. A crude characterization of these west-to-east topographic trends defines a wedge-shaped form whose steepest side to the west faces the Minches, Irish Sea and Celtic Sea. The remaining portion slopes more gently eastwards towards the margins of the North Sea. These broad trends are explicable if large-scale west-to-east crustal tilting has taken place in the recent geological past. A favoured culprit for this is the addition

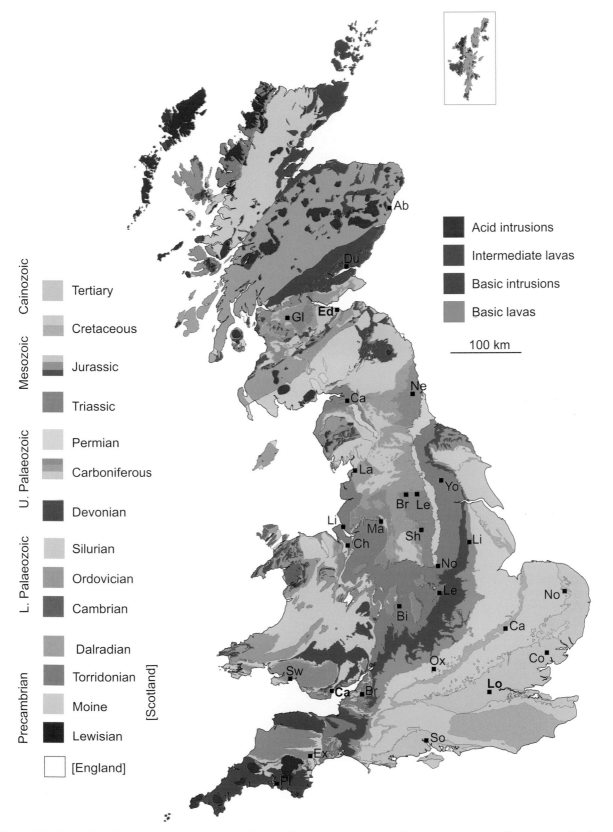

Figure 5.6 Generalized geological map of the Island with the major eras divided into their constituent periods. Data generalized from British Geological Survey (2007a, b) sources.

of igneous rocks into the deep crust and mantle under Palaeocene (*c.*60Ma) volcanic centres from Northern Ireland to the Inner Hebrides (Chapter 7). As these solidified, their relative buoyancy with respect to the lower crust and mantle caused permanent uplift. The general eastward decline in elevation has been accentuated by the persistence over some 200Myr of crustal subsidence ('sagging') in the North Sea sedimentary basin. This now saucer-shaped area was formerly a rift valley (like that of modern Kenya). As the hot mantle underneath it gradually cooled it shrank and subsided to form the basin into which the rocks of eastern England consistently tilt.

Superimposed upon these general trends are regional uplands: the Scottish Highlands, Southern Uplands, Lake District, Snowdonia, Cambrian Mountains and Pennines. Within these, local variations in erodibility can cause severe local gradients as strata of contrasting hardness crop out, as in Pennine gritstone 'edges' (Fig. 5.7A). Larger massifs often comprise blobby masses of more resistant granitic rocks formed as intrusions into less-resistant metamorphic or sedimentary rocks. Such are the Cairngorms, isolated mountains in the Southern Uplands and Galloway (Cheviot, Merrick, Fleet and Criffell), Lake District (Carrock Fell (Fig. 5.7B), Shap) and south-west England (Dartmoor, Bodmin Moor, Carnmenellis, St Austell, Land's End). Ridges formed by ancient volcanic flows interstratified with sedimentary rocks define the most rugged parts of the formerly volcanic Hebridean Islands (Skye, Mull, Arran), Lakeland and Snowdonia, and account for the greatest relief in the Ochil and Sidlaw Hills of Midland Scotland.

Lowland areas also have isolated rocky ridges and peaks set amongst otherwise pastoral vales. These often lie along major fault lines bordering former rift basins, the more resistant older igneous and metamorphic rocks having been uplifted along them. Examples are the Long

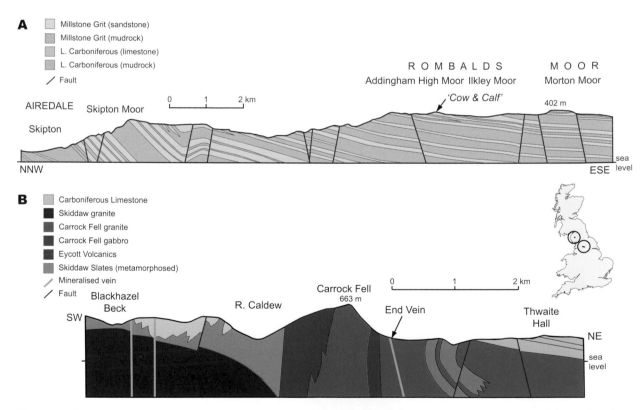

Figure 5.7 Geological sections to illustrate the influence of rock-type on weathering and landscape forms. **A** South Pennines Millstone Grit landscapes are dominated by the resistance to weathering of well-indurated sandstones compared to softer mudrock interbeds. **B** Lakeland contains numerous igneous intrusions like Carrock Fell, whose harder and more resistant crystalline rocks stand out proud from the surrounding 'country rock'. The fells to the west are high because they are underlain by the Skiddaw Granite whose intrusion baked and hardened the Skiddaw Slates. Data generalized from British Geological Survey sources.

Mynd in Shropshire, the Malverns in Worcestershire, and Charnwood in Leicestershire. Particular landforms in lowland Mesozoic southern and eastern England are linear scarplands and downlands with chalk, limestone, ironstone or sandstone bedrock, that define locally hilly terrains like the Cotswolds, Chilterns, North and South Downs, Kent and Sussex Weald, Lincoln Edge and the Cleveland Hills. Such linear scarps, though subtle compared to the mountainous ridges of the western and northern uplands and mountains, are made more obvious by their juxtaposition against dead-flat vales and dales grounded in softer mudrock and through which flow major lowland rivers. The linear arrangement of scarps and vales is due to the outcrops following the trends established by regional tilting and folding.

All the above-mentioned linear ranges, mountain massifs, upland ridges and scarplands, are explicable in terms of their different erodibility. Hillslope erosion, river and glacier erosion has emphasized the contrasts between adjacent rock types previously brought together by simple succession or by tilting, folding, faulting and the intrusion or extrusion of magma. In general, igneous and metamorphic rocks, well-cemented sandstone, limestone and permeable/porous chalks are more resistant to erosion than softer mudrocks, clays, marls and slates.

Chapter 6

'Deep Time' and the 'Mobile Earth'

It is perhaps a little indelicate to ask of our Mother Earth her age, but Science acknowledges no shame and from time to time has boldly attempted to wrest from her a secret which is proverbially well-guarded.

The first few eloquent lines of Arthur Holmes's preface to his epoch-making book of 1913 (see discussion below). He goes on to reference his 1911 seminar at Imperial College, London (then the Royal College of Science) in which he first publicly outlined his revolutionary findings: the first reliable radiometric dating. What it must have been like to be in the audience at that seminar!

A Procession of Successions

As in human history, geological history is concerned with the timing of events, their succession and linkage to contemporaneous events. Hence, a succession of sedimentary layers, strata or beds may be established whose deposition (as Steno first realized) proceeded from oldest below to youngest above. Take a specific example from Dyfed, south-west Wales (Fig. 6.1). The featured strata are from the Lower Carboniferous period of the Upper Palaeozoic era. The name 'Carboniferous'

Figure 6.1 The 25m high natural arch of the 'Green Bridge of Wales' (51.610416, –4.998500), south of Castlemartin, Dyfed, SW Wales. Photo: Wikimedia Commons©W.Lloyd MacKenzie, via Flickr. Saffron Blaze. The cliffed outcrops of north-tilted light-grey Carboniferous Limestone define a prominent 40m elevation coastal platform. Its entire coastal stretch from Linney Head to St Govan's Head forms part of the southern limb of a downfold (syncline) formed by crustal compression and shortening towards the end of the Carboniferous.

celebrates the abundant coal in somewhat younger strata of this period in Europe, Russia and North America. The fossils (brachiopods, corals) found here enable correlation with far-flung deposits of the same age as far apart as Kentucky and the Kun Lun terrane of northern Tibet. As discussed in Chapter 5, the overall stratal succession across the Island has been patiently put together by geological fieldwork, the strata divided up into periodic chunks (Chapter 3) and named after locations, salient features and their fossil content.

The succession-led division of geological time was confirmed when independent evidence for the younging direction of strata (their 'way-up') was found. This came from what are known as 'sedimentary structures'. These are often caused by sediment deposition from ancient water or wind currents, like ripples or dunes visible on modern tidal flats or beaches, and whose triangular shapes were preserved on the tops of beds. Others show distinctive patterns of what is called cross-stratification: declining downward-sloping laminations formed by the migration of the same ripples and dunes. Downward-tapering mudcracks due to the surface desiccation of muddy sediment also occur. Some strata/beds exhibit graded bedding where grain size changes from coarser at the base to finer at the top, implying that the depositing current waned with time.

All such sedimentary structures independently demonstrate the correct younging ('way-up') of the strata that contain them. But in the early twentieth century, field evidence from strata in the French Alps established that overfolding had occurred during the severe tectonic compression that caused the mountain-building there. This kind of folding occurs when strata are not only bent into wave-like forms with culminations (upfolds/anticlines) and depressions (downfolds/synclines) but are folded over on themselves (Fig. 6.2), rather like a crumpled sheet or a folded omelette. Such overturned folds (*aka* recumbent folds) may be pushed along a faulted slide-surface over their overturned limbs. They were named 'nappes' by M. Bertrand, a pioneer French Alpine geologist, the word meaning literally a folded tablecloth. The faulted slide-surface involved in such tectonics is a thrust fault of the type spectacularly developed in north-west Scotland as the Moine Thrust (Chapter 20). Giant overfolded structures discovered during fieldwork by solicitous use of 'way-up' criteria were later discovered in the Highland Border Zone of Scotland (most spectacularly the Tay Nappe; Chapter 20) in the 1930s–50s.

Figure 6.2 An overturned anticlinal fold at Sandgreen, near Gatehouse of Fleet, Scotland (54.829059, −4.217340). Compression has squeezed and buckled the well-bedded sandstone strata into an upfold, pushing them over to the left (north) so that the strata on the left side of the fold are now inverted (turned 'upside-down') by about 15 degrees.

Unconformities

Hutton's *Theory of the Earth* (Chapter 3) envisaged great cycles of mountain formation, their destruction by erosion, subsequent deposition on top of the eroded remains, then renewed uplift and so on. Hutton realized that the erosion of former mountains reveals the stratal disruption (tilting, folding, faulting) caused by their formation. Over later years of arduous fieldwork (1785–1795), by now in his sixties, he discovered examples of disrupted strata overlain by younger, less-tilted strata at three key localities. First discovered was that on the Isle of Arran with two more later (conveniently closer to Edinburgh) at Jedburgh and Siccar Point in the Scottish Borders (Fig. 6.3). Each featured older, steeply tilted strata (his 'schistus') overlain by younger, only slightly tilted, red-brown sandstone strata whose lowermost parts just above their junction with the schistus contained eroded fragments of that rock. Hutton

Figure 6.3 The Siccar Point unconformity (55.9317, −2.3009). Photo: Wikimedia Commons ©Dave Souza.

and friends (including John Playfair; see Chapter 3) approached Siccar from the north by boat at high tide. He wrote:

> at Siccar Point, we found a beautiful picture of this junction washed bare by the sea. The sand-stone strata are partly washed away, and partly remaining upon the ends of the vertical schistus; and, in many places, points of the schistus are seen standing up through among the sandstone, the greatest part of which is worn away. Behind this again we have a natural section of those sand-stone strata, containing fragments of the schistus …

Such discoveries defined the beginning of modern geology. 'Unconformity' was the term Hutton's successors (John Phillips, Roderick Murchison and others) chose to call the arrangement of younger, less-tilted 'sand-stone' strata resting with erosion upon older, more-tilted strata. To Hutton they were proof of the occurrence of 'revolutions' in the evolution of the earth. As he saw it, the older strata (now known to belong to the Lower Palaeozoic era) were originally deposited underwater on the very-gently sloping sea floor. As the loose sediments were gradually buried they lithified (converted into rock), chiefly by elevated temperature in the deeper crust. The now-solid sediment was subsequently tilted, folded and uplifted above sea level by tectonics during the phase of mountain-building now known as the Caledonian orogeny. It was subsequently eroded down and the remaining landscape overlain by the deposits of younger waterlain sedimentary strata (now known as the Devonian 'Old Red Sandstone'). Hutton thus viewed the geological past as having been broken by periodic such upheavals, when mountains were born from oceans where sedimentary deposition had previously occurred. An unconformity thus marks a period of 'missing' time at the junction of the unconformity, analogous to a historical interregnum, a time-gap between successive governments or monarchs.

Subsequent to Hutton's discoveries other unconformities were found by nineteenth-century fieldworkers. These enabled the identification of further geological interregna. Several were discovered by John Playfair and Lord Webb Seymour in the years immediately after Hutton's death. The most spectacular, in North Yorkshire, features Carboniferous Limestone resting unconformably upon Siluro-Ordovician sandstones and slates (Fig. 6.4). Then it was discovered that in South Wales and around the Mendip, the same limestones had themselves been pushed up on their ends by tectonics, eroded and unconformably overlain by near-horizontal Mesozoic strata (Figs. 3.1, 3.6). This led to the recognition of the Hercynian (Variscan) tectonic mountain-building event towards the close of the Carboniferous period.

Arthur's First Date

100 years or so after Playfair rewrote the Huttonian message (Chapter 3), geological time was calibrated: 'real' ages were established for ancient rock successions that had hitherto been only relatively determined, i.e. older or younger. This fundamental breakthrough followed from the discovery of radioactivity generated by the spontaneous decay of uranium isotopes in certain rock-forming minerals and the elucidation of the physical laws that govern the isotopic rate of their decay to stable lead. Previous to this many indirect (and incorrect) estimates for the age of the Earth had been made.

Figure 6.4 The Playfair–Seymour unconformity near Horton-in-Ribblesdale, North Yorkshire (54.123500, –2.310261). **A** East-facing outcrops at Moughton Nab and in the quarries beneath (view is *c.*1km across) show horizontal Carboniferous Great Scar Limestone overlying north-tilted Silurian sandstone and cleaved slaty mudrock clearly visible in Dry Rigg quarry to the left. **B** View underground in White Scar Cave, Ingleton (54.165626, –2.441345) of the same unconformity. The walls of the *c.*2m high passage are formed from brown-stained Silurian slaty mudrocks whose cleavage dips steeply leftwards. The roof rock is horizontal Great Scar Limestone.

The measurements were by a Ph.D student who was to become arguably the greatest of twentieth-century geologists, Arthur Holmes of Gateshead.

Holmes' first result, made in the chemical laboratories of his supervisor, R.J. Strutt, at Imperial College, London was from a carefully chosen rock sample from south Norway. It came from part of an igneous intrusion into fossiliferous sedimentary strata that contained Lower Devonian fossil fish. The igneous intrusion therefore had to be younger than these fossils and their entombing strata. After crushing and chemical treatment of its powdered minerals Holmes determined the rock's age as 370Ma (within some small error margins). This is within the span nowadays recognized as the time limits of the Devonian period, 420–360Ma: appropriately in its uppermost part. Holmes' pioneering result was wonderfully accurate and a tribute to his painstaking analytical techniques. It truly helped to change the world view: not a bad outcome for a young research student.

Holmes's earliest radiometric timescale incorporated further age results that he obtained from Carboniferous, Ordovician and Precambrian igneous rocks. It was published in 1913 (Fig. 6.5) whilst he was in southern Africa employed in mineral prospecting. His date for the Precambrian samples was in the range 1400–1600Ma. So, one year before the cataclysm of the First World War, humankind learnt that their planetary crust was unimaginably older than ever realized. Such talk of hundreds or thousands of millions of years bewildered and alarmed many people. Here was the response of geologist and author A.E. Trueman in 1949:

> And if there are readers who would say that scientists have no right to talk in terms of those inconceivably long periods we may point out that the geologist is as well able to appreciate the hundreds of millions of years representing the age of the earth as the Chancellor of the Exchequer is able to appreciate the total of his budget.

Geologists nowadays have four and a half billion years to play with, courtesy of the sophisticated techniques of modern geochemistry; modern Chancellors (still male at the time of writing) talk in their trillions, or, rather, our trillions.

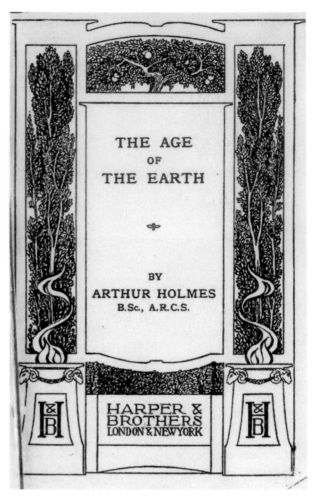

Figure 6.5 The frontispiece to Holmes's pioneering book of 1913, the Art Nouveau design suitably *avant garde* for such a revolutionary volume.

Alfred's *'Verschiebungen'*

The Island's geological story cannot be divorced from further startling discoveries in what came to be known as 'deep time' (see further below). Charles Lyell himself had toyed with the idea that the distributions of land and sea might have changed over time due to surface uplift and subsidence during Huttonian 'revolutions'. He proposed that there were former 'land bridges' between continents that periodically did this. The idea was widely used to explain the distribution of fossil and extant species and the existence of distinctive faunal provinces, e.g. the uniqueness of certain island faunas and floras, like those of Australasia.

This all changed at the same time as Arthur Holmes was compiling and gleaning his radiometric dates for publication. Alfred Wegener, a German meteorologist, polar explorer and brilliant polymath, was proposing in lectures and papers a radically new and coherent theory. This involved not just the ups-and-downs required for 'land bridges' but for more fundamental horizontal crustal displacement; continental 'Verschiebung' in German, widely translated into the more subtle English as 'drift'. His theory came from two strands of evidence. Firstly, that the present continents could be rearranged by eye to fit snugly together into a former late-Palaeozoic supercontinent that he named Pangea. Secondly, that this rearrangement enabled many geological and fossil similarities and trends to be recognized across the formerly conjoined terranes. The corollaries were startling: that the continents had not only drifted apart since the relatively young geological age of around 200Ma but that the world's oceans were therefore geologically young, and of necessity, quite distinct.

Wegener's theory was first published in book form in 1915 (he was a wounded soldier in Belgium at the time) and quickly ran through several revised editions. Some well-travelled and well-read geologists (most notably A. du Toit in South Africa) strongly supported these conclusions; many did not. A great problem was why, mechanically-speaking, should the continents move around in the way that Wegener envisaged? Another Arthur Holmes masterpiece published in 1929 gave a thermo-mechanical answer.

Holmes knew from his work on radioactive decay that the Earth was not just a cooling body losing heat by conduction. Heat produced by radioactive decay could not be dissipated so: the process was too slow. It could only dissipate if there was another heat-loss mechanism at work. Holmes argued that this could only be the action of convection currents in the Earth's mantle. These could push and pull Wegener's drifting continents around during their great Verschiebungen. Holmes envisaged a continental crust underlain everywhere by a continuation of the basaltic layers of the oceanic crust, both resting on slowly-convecting upper mantle that rose under the mid-ocean ridges and descended to melt under oceanic trenches. Thus were old oceans closed up and new oceans expanded in tandem as great supercontinents like Pangea formed, broke up and reformed.

Wegener's and Holmes's proposals concerning continental mobility were controversial for decades (see the nice review by Hallam, 1989). They eventually gained widespread acceptance by the ground-breaking discoveries of post-Second World War continental and marine geophysics. From these were born the theory of plate tectonics and the acceptance of the twin concepts that heralded the twentieth-century geological revolution – 'Deep Time' and 'Mobile Earth'. The former is an evocative phrase for the great antiquity of the Earth, first used by author John MacPhee in his 1981 book Basin and Range that charted his geo-journey of discovery in the western USA. The latter is R.A. Daly's eponymous and influential book of 1926 in which he cautiously explored the implications both of Holmes's timescale and Wegener's theory of continental drift. Today, few of us are unaware of plate tectonics: its tenets underpin our entire understanding of how the outer Earth works and, as we wrote at the beginning of our Preface, its name and actions have entered into the spin of everyday communication worldwide.

Cool Pacemaker

The final peg upon which we hang modern geological explanation is the orbital theory for periodic climate change. Lyell had originally used the Earth's 'wobble' about its spin-axis to help explain what he saw as the undoubted geological fact that climate change had occurred in the past. The French physicist d'Alambert had previously proposed that this 'wobble' caused the 'precession of the equinoxes' every twenty or so millennia. Lyell cautiously wrote in the first edition of his Principles of Geology that, as a consequence of this: 'The two hemispheres receive alternately, each for a period of upwards of 10,000 years, a greater share of solar light and heat. This…must sometimes tend to increase the extreme of deviation which certain combination of causes produce at distant epochs…' In later editions (from 1834) he also discussed the possible influence on climate of the planet's changing orbital path around the sun (known as eccentricity: the changing elongation of orbit), a link first mentioned by John Herschel and developed by Jean Adhémar. James Croll considerably furthered the argument in the 1860s as a possible cause of the Ice Ages that Louis Agassiz had proposed from geological evidence in the Alps and elsewhere.

By the twentieth century, orbital theory had been refined by the additional discovery of variations in the tilt of the Earth's axis. This enabled Milutin Milankovitch to calculate the combined orbital contribution (from precession, eccentricity and tilt) to the intensity of incoming solar radiation during northern hemisphere summers over time. His results were published in the 1920s, when Wegener and the climatologist Koppen, his father-in-law, also became interested in the problem.

Milankovitch suggested a strong influence by the combined orbital phenomena as a driver for the Ice Ages.

In 1975 J. Hays, J. Imbrie and N. Shackleton published geochemical results on oxygen isotopic composition from deep-sea cores containing fossil calcareous microplankton that confirmed the periodic timing predicted by Milankovitch's orbital mechanisms. In their own evocative phrase, fit to accompany 'Deep Time' and 'Mobile Earth' as one of the iconic mottoes of modern earth sciences, they had determined the 'Pacemaker of the Ice Ages'.

We can now see more clearly something that Wegener first broached: that major earth cycles must proceed in parallel. The moving plates of the solid outer Earth create new oceans and destroy old ones to form mountain ranges. These in turn must interact with the global weathering cycles of atmosphere and ocean controlled by orbital events. Historical geology now uses (since the 1980s) such causal explanations widely: past plate tectonics, climate, erosion and deposition all determine the geological history of crustal rocks and the chemical evolution of the atmosphere and oceans. It is a development and an outcome that one feels James Hutton would have been thrilled to witness. It remains controversial as to whether James Lovelock's Gaia hypothesis, that of an entirely self-governing planetary system, can be upheld by the evidence from geology's 'Deep Time'.

Coda

So, with more than a century of radiometric dating and geological enquiry behind us, we can revisit Hutton's Siccar Point unconformity (Fig. 6.3) in the company of Clarkson and Upton's (2010) guide to the history of the vanished Iapetus Ocean. The older, folded/tilted 'schistus' strata are of Middle Silurian age, deposited under oceanic currents that also collected swirls of dead planktonic graptolite creatures adjacent to an oceanic trench around 430Ma. The younger, upper strata are the Upper Old Red Sandstone of late-Devonian age, around 370Ma. They were deposited as sands and gravels in shallow shifting river channels with primitive jawless and boney-scaled fish swimming amongst migrating bars, ripples and dunes. The channels flowed between inundated and periodically desiccated muddy floodplains.

The geological interregnum at Siccar is thus around 60Myr, roughly the length of time since dinosaurs last roamed our planetary surface or since the Hebridean region exploded with fiery volcanism. During this long interval the future Southern Uplands formed as plate subduction scraped off heaps of sediment against the ocean margin to form an accretionary prism, as in modern Sumatra. This was further scrunched together as the Anglo-Welsh continental crust and Scandinavia shuddered into the Scottish landmass during Middle Devonian mountain-building. The whole lot was then indurated and pushed above sea level as part of the Caledonian mountain chain. After some 35Myr of erosion, the mountain chain had been reduced to near sea level once more as a rift province established itself in southern Scotland and northern England: swift-flowing rivers depositing their eroded sediment over the riverine plain of the Scottish Border basin. Meanwhile, far away in polar Gondwanaland, a continental ice sheet was beginning to grow upon the southern continent. This would soon cause major sea level variations and global climate change – part of the story of the whole Island explored in our next theme.

PART 3
Remembrance of Things Past

We all benefit from some knowledge of human history and the way it has run out: it prepares and warns us of dangers that might arise if history's lessons are not learnt. It is the same with geological and landscape history: the record of how our crustal foundations have been laid down over time can lead to wisdom concerning the impact of present and possible future scenarios. Chapter 7 takes on the entire history of the Island, from its murky beginnings detected in our oldest rocks in Assynt (*c.*3Ga), right down to those first Pleistocene ice invasions of East Anglia almost 500ka and our current ascent in the warm Holocene. It is a story of lost mountains and oceans, extinct fiery volcanic island arcs, former blistering hot deserts and warm sunlit semi-tropical carbonate platforms. Then the violent spasms of new oceanic rifting in the Iceland–Faroes gap, accompanied by a positive storm of Hebridean volcanism. Much of the information is presented in global and regional maps and reconstructions.

It was the rifting of Avalonia and Baltica terranes from Gondwana in the Cambrian that led to the creation of the Iapetus Ocean.

The volcanic ashes of late-Precambrian Charnwood have yielded what is arguably one of the single most important fossils ever found in the Island's rocks: the sessile marine creature *Charnia*, the first discovered anywhere. It is Europe's sole representative of the earliest multicellular life assemblage later discovered in greater abundance in a far-off part of Gondwana – modern South Australia. Its discovery, by two Leicester sixth-formers in the 1950s, is a fascinating example of the serendipity of both discovery and happy endings. Look it up online.

The Iapetus and the Caledonian Mountains (600–390Ma)

During the Iapetus' lifetime its salty waters hosted multicellular life -forms that evolved into trilobites, graptolites, corals, molluscs and brachiopods. Along eastern Laurentia (Fig. 7.1), across what are now the Grampians and Argyll, Dalradian sedimentary rocks were deposited. In the Inner Hebrides sedimentation of the thick Jura Quartzite by tidal shelf sand streams was followed by rapid subsidence. Later eruption (around 600Ma) of the Tayvallich Volcanics signalled formation of the juvenile Iapetus, its deepening waters soon receiving floods of detritus introduced as turbidity currents.

A notable interval within the oldest Dalradian is the Port Askaig Group on the island of Islay. Close comparison of its mixed-up boulders, pebbles, granules and silty mudrocks can be made with the modern deposits of grounding tidewater glaciers. The varied clasts were glacial erratics, gleaned from far and wide and dumped on Iapetus' low-latitude coasts by ice streams and icebergs. Similar deposits occur worldwide in rocks of this age. The 'Snowball Earth' theory attributes them to a series of global glaciations and deglaciations. The youngest Dalradian of the Southern Highland Group contains limestones with Lower Cambrian-age trilobites whose affinities lie with North American examples in Newfoundland. In Assynt the shallowest waters of the Dalradian shelf deposited the Cambro-Ordovician Durness Group, including well-preserved limestones, again with fossils of New World affinity.

By Cambrian times the western side of Gondwana, over today's central England and Welsh Borderlands, was a shallow-water shelf whose faunas were strikingly Gondwanan rather than Canadian types. In the Harlech Dome of North Wales a deeper water basin with turbidite deposits was surrounded by land on three sides. During the early Ordovician plate consumption by subduction began along Iapetus' northern continental margin (Fig. 7.2). The Dalradian succession began to deform as smaller continental terranes were pushed by plate forces to dock onto its oceanwards margin. The result was intense compression and the formation of the Grampian mountain belt, which welded onto and re-deformed the Moine mountains. The deformation took the form of gigantic overthrust folds (nappes; see Chapter 5) witnessed by large areas of 'upside-down' strata in the Highland Borders.

Meanwhile, across the other side of Iapetus, Avalonia/Baltica had rifted away from Gondwana. Deep-water muddy sedimentation formed the Skiddaw Group of the northern Lakes and the Isle of Man. Uplift in the mid-Ordovician saw initiation of massive submarine slides as Avalonia's margin became a site of active subduction of the Iapetus plate. This generated the arc and back-arc volcanics that largely define the rugged modern scenery of the central Lakes and of Snowdonia. Initially came prolonged and thick (> 2km) accumulations of lavas. Then more silica-rich magmas caused explosive eruptions from giant circular depressions (calderas) bounded by arcuate faults. These caused statosphere-high plumes of ash, eruptive surges and flows of hot ash that filled up large crater lakes. Below individual calderas (half a dozen have been mapped out in the Lakes) great volumes of magma remained in place. These slowly crystallized into granitic rocks that comprise the Lake District batholith. Associated circulation of heated groundwaters scavenged metallic elements like copper and lead and slowly precipitated them in cracks as mineral veins during their briny passage upwards.

Magmatism ceased in later Ordovician and Silurian times. Along the northern Iapetus margin slow northward subduction scraped off sedimentary detritus to form a linear wedge of thrust and folded sediments, an 'accretionary wedge' that is now the Southern Uplands of Scotland (Fig. 7.3). The final elimination of Iapetus was accomplished between the late Silurian and middle Devonian as Baltica/Avalonia docked into Laurentia. It was accompanied by widespread compression and thrust faulting (notably the Moine Thrust, see below) and folding. Heating under northern England and much of Scotland south of the Great Glen led to widespread magma production, forming many granitic intrusions (Cairngorms, Galloway) and explosive lava eruptions (Cheviot, Midland Valley). The Caledonian mountain

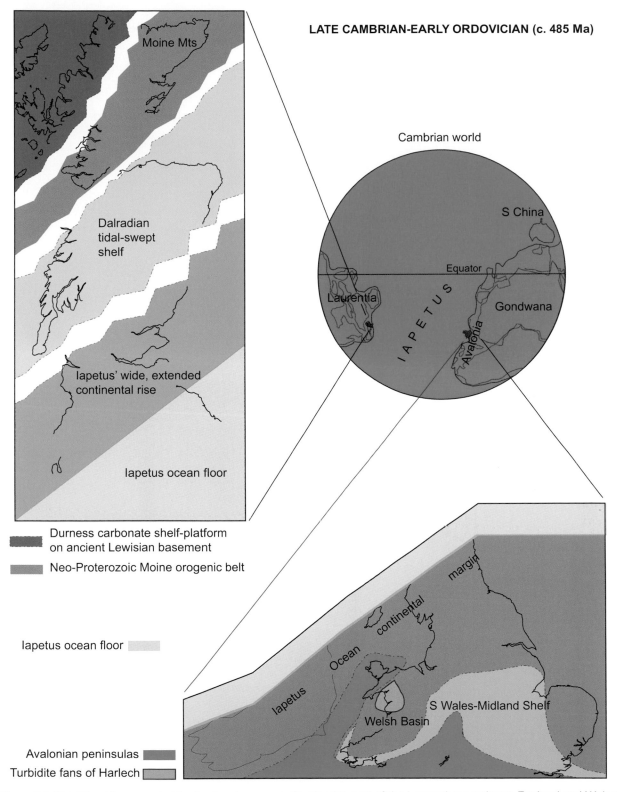

LATE CAMBRIAN-EARLY ORDOVICIAN (c. 485 Ma)

Moine Mts

Dalradian
tidal-swept
shelf

Iapetus' wide, extended
continental rise

Iapetus ocean floor

Cambrian world

S China

Equator

Laurentia

I A P E T U S

Gondwana

Avalonia

Durness carbonate shelf-platform
on ancient Lewisian basement

Neo-Proterozoic Moine orogenic belt

Iapetus ocean floor

continental margin

Iapetus Ocean

S Wales-Midland Shelf

Welsh Basin

Avalonian peninsulas

Turbidite fans of Harlech

Figure 7.1 The Island is separated by the Iapetus ocean; Scotland is part of the Laurentian continent, England and Wales are at the western margin of the Gondwanan continent. The Durness Shelf features shallow-water sedimentation. We have little idea how the Moine mountains relate to either Dalradian sedimentation or to the Assynt foreland. Data from many sources, chiefly Anderton et al. (1979), Woodcock and Strachan (2000), Torsvik et al. (2002), Trewin (2002), Brenchley and Rawson (2005).

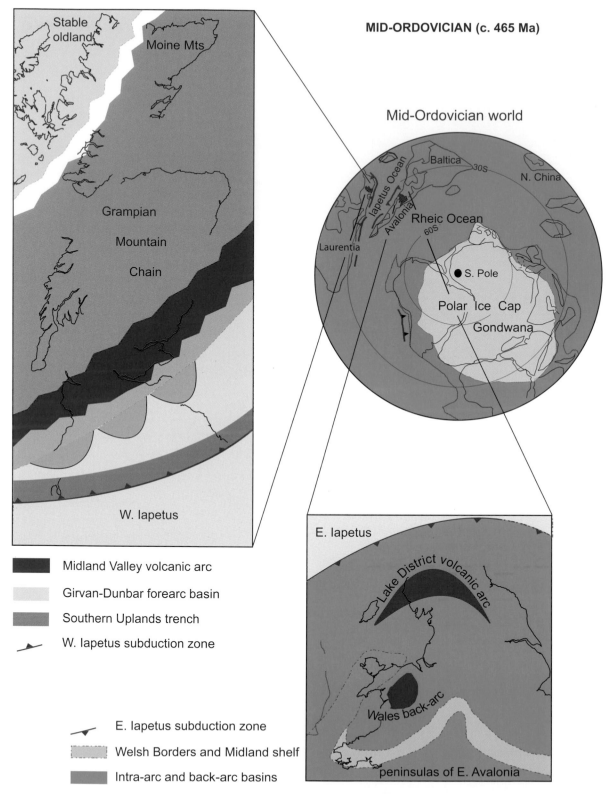

MID-ORDOVICIAN (c. 465 Ma)

Stable oldland

Moine Mts

Grampian

Mountain

Chain

W. Iapetus

Mid-Ordovician world

Baltica

30S

N. China

Iapetus Ocean

Avalonia

Rheic Ocean

60S

Laurentia

• S. Pole

Polar Ice Cap

Gondwana

■ Midland Valley volcanic arc

Girvan-Dunbar forearc basin

Southern Uplands trench

W. Iapetus subduction zone

E. Iapetus subduction zone

Welsh Borders and Midland shelf

Intra-arc and back-arc basins

E. Iapetus

Lake District volcanic arc

Wales back-arc

peninsulas of E. Avalonia

Figure 7.2 The Iapetus and Rheic oceans border the Baltica terrane whose extraordinary peninsula, Avalonia, is rooted in the late-Precambrian crust of England and Wales. The Iapetus is closing, with subduction zones on both margins, and with the Grampian mountains bordering the Midland Valley volcanic fore-arc. The Southern Uplands are just beginning to form as off-scrapes along a deep trench. Lakeland and Snowdonia are the sites of explosive volcanism. Data from many sources, chiefly Anderton et al. (1979), Woodcock and Strachan (2000), Torsvik et al. (2002), Trewin (2002), Brenchley and Rawson (2005).

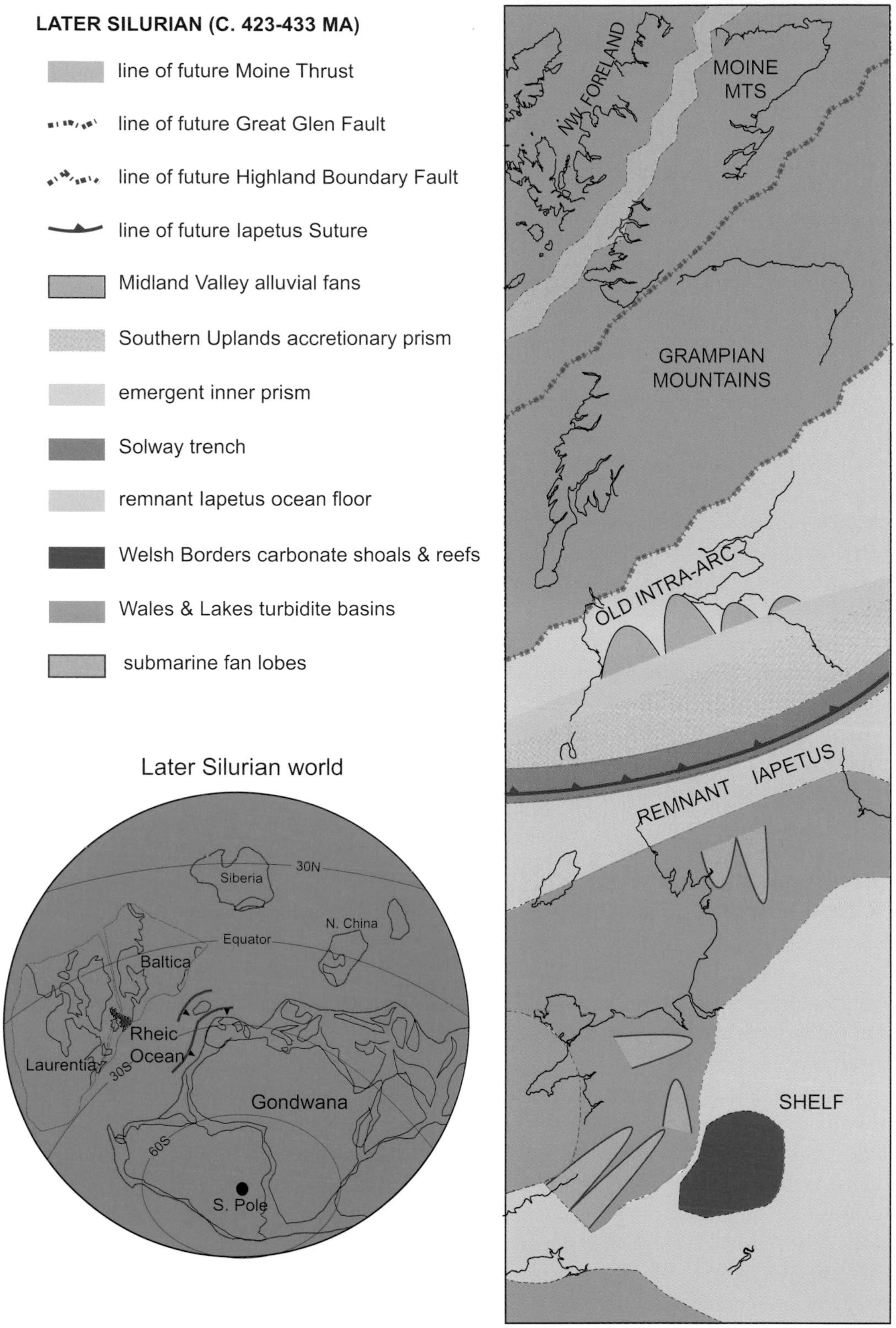
LATER SILURIAN (C. 423-433 MA)
line of future Moine Thrust
line of future Great Glen Fault
line of future Highland Boundary Fault
line of future Iapetus Suture
Midland Valley alluvial fans
Southern Uplands accretionary prism
emergent inner prism
Solway trench
remnant Iapetus ocean floor
Welsh Borders carbonate shoals & reefs
Wales & Lakes turbidite basins
submarine fan lobes
Later Silurian world
Siberia
30N
N. China
Equator
Baltica
Rheic
Ocean
Laurentia
30S
Gondwana
60S
S. Pole
NW FORELAND
MOINE
MTS
GRAMPIAN
MOUNTAINS
OLD INTRA-ARC
REMNANT
IAPETUS
SHELF

Figure 7.3 Subduction grinds to a halt. The Midland Valley contracts over a deep thrust fault that hides all trace of its old fore-arc. The northern edge of the Southern Uplands is emergent above the remnant of Iapetus, sending riverine sediments northwards. The deep, murky remnant oceanic waters of Lakeland, North and Central Wales are being infilled by turbidity currents. Westwards, clear, shallow waters on the Welsh Borders shelf are host to the exquisite coral reefs of the Wenlock Edge. Data from many sources, chiefly Anderton et al. (1979), Woodcock and Strachan (2000), Torsvik et al. (2002), Trewin (2002), Brenchley and Rawson (2005).

belt was the result. This amalgamated with the earlier Grampian belt and formed part of a linear mountain chain from northern Norway, stretching south-west across the Island into Ireland and eastern North America.

The Moine Thrust (Figs 7.1, 18.1) was discovered in the 1880s when Charles Lapworth and John Calloway observed Moine metamorphic rocks resting on top of the fossiliferous Durness Cambro-Ordovician sedimentary succession. They rejected the then-current Geological Survey view that it could be a normal stratigraphic relationship, proposing instead that the Moine rocks had been pushed from the east onto the Assynt foreland by an east-inclined thrust fault, their Moine Thrust (Lapworth apparently had nightmares of being trapped under the great grinding mass of rocks trundling westwards on his proposed thrust). It produced the characteristic platy-to-flaky textured and finely crystalline rock seen everywhere at the fault surface. Lapworth suggested that this had originated by the intense deformation caused by fault motion: a product of the immense kinetic energy involved. He christened the fault rock 'mylonite' from the Greek for 'mill', μύλος, *mylos*. Today, at Knochan Crag on the Sutherland/Ross-shire border, the Moine Thrust has its very own visitor centre in the North West Highlands 'Geopark'. The fault and its zone of subsidiary thrusts is widely visible around about. It is an iconic and spectacular geological sight – a tectonic redaction of time spanning almost 2.5Gy.

Final Caledonian Contractions and Rhea's Rifted Margins (390–250Ma)

Upper Palaeozoic times were extraordinary intervals in the Island's geological history. In the early Devonian, Rhea lapped its shores across southern Wales and the Borders as primitive jawless fishes swam in rivers along its coastal riverine plains (Fig. 7.4). The climate was hot and semi-arid, calcrete soils developing in the river flood-plains, just like those that form in the New Mexico desert today. Sedimentation was widely interrupted in Middle Devonian times south of the Scottish Highlands by a final violent spasm of compressive deformation within the Caledonides, the Acadian event, named after its type locality in Maritime Canada.

The most spectacular consequence of the Acadian was the formation of the Great Glen Fault. In a paper read to the Geological Society in 1939 (the Second World War delayed publication till 1946) W.Q. Kennedy proposed that the structure had a 100km transverse displacement along it, as a leftwards sense of motion. The evidence

he carefully marshalled for this startling proposal had three chief elements: that the linear fault line had cut pre-existing granitic intrusions to produce the now-separated Foyers and Strontian intrusive masses; that it had ground up and mylonitized (à la Lapworth) the Lower Old Red Sandstone; and that it displaced distinctive zones of Dalradian metamorphism. Kennedy saw it as a fundamental vertical structure rooting deep below the crust:

> It is tempting to identify the plane of weakness with the level (about 80 kilometres) at which Daly and Gutenburg have postulated a change in the condition of the earth's material from a crystalline to a vitreous state…

We see it today as Scotland's very own continental transform fault, an ancient example analogous to California's San Andreas or Turkey's Anatolian faults. A consequence was the subsidence of a giant Middle Devonian lake basin over its northern splays, itself analogous to the Ridge basin of California. The sedimentary deposits of this wider 'Lake Orcadia' include finely laminated flagstones famed for their fossil fish faunas. These were preserved *post-mortem* because foul waters in the stratified lake depths lacked oxygen, so allowing preservation of the fishy carcasses as imprints of skin and hard parts. The remains are now seen in the famous flagstone lithologies of Orkney (at Sandwick, Mey) and Caithness (at Achanarras, Rousay, John O'Groats).

Late-Devonian to early-Carboniferous times saw the onset of pervasive plate stretching that caused formation of an Island-wide rift province adjacent to the Rheic ocean basin to the south. The stretching fractured the crust into tilted blocks with widespread volcanism in the north (Fig. 7.5). Riverine sedimentation re-established itself in northern basins (the Orkneys, Caithness, Midland Valley, Borders) with marked unconformity between the Upper Old Red Sandstone and older deposits (Hutton's unconformities). Shallow saline lagoons and lakes developed in the early Carboniferous, with clayey limestones ('cementstones') and oil shales widely deposited. Over many central and southern areas of England the rift basins and their tilted bounding flanks were covered by open-water shallow seas. In these were deposited calcareous sediments derived from the calcitic shells and frameworks of brachiopods, crinoids and corals. After burial and hardening by interstitial precipitation of calcite these became the Carboniferous Limestone.

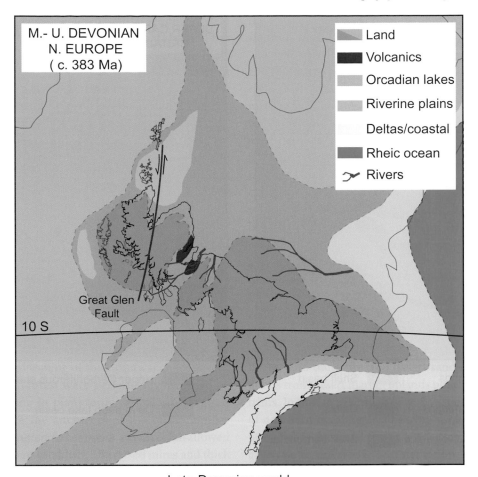

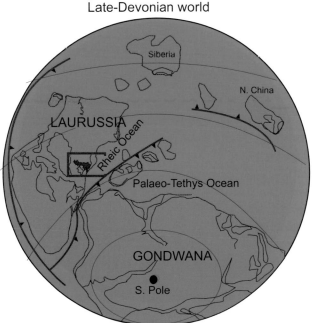

Figure 7.4 The 'Old Red Continent' borders the western margins of the Rheic Ocean. Shallow-marine sediments fringing its coast are well developed as coralliferous limestones in Devon. Elsewhere the record is mostly of riverine sediment, with large freshwater lakes in Orkney and the offshore Hebrides. Data from many sources, chiefly Anderton et al. (1979), Woodcock and Strachan (2000), Torsvik et al. (2002), Trewin (2002), Brenchley and Rawson (2005).

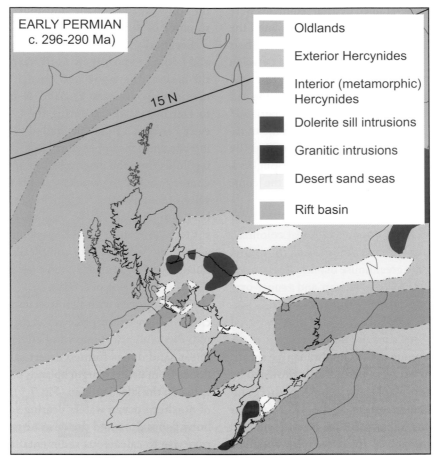

Early Permian world

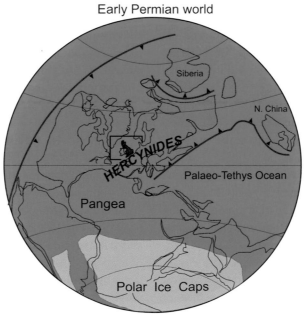

Figure 7.6 Complete closure of the Rheic Ocean creates the Hercynian (Variscan) mountains, part of a wider belt that eventually stretches from Texas to the Tien Shan. Rocky plateaux to the north are eroded and subside once more to become desert basins: the southern North Sea prominent. Data from many sources, chiefly Anderton et al. (1979), Woodcock and Strachan (2000), Torsvik et al. (2002), Trewin (2002), Brenchley and Rawson (2005).

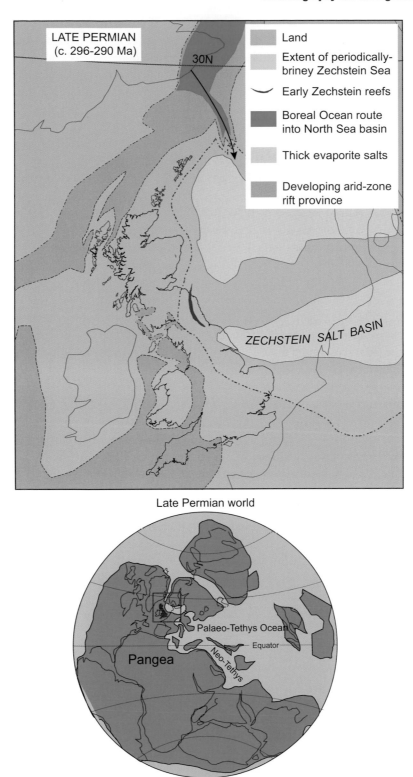

Figure 7.7 The western end of the great Pan-European Zechstein salt basin stretches from Poland to Hartlepool. There is a tenuous connection into northern seas whose deposits may be found on Spitsbergen today. Further west, a strait connects to the Palaeo-Tethys ocean. Data from many sources, chiefly Anderton et al. (1979), Woodcock and Strachan (2000), Torsvik et al. (2002), Trewin (2002), Brenchley and Rawson (2005).

During the early Mesozoic the Island was situated in the interior of northern Pangea, later bathed by tide and wave from three sources: the young central Atlantic, a Boreal ocean to the north, and the dominant swirl of Tethys to the south (Fig.7.8). Subsequently Pangea fragmented into a series of archipelagos and shoals on the eastern flank of the developing north Atlantic. The Island's rock successions enable us to follow both the birth and growth of the Atlantic and the demise of the Tethys ocean as it morphed into the Alpine–Himalayan mountain chain.

The Triassic period heralded the onset of rift tectonics as the nascent central Atlantic developed and crustal stretching propagated normal faults widely over the Pangean landmass (Appalachia, north-west Europe, north-east Africa, Biscay). Detritus flushed by rivers into the rifts of Midland England came partly from erosion of the Hercynian mountains of central France. Other detritus in the Welsh Borders, northern England and SW Scotland came from local range fronts.

Pangea was still intact in the Jurassic, though oceanic rifting had begun in the far southern central Atlantic. The growth of new mid-ocean ridges caused a global marine flooding event over the former desert rift basin province of north-west Europe as it slowly subsided as shallow-water shelf basins. Periodic plumes of muddy effluent poured into the basin floors from distant river systems, depositing fine-grained sediment that entombed the remains of abundant fossil faunas (ammonites, molluscs and marine vertebrates) forming deposits like the Lias and the younger Oxford and Kimmeridge Clays. These were sometimes rich in dissolved and particulate iron minerals and the organic blooms of planktonic organisms. After burial and heating up, some later became fertile sources of ironstone (the Lias) and hydrocarbon (the Kimmeridge Clay).

Active faulting resumed in the Middle Jurassic. Partial melting of uprising mantle caused domal uplift and basic volcanism over the Viking rift that ran along the north–south axis of the North Sea. River deltas gradually impinged on Skye and Yorkshire. In central and southern areas away from the sandy influx, shallow and clear warm waters driven locally by strong tidal currents enabled deposition of the marine calcareous sediments that, after burial, became shelly and oolitic limestones.

Sediments deposited in the Island's subsiding Cretaceous basins enable us to chart the course of the break-up of Pangea (Fig. 7.9). The split occurred in separate pulses. In the early Cretaceous, active subsidence in the North Sea and Scotto-Greenland rifts spread seawater from the proto-Atlantic northwards. A strait opened across eastern and southern England, swept by tidal currents that deposited the widespread Lower Greensand. Rifting then jumped westwards to begin the Rockall and Porcupine rifts west of Ireland. This left the North Sea stranded as a 'failed rift'. It and much of the adjacent Island subsequently underwent subsidence triggered by cooling of the mantle under the old rift. A global ocean deepening event in the mid-Cretaceous was coincident with ocean crust formation in the south Atlantic. This led to the drowning of the majority of upland relief left in the rift province, chiefly the oldlands of western parts and the London–Brabant uplands. Supply of river-borne detritus ceased, replaced by organic calcareous sedimentation from its well-lit and warm near-surface waters. This comprised platelets derived from the carapaces of minute coccolith plankton; their seasonal blooms caused prolonged but slow sedimentation of soft calcite oozes on the sea floor. After burial and partial hardening they became the extraordinary sedimentary formation known as the Chalk.

Early Cenozoic times saw new tectonic forces set up, caused by the onset of sea-floor spreading in the NE Atlantic (Fig. 7.10). Widespread uplift occurred in response to a mantle plume under the Iceland–Faroes ridge. It caused eastward tilting of northern Britain along a magmatic upwelling that caused volcanic eruptions and plutonic intrusion from the Faroes southwards to Ulster and the coast of Ayrshire (Ailsa Craig). As noted in Chapter 5, today's general increase in elevation westwards and the dramatic line of extinct Cenozoic volcanic centres in the Hebridean chain bear witness to the event. Associated stretching cracked the Island's northern crust far and wide as basaltic magma intruded from Mull to the North York Moors as the Cleveland Dyke. The uplift caused erosion and stripping off of Mesozoic strata from the western oldlands: a 'Great Western Rejuvenation', revealed by analysis of the cooling of crystals of the mineral apatite in today's sedimentary rocks in NW England and elsewhere. In all these areas ancestral rivers incised deeply into the Mesozoic sedimentary cover, often removing it completely, cutting down to the Palaeozoic core beneath. It was these elevated lands that much later were able to source the indigenous ice caps in the Quaternary (see below).

Meanwhile, in southern England sedimentation and subsidence continued in the North Sea basin, unaffected

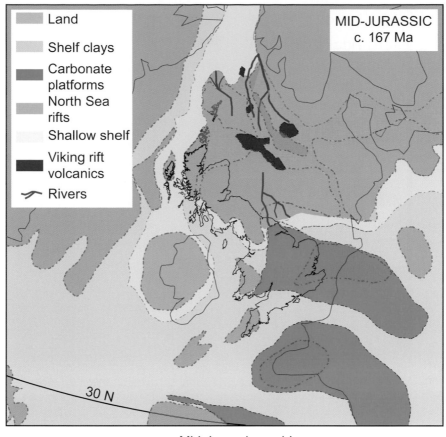

Mid-Jurassic world

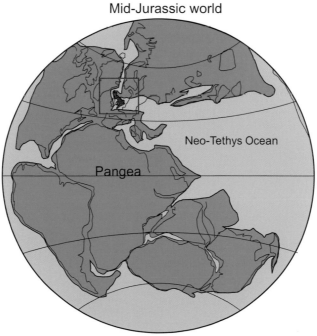

Figure 7.8 Active rifting and basaltic volcanism feature in the North Sea, with rifting along the Hebridean margins. Rivers flow both along the rift basins and outwards from the uplifting rift flanks, notably in the Yorkshire and Outer Hebrides basins. Further south, away from the influx of clastic sediment, carbonate platforms develop, swept by strong tidal currents and with iron-rich sediments locally developed. Data from many sources, chiefly Anderton et al. (1979), Woodcock and Strachan (2000), Torsvik et al. (2002), Trewin (2002), Brenchley and Rawson (2005).

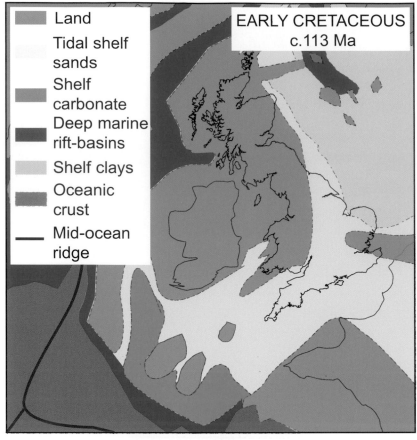

Early Cretaceous world

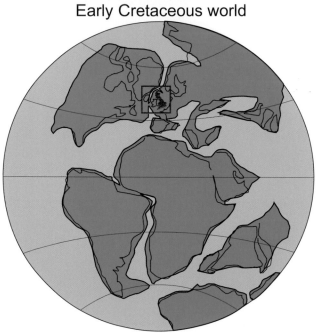

Figure 7.9 In the Early Cretaceous a tidally swept seaway joins the North Sea basin and the developing North Atlantic ocean to the south-west. In these straits tidal sand waves develop, which today define the Greensand deposits of southern England. The Viking rift continues active in the far north. Data from many sources, chiefly Anderton et al. (1979), Woodcock and Strachan (2000), Torsvik et al. (2002), Trewin (2002), Brenchley and Rawson (2005).

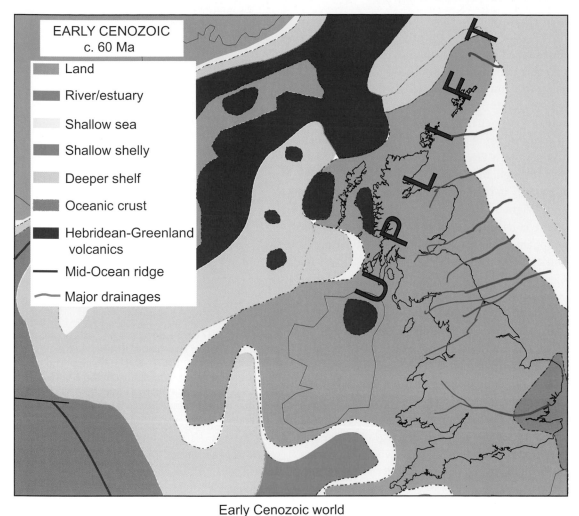

EARLY CENOZOIC
c. 60 Ma

Land
River/estuary
Shallow sea
Shallow shelly
Deeper shelf
Oceanic crust
Hebridean-Greenland
    volcanics
Mid-Ocean ridge
Major drainages

Early Cenozoic world

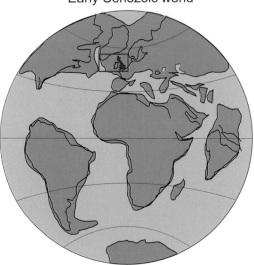

Figure 7.10 Palaeocene rifting, volcanism and uplift herald the opening up of the NE Atlantic through Eastern Greenland and along the future Iceland–Faroe gap. Data from many sources, chiefly Anderton et al. (1979), Woodcock and Strachan (2000), Torsvik et al. (2002), Trewin (2002), Brenchley and Rawson (2005).

by this northern and western mayhem. The Eocene London Clay formed as a thick deposit of estuarine muds introduced by a proto-Thames possibly sourced in the distant rising Welsh mountains. The deposit is notable for its abundant tropical fossil flora: an indication of a temperature maximum that reached its peak early in the Paleogene. By the late-Oligocene epoch, after intervals of younger coastal and river sedimentation over southern England, global temperatures had generally declined, but with seasonal aridity indicated by salt deposits in the Paris basin and calcrete soils in the Isle of Wight.

Now the consequences of the Tethys' final demise became apparent. Northward movement of the African plate put the thick accumulations of Mesozoic and early Cenozoic sediment across Europe under strong compression. As a consequence the intricate loops and recesses of the Alpine–Himalayan mountain belt gradually arose. No grand mountains were set up in the Island, but over southern England impressive regional folds (Weald–Artois anticline/London Basin syncline) and thrust faults gradually formed in the late-Oligocene epoch as uplands emerged and rivers cut deeply through into the Mesozoic substrate beneath. Thus was formed the foundation of the modern drainages that we see today, cutting through the Chalk Downs escarpments through their many gaps (Dorking, etc.) in southern and eastern England.

Few strata other than remnants of terrestrial and lacustrine strata in the Midlands exist from the Cenozoic interval between 23 and 5Ma, i.e. most of the Miocene epoch. Globally, temperatures were in slow decline. Here the whole Island suffered net erosion, detritus being transported down ancestral rivers into the subsiding coastlines and shelves of surrounding shelf sea basins (Celtic, Irish and North Seas) whose subsidence continued unabated.

The Great Cooling and the First Record of Hominid Inhabitants (5Ma–21ka)

We have now reached the late-Neogene (Pliocene) to late-Quaternary (most of the Pleistocene) interval. To begin with, climatic changes were not extremwe. Sea level fluctuated by a few tens of metres every 20–40kyr in response to the waxing and waning of polar glaciers (Fig. 7.11); the West Antarctic and Greenland ice caps were probably absent or diminished compared to the later Pleistocene. During

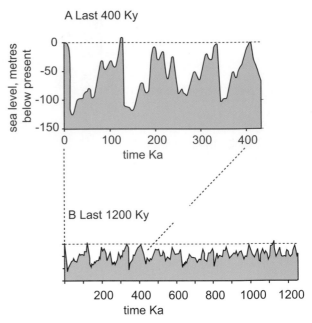

Figure 7.11 A The repeated rises and falls of global sea level during glacial and non-glacial times over the past 1.200My (after Shackleton 1987). **B** Detail from the last 400ky illustrates the saw-blade nature of the sea-level changes, with rapid rises culminating in interglacial highstands of sea level and slower punctuated falls leading to full-glaciation lowstands.

warm phases, mean global sea levels were 30–40m or so higher than today: this is when the coastal platforms visible today around much of the southern Island's unglaciated coastline were cut by marine erosion. The Pliocene rock record on the Island is scant, the majority of its land area still undergoing erosion by rivers and mass wasting. Only on the extreme borders of the southern North Sea basin do we see significant deposits: the shelly sands and gravels of the East Anglian 'Crags'. These accumulated as offshore shoals along the tidal-swept margins to the North Sea (Fig. 7.12). They rest unconformably upon slightly tilted London Clay and Chalk.

From around 800ka, the later Pleistocene epoch, climatic fluctuations became more severe, sea level changing by as much as 120m every 100kyr or so. The youngest deposits that pre-date glaciation comprise freshwater to estuarine deposits of the Cromer Forest Formation. These are very special deposits indeed because of recent fossil finds. Firstly in the 1990s a whole fossil steppe mammoth was excavated at West Runton, north Norfolk, evidently from a cooler interval in the succession. Then, more recently, a superb flint hand axe was discovered further east at Happisburgh

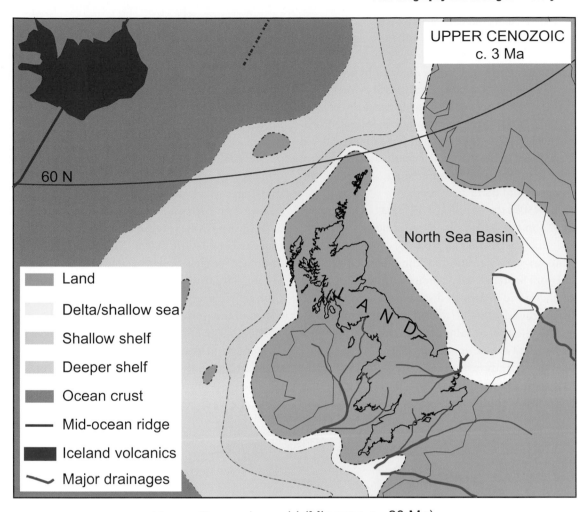

UPPER CENOZOIC
c. 3 Ma

60 N

North Sea Basin

L A N D

Land
Delta/shallow sea
Shallow shelf
Deeper shelf
Ocean crust
—— Mid-ocean ridge
Iceland volcanics
∿ Major drainages

Upper Cenozoic world (Miocene c. 20 Ma)

Figure 7.12 For much of the later Cenozoic period the Island is an uplifted neck of land, a promontory of NW Europe, legacy from Palaeocene rifting 30 million years earlier. Only East Anglia has late-Cenozoic deposits. Data from many sources, chiefly Anderton et al. (1979), Woodcock and Strachan (2000), Torsvik et al. (2002), Trewin (2002), Brenchley and Rawson (2005).

(Fig. 7.13): a *Homo heidelbergensis* owner walked the Cromerian wetlands at least 500 millennia ago. Coeval deposits further south at Pakefield contain worked flint fragments (microliths) perhaps chipped by members of the same species. The only direct fossil evidence for these first hominids recorded in the Island are in coastal deposits of the very youngest Cromerian interglacial at Boxgrove, Hampshire. Spectacular finds of teeth and a shin bone serve to identify that a family or tribal group of *Homo heidelbergensis* manufactured or finished off and trimmed flint hand axes here.

Around 450ka the merciless onslaught of Quaternary glacier ice on the Island's landscapes began. The first evidence for lowland ice is from East Anglia, hence the name Anglian Glaciation for the event. Ice streams sourced from Scotland and Scandinavia ground in over a dry North Sea bed to deposit distinctive and widespread detritus as ground moraine (till). These Anglian deposits also include the end-moraines of the Cromer Ridge and widespread meltwater outwash deposits. This glaciation extended significantly further southwards than any of the three subsequent ones. The oldest of three Anglian tills in NE Norfolk yields plentiful examples of Norwegian erratic pebbles (Fig. 7.14). The erosive power of the ice streams entering East Anglia from the north was awesome, sufficient to bulldoze away an entire 40km stretch of the former *c.*100m high Chalk escarpment of north Suffolk, Norfolk and south Lincolnshire. The remains were distributed over much of East Anglia as a thick (up to 10 metres) chalk-rich till, formerly aptly named as 'Chalky Boulder Clay'.

Eventually the Anglian ice retreated back to its sources and was gone. It left utterly changed upland landscapes – the glaciated peaks and valleys of northern and western Britain. Eastern meltwaters joined those of the proto-Rhine/Meuse in the southern North Sea. Dammed against the Dover–Boulogne chalk ridge, the combined weight of the discharge eventually breached it as a 'megaflood' whose deposits were imaged by Imperial College geophysicists (Gupta *et al.* 2007) below the waters of the present-day English Channel.

The Anglian and the two subsequent glacial periods divided Island landscapes into glaciated and non-glaciated parts, the latter a minority (Fig. 7.15), but including all of southern and south-west England. Here the landscape was permafrosted tundra during peak glacial

Figure 7.13 The superb flint hand-axe from Happisburgh, North Norfolk found protruding from the topmost Cromerian clay (but still *in situ*) after severe winter storms in 2007. Photo: ©Norwich Castle Museum. It was at least half a million years since the implement was last held snug in the warm grip of a hominid hand.

Figure 7.14 The oldest direct record for the existence of land ice on the Island comes from the Happisburgh Till of NE Norfolk. Here is a water-worn pebble of rhomb porphyry, a distinctive glacial erratic sourced from Permian volcanic rocks in the Oslo rift.

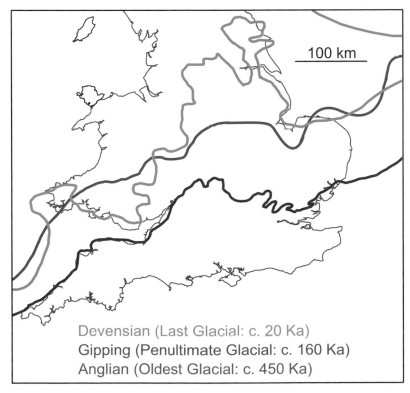

Devensian (Last Glacial: c. 20 Ka)
Gipping (Penultimate Glacial: c. 160 Ka)
Anglian (Oldest Glacial: c. 450 Ka)

Figure 7.15 At least three major ice advances have overwhelmed the north and midland parts of the Island over the past half million or so years (data after Gibbard and Clark 2011). Each time the ice retreated from a more-or-less well-defined southern limit, shown on this map as coloured lines. The parts of England below the red line have never been glaciated.

periods. As climate ameliorated, seasonal freeze-thaw produced expansion-contraction features: ice wedges in polygonal and linear ground cracks, mounded pingos of soil-covered ground-ice, and deposits from slow downslope soil and sediment movement (traditionally called 'head'). All this happened in ice-margin environments that effectively churned and softened all pre-existing near-surface rock. It was never scoured away by glacier ice streams, and so much remains today.

In formerly glaciated lowlands, temperate interglacial lake deposits are widely preserved. Perhaps most notable for their fossil and artefact content are those that succeed Anglian glacial deposits at Hoxne in north Suffolk. Foremost amongst its varied warm temperate fauna and flora are horse remains, presumably prominent on the local menu. They were efficiently butchered by hominids using exquisite pointed/tapering flint hand axes. Neanderthal remains with a plethora of flint handaxes

feature from ancestral Thames river gravel deposits at Swanscombe, Kent. The earliest such skeletal remains, dated to around 230 millennia ago (the penultimate interglacial), come from cave deposits at Pontnewydd in Denbighshire, North Wales.

The last interglacial sea level around 130ka was 2–4m higher than that of today; clear traces of its clifflines and deposits are recorded around our present coastline (Figs 7.16, 7.17). Since that time, during cold intervals with low sea level, Neanderthals and latterly our own ancestors could walk to and fro between here and Europe. Evidence from the continent suggests Neanderthals and *Homo sapiens* overlapped by up to 5 millennia around 40ka or so. The oldest *Homo sapiens* remains on the Island (dated to 33ka) are from the Paviland sea-cave burial in South Wales (Chapter 4), the remains interred at a time of slow climatic deterioration that heralded the last Ice Age.

Figure 7.16 Around Start Point, South Devon, a prominent coastal bench cuts across Devonian schists 2–3 m above present mean sea level. It is visible here (50.222187, −3.681496) as the flat-topped band of dark algal-encrusted rocks beyond and above the horsewoman. It formed by wave abrasion during the last interglacial highstand of sea level around 132 millennia ago. Resting directly on it are brownish-weathering deposits: glacial borderland scree and landslides of the last glaciation. They are the same age as the glacial till depicted in Figure 7.17.

The Last Ice Age and Meltout (21–11.5ka)

The Last Glacial Maximum was the climax of the last Ice Age when glaciers reached their southernmost limits – advance exactly balanced by meltout at the terminus. With sea level some 120m below present day levels the Island once again became an ice-covered promontory of the Eurasian continent. The limits to ice advance are confidently mapped from heaps of glacial detritus, terminal moraines, most prominent in east Yorkshire, but with distinctive tills elsewhere (Fig. 7.17). South of the ice front was tundra. The northern Atlantic Ocean lapped coldly around an iceberg-infested coastline to form a gulf between SW England and Brittany.

Subsequently, Earth's mid- to high-latitude regions warmed up as more and more solar radiation fed into the atmosphere. The great ice caps retreated as the outflow of ice from their centres became insufficient to compensate for that lost by melting at the margins. Global sea level rose slowly at first, accelerating around 16ka (Fig. 7.18). Large ice-dammed lakes formed in the lowlands of central and north-eastern England and central Scotland. Sea level continued to rise, the climate ameliorating from sub-polar to cool-temperate, then to warm-temperate. The ice vanished from the lowlands and then the highlands. A permafrost desert changed to steppe and then inexorably into birch, and then mixed birch/pine forest. Humans and a succession of life-forms re-invaded the Island landscape.

Around 13ka a severe worldwide climatic reversal occurred that lasted some 1300 years. This was the 'Younger Dryas' cooling event noted previously (Chapter 1). Scottish ice from mountain corries ground back down pre-existing glacial valleys and

Figure 7.17 The glacial deposits of the last Ice Age include examples of till (boulder clay) like this example of brownish-red weathered Hunstanton Till (the upper half of the image) from near Morston, North Norfolk (52.955841, +1.000234). The fingers rest on well-rounded flint pebbles in older beach gravels of the Last Interglacial. The site is a low, degraded cliff a metre or so above present-day high-water mark; it is probable that ongoing post-glacial subsidence is occurring here.

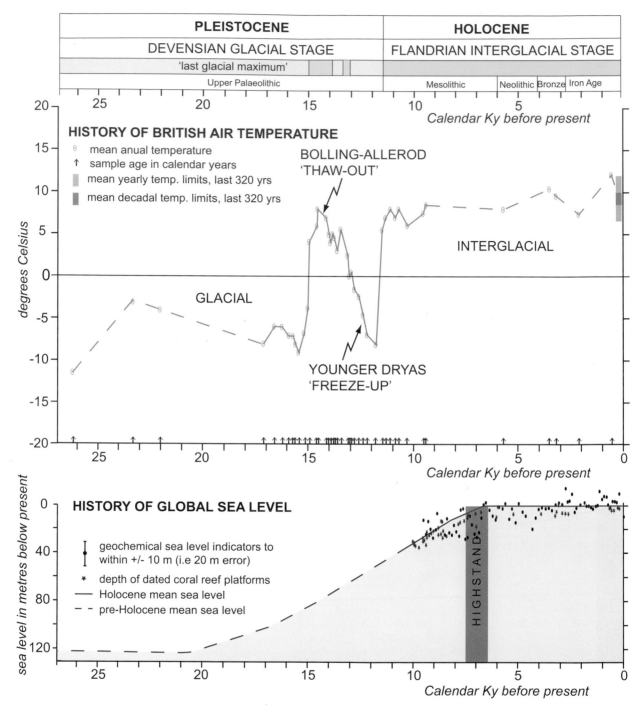

Figure 7.18 Land temperature and sea-level changes in the past 20 millennia. The temperature curves are calculated from radiocarbon-dated fossil beetle remains (data of Atkinson et al. 1987). See Brown (2008) for an accessible review of the arguments for late-Bronze Age climate change. The sea-level curve is smoothed from data provided by calcareous plankton and the depth of submerged coral reefs (data of Siddall et al. 2003). Modern sea level is extrapolated back in Holocene time. It is constant to about 6.5–7.5ka, the date of Holocene highstand.

formed a small ice sheet over local peaks, an event known as the Loch Lomond Readvance. More widespread was intense frost-shattering that caused abundant scree to accumulate around glacier fringes on the high plateaux of the Cairngorms, Snowdonia, the Lakes, North Pennines and elsewhere. We see its remains clearly in these high places today. Then global climate ameliorated again around 11.7ka: the glaciers retreated to their sources and finally disappeared. The Holocene epoch began.

Holocene (11.7ka–present)

The beginning of the Holocene heralded profound environmental changes. New mountain landscapes of former ice-filled U-shaped valleys now showed off their corries, cwms, arête ridges, hanging valleys, moraines, kame terraces and scoured-out lake basins. The piedmont and lowlands revealed their own terminal moraines, drumlin fields, eskers and flat, former pro-glacial lake floors. All exposed rocks and glacial sediments began the long process of chemical, biochemical and physical breakdown to form ever-thickening soil horizons: firstly as flakes of eroded sediment accumulating in hollows, the deposits thickening as plant roots took hold; then brown forest-soils slowly developing under leaf fall from tree canopies and by the subterranean effects of penetrative root systems.

New coastlines advanced, cutting into previous interglacial cliffs, renewing rapid coastal retreat in the many localities with softer Mesozoic and Cenozoic mudrocks and friable older Quaternary glacial deposits. Where rivers met already low-lying and low-sloping *terra firma*, extensive coastal wetlands (Trent-Ouse Levels; Lincolnshire Fens; Somerset Levels) developed as the incoming river systems spread their flow through numerous delta distributaries. At first there was little sediment brought in, for much of the Island's uplands were tree-covered. Freshwater peats accumulated in the floodplains and lakes that separated the wandering river channels. Coastal estuaries cut in former Ice Age valleys started off as narrow exits to the sea. The extensive freshwater wetlands advanced seawards in uneasy equilibrium with the ever-present counter-threat of winter storm-surge flooding around the North Sea fringes.

Mesolithic hunter-gatherers soon became permanent inhabitants of the Island. Post-holes and hearths of fixed roundhouse homes are dated as early as 11ka in Northumberland. The Island's oldest known village is at Starr Carr on the margins of what was formerly glacier-dammed 'Lake Flixton' in the eastern Vale of Pickering, North Yorkshire. Its two hectares of prime foraging territory featured a lakeside jetty of pile-wrought timbers. After 10ka such settlements must have been a common feature in many post-glacial lowland landscapes; examples currently under excavation include a group of structures at Sefton, north Merseyside. The earliest Holocene human remains date from around nine millennia ago: a complete skeleton from a cave-shelter close to Cheddar Gorge, Somerset.

The Holocene climatic amelioration saw oak and other temperate species (elm, lime, elder) appear by the time of the late Boreal, six millennia ago. This was preceded by a brief but severe cold snap thought to have been caused by a diversion of the Gulf Stream for a few hundred years around 8.2ka. These were already trying times for our Mesolithic ancestors when what must have been a coastal catastrophe occurred along the northern North Sea margins and over what remained of Doggerland. This was the Storegga tsunami whose deposits occur today at inland sites from Orkney to Lothian (Fig. 7.19). Caused by massive collapse of Norwegian shelf margin sediments, its wave would have radiated across the northern North Sea, growing to many metres in height along the shallow-shelving Scottish coastlines.

Global sea level finally rose to reach present levels ('highstand') between 6.5 and 7.5ka as a final equilibrium was established between polar ice accumulation and melting. But though Island climate and global sea level were generally stable, the Holocene witnessed further environmental changes. In highland areas, particularly over much of Scotland, land that had been glaciated rose upwards several metres relative to the now-stationary global sea level. The raised beaches we featured earlier (Fig. 1.2) were the result. In south-east England local sea level rose somewhat, helped by longer term North Sea basin subsidence: the result was marine flooding of coastal estuaries.

The story of modern humans farming the Island landscape begins with the 'Neolithic Agricultural Revolution' around 5.8 millennia ago. It took all of the five previous millennia to reach the islands off NW Europe from the Near East. Studies of skeletal DNA from dated cultural grave-sites in Ireland (Cassidy *et al.* 2016) indicate that it was an influx of Near East agriculturalists via the European mainland that was responsible and not just the passage of a cultural adaptation copied along the way by indigenous folk (the Irish results must also apply to Britain). The entire agrarian culture they brought with

Figure 7.19 A tsunamite: the pale-coloured sand deposit (centre) left by the Mesolithic Storegga tsunami of 8.2ka. It covered and was then succeeded by 'normal' coastal marsh sediments. Maryton, Montrose, Angus, Scotland (56.693834, –2.516866) Photo: Wikimedia Commons©Stozy 10.

them was radically different from that of the indigenous and settled Mesolithic hunter-gatherers. It involved ploughing, planting, tending and harvesting a variety of nutritious crops, careful seed storage and the breeding, seasonal slaughter and preservation of domesticated animals and birds. All this required clearance of late Boreal–early Atlantic forested landscapes of oak, elm, lime and alder. The new way of life also included engineering skills (stone quarrying and transport chiefly), astronomical prediction and communal ventures like monument building by local and wider kinship groups. The indigenous Mesolithics must have eventually integrated with this new culture, slowly learning required skills down the generations. Or perhaps they remained marginalized, like countless colonized folk ever since.

The above-quoted DNA studies also reveal a further influx of peoples around 4.3ka, this time descended from Bronze Age herders of the Pontic Steppes (the Yamnaya peoples), in common with Steppe genetic heritage transmitted during trans-European population upheavals from southern Siberia to the Atlantic margins. These immigrants made a major contribution to the indigenous genetic heritage of the 'Celtic' peoples in modern Irish, Scottish and Welsh populations. Long-named as the 'Beaker Folk' they had practical metallurgy and mining skills (gold, copper, tin) first developed in Anatolia three to four millennia earlier. It had been a long, slow migration; a different culture with distinct pottery, burial and ceremonial forms that soon permeated the long-established Neolithic farming (and flint-mining) communities of the Island. It kick-started the Bronze Age here, mining both indigenous copper and tin deposits and fostering a great expansion in agricultural land use.

The direct evidence for agricultural expansion is most obvious in upland areas like the Dartmoor granite massif. Spread over thousands of acres in the lower and middle valley slopes of the modern peaty moorlands are the still visible settlements of permanent stone roundhouses and grids of small rectangular fields with low drystone-cored hedge boundaries (reaves). (Fig. 7.20) What a sight they

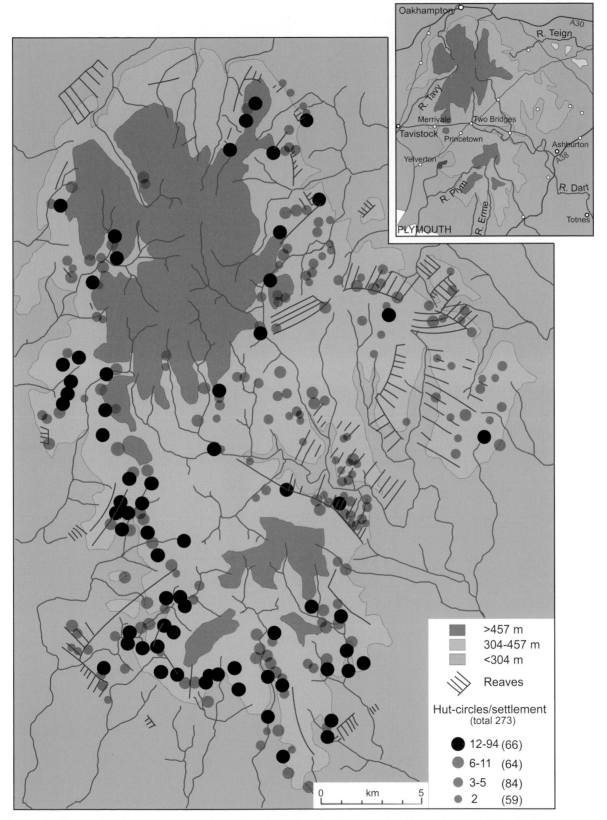

Figure 7.20 Bronze Age farmer-tinners colonized Dartmoor's uplands up to elevations of around 450m. Data redrawn after Price (1985). See Brown (2008) for an accessible review of the arguments for late-Bronze Age climate change.

must have been with their annual blooms of blackthorn and whitethorn in the five hundred or so Bronze Age springs of their existence. Though only a presumption at present, for there is no direct evidence, the Dartmoor farmers may also have sought out stream-tin deposits to smelt and mix with the abundant copper they mined at several localities across the Island (Chapter 11).

In the late Bronze Age around 3ka came what seems to have been a global climate-change event. Drought conditions spread across the Middle East and the Aegean where the late Bronze Age 'Palace-Cultures' collapsed. On the Island a change to a more north Atlantic/sub-Polar-influenced climate occurred; rainier with cooler winters. The previously cultivated uplands and their villages became first marginal and then unfarmable, the kept hedges of the Dartmoor reavelands overrun with brambles and gorse. Waterlogging eventually spawned peat and gley soils above the ditches cut through former native forest brown-earth soils. At about the same time, *c*.800 BC, Iron Age immigrants with aggressive tribal cultures advanced in from Belgic and Germanic Europe, with particular tribal groups prominent in the Yorkshire Wolds and southern England. Neolithic and Bronze Age monuments were abandoned forever, replaced by the cultural ferocity implied by the hundreds of Iron Age fortresses that spread over every habitable eminence of the Island's uplands and coastal highlands. Large-scale iron working began in the Sussex Weald and the Forest of Dean.

By the time of the Roman invasions the Iron Age settlers, farmers, charcoal burners, miners, potters and builders had steadily reduced the protective upland forest cover. They undoubtedly practised renewable coppicing but probably only on selected areas of lowland willow, hazel and ash. The uplands became vulnerable to erosion, particularly during stormy intervals (as in the later-medieval thirteenth and fourteenth centuries). Lakes and downstream floodplains became sites of increased sediment deposition. The flat-bottomed lower alluvial valleys and estuary margins developed into a new human landscape, that of managed water meadows whose channel sluices, levees and wind-powered pumping mills controlled water levels and enabled all-year grazing.

And the rest, as they say, is history.

PART 4
Material GeoBritannica

Our material needs down the millennia have been almost entirely satisfied by geological materials: hand-held hammer tools of quartzite and igneous stones; cutting implements of flint; freestone for building durable and fireproof structures; lime and phosphate for fertilizers; lime for cement and mortar; ores for metals; carbon-based sources of fuel, heating and locomotion; and so on – the list is long. Chapter 8 examines how settlement and communication were and are governed by the physical arrangement of landscape. Chapter 9 looks at natural resources in general. Chapter 10 posits that since we all need warm and weather-proof shelters, places to congregate or shop in, the nature and source of building materials is of prime concern. They control the appearance, feel and appeal of any urban or rural settlement. Local stone, brick and slate blend timelessly into natural landscapes, defining our self-built living spaces.

Chapter 8

Settlement and Communication

A cellar underneath the house, though not lived in,
Reminds our warm and windowed quarters upstairs that
Caves water-scooped from limestone were our first dwellings,
A providential shelter when the Great Cold came…

W.H. Auden, *Down There* in *About the House* (1966)

Choice of settlement is never random, it involves practical possibilities as nuclei of migrating or invading settlers sample their new landscape. They look for the best to suit their purpose from the perspective of previous experience and current opportunities. The former is their 'cultural package', in the *argot* of socio-historians and archaeologists. The latter, inevitably, means that each potential site is a compromise between physical reality and the ideal.

The Soft Stuff

The Island's maritime climate ensures plentiful rainfall, hence an abundance of runoff. Although periodic droughts affect the drier south-east, sustained periods of 'water stress' are rare. It was probably ever so, even through the vicissitudes of changing Holocene climate. Water availability close to settlements was always a key factor. In more drought-prone areas, surface supplies were augmented by local wells dug to the water table and its groundwater. Springs issuing from the junction of permeable and impermeable bedrock were prime settlement sites everywhere.

Our Bronze Age ancestors, probably earlier cultures also, regarded water sources as sacred, perhaps hosting individual deities, like their Aegean contemporaries dramatized by Homer. The Romans needed plentiful supplies for sanitation, home consumption, relaxation and exercise. This was enabled by their unrivalled engineering skills that delivered water from often distant natural sources. Later monastic foundations were carefully sited so that ample amounts ran through or close

by their kitchens, toilet blocks, grounds and gardens. An accompanying aesthetic of pleasant valley scenery and the practicalities of mixed farming between upland and lowland settings is often evident, as in Yorkshire (e.g. Rievaulx, Fountains, Jervaulx), the Scottish Borders/Galloway (e.g. Dundrennan, Sweetheart, Jedburgh, Dryburgh, Melrose, Kelso) and elsewhere.

During the nineteenth century, population growth rates soared, industries developed, medical science advanced, and the demand for clean water supplies burgeoned. Subsurface extraction of groundwater began from deep wells drilled into more porous and permeable bedrock; slowly the science of hydrogeology developed. Major aquifers in the urban districts of Midland Scotland, Northern England and South Wales tapped extensive Old Red Sandstone and Carboniferous aquifers; the Midlands pumped from Triassic sandstone while London and the South-East utilized the Chalk. Such borehole sources, though often individually prolific, were unable to satisfy metropolitan demand. Major cities requisitioned (sometimes controversially) neighbouring, or sometimes far-distant, deep-glaciated upland valleys for reservoir construction. So the inhabitants of modern Manchester drink Lakeland water and those of Birmingham and Liverpool the waters of upland Wales.

Home Sweet Home

Settlements have grown, stagnated or decayed over time. We see this interplay from Iron Age and Roman times to the present day in the history of individual houses, hamlets, villages, market towns and regional cities. Decline has had many reasons. Populations have perished or shrunk below viability through disasters such as pestilence (e.g. deserted medieval plague villages). Medieval windstorms brought destruction to treeless croplands by great sand blows, most notably forming the East Anglian Brecklands in late-medieval times. Marine flooding has periodically inundated reclaimed coastlines like the pre-modern Thames estuary, the East Anglian

Fenlands and Broads and the Somerset Levels. Coastal erosion by cliff-retreat destroyed the rich medieval town and port of Dunwich in Suffolk. Settlements have also been abandoned for non-environmental reasons: familiar is the once widespread monastic culture destroyed by Tudor legislation. Wider civil communities have been demolished by conquerors (Norman destruction) or by diktat (Army training camps and ranges).

Rock outcrops themselves served as fortifiable foundations and as convenient sources of economically extractable and tradeable resources like metallic ores, building stone and fuel. They also helped to provide safety from flooding, a means of shelter in caves, and the foundation for dependable ridgeway communication routes away from riverine or coastal wetlands. Defensive sites ranged from Iron Age hill forts and settlements built upon isolated mounds and crags formed by volcanic rocks (e.g. Burnswark, Dumfriesshire – also spelt Birrenswark in older BGS literature, as in 'Birrenswark Lavas' – Eildon Hills, Roxburghshire; Traprain Law, Lothians), scarpland rock outliers over the clay vales of SW England (e.g. Glastonbury Tor; Maiden Castle; Hod Hill) and resistant Precambrian inliers along major faults (Malverns).

Yet the Romans famously eschewed building their new forts and towns on such sites; no true-blooded Roman would hide behind walls or look to precipices for deliverance. Medieval castles, though, are frequently located thus: well-known examples are Edinburgh Castle on a volcanic plug; Stirling Castle on a dolerite sill and, perhaps most spectacular of all, Bamburgh Castle (Fig. 8.1) on the Whin Sill of Northumberland's rocky Carboniferous coastline. Earlier than any of these, and perhaps as impressive in their own way, were the entirely artificial mounds built of the rubble and earth from the Norman demolition of parts of Saxon cities like Norwich.

Figure 8.1 Bamburgh Castle on precipitous black dolerite coastal outcrops of the Whin Sill in Northumberland (55.608960, –1.709900). Photo: Shutterstock 186024122 ©Philip Bird LRPS CPAGB. The castle itself is constructed with the more easily crafted Lower Carboniferous sandstone into which the sill originally intruded.

In these lowland settings, in the most populous parts of their new colony, the Royal Castles were emphatic statements of absolute power (Fig. 8.2).

The many settlements that grew up adjacent to worked mineral deposits declined as these were exhausted or regarded as no longer economically viable. Sometimes new reserves kept a working population together or led to regional migrations, such as the spread of pit villages across the verdant countryside above deep 'concealed' coalfields (notably Yorkshire, Nottinghamshire and Kent) from late-Victorian times to the 1970s (North Yorkshire). Metallurgical breakthroughs (Chapter 11) enabled the large-scale mining of low-grade sedimentary iron ores and their reduction to iron and steel in eastern England (Middlesbrough, Scunthorpe, Corby) for a hundred years.

But all this is now gone, testing to the limits the resilience of populations to take up new trades and use newly-acquired skills. They endeavour to preserve their landscapes *in situ* as it were, to renew their culture and history without forced or assisted emigration. Most try to gain financial help from central government and (formerly) the EU to ameliorate the effects of home-grown economic dogma and misfortune of one sort or another.

Through the Wetlands

For millennia the Island's major rivers and estuaries were a key means of regional communication, made use of by both peaceful traders and hostile incomers. Exported goods and expeditionary forces went out and imported

Figure 8.2 Norwich's Royal Castle (52.628729, +1.296386). Photo: Shutterstock 1017824 ©Darren Pierse Kelly. The original Caen limestone cladding around its flint core was completely replaced in the 1830s by Bath Jurassic oolitic limestone. The architect faithfully reproduced the much-decorated Norman exterior work.

goods, traders and invaders came in. In upland Britain, glaciation had created the very deep estuaries and sea lochs that are such a prominent feature of western seascapes. Those like the Clyde, sited close to supplies of raw materials, became prime ship-building sites during the Industrial Revolution. Major lowland rivers in England (Severn, Thames, Trent, Yorkshire Ouse) usually follow outcrops of weaker bedrock, the navigable channels surrounded by riverine wetland and fen, as they pass downstream to estuaries and open coasts.

Since sea-level highstand, many lowland estuaries are now shadows of their former selves. Sediment usually infills inner-estuarine environments to give the distinctive flat-bottomed vales made famous in Constable's landscapes (see Fig. 15.7). These were drained and the floodwaters controlled to nurture floodplain water meadows. Such pasture traditionally provided lush grazing for sheep and overwintered herds of Scottish Border and Galloway cattle ('runts') destined for metropolitan markets.

Communications in the Island's lowlands has always had to cope with watery landscapes: periodically flooded or permanently saturated environments of floodplain, fen, marsh and carr. According to some estimates they occupied up to a quarter of the Island's lowland acreage in prehistory. Riverine wetland was luxuriant in reedbed and wildfowl. It lay adjacent to settlements, pasturage and arable land and bordered a subaqueous realm rich in fish and eels, dotted with often sizeable cultivable islands, like the sometime Isle of Ely in the Cambridgeshire Fens. Wetland drainage has had a long history, beginning with the Romans, skilled drainers of their native Pontine marshes, in our eastern and northern fenlands and those bordering the Somerset Avon and Severn. The majority have been artificially reclaimed by later drainage schemes: sucked dry by windmill-driven pumps, and in some cases squeezed metres below sea level by compaction (see Fig. 34.5). On the rich, drained earth have grown villages, market towns and regional administration centres with agro-industrial factories dotted across their flat hinterlands.

Safe and secure communication was essential for both humans and animals within wetland environments. Major archaeological discoveries over the past few decades have laid bare the remains of buried wooden causeways – wetland's A-roads in prehistory. These had firm lateral and vertical foundations provided by closely spaced pile-driven logs, in between which were spread planks and branches to provide non-slip walking surfaces.

A fine example, Sweet Track, nearly 2km long and about two metres wide, with a planked upper surface resting on pegged 'rails', has been excavated in the Somerset Levels. The oldest timber is dated from tree-ring analysis to the early Neolithic, precisely 5807-5806 years, making it the oldest known wooden trackway. Others include Bronze to Iron Age Flag Fen near Peterborough and the Iron Age causeway at Fiskerton in the Witham valley near Lincoln. Constructed wide enough to take carts and pack animals, the trackways provided vital local communication around marginal wetlands and to isolated islands.

Flag Fen causeway is especially well documented, comprising an estimated sixty thousand or so sharpened stakes. The route cuts across pre-existing field boundaries, suggesting to archaeologist Francis Pryor that it was constructed in an effort to regain access to 'island' farm sites on slight rises as marshland encroachment occurred as groundwater levels rose. The causeway also became the launching place for votive offerings and, like sites in ancient Greece, these may have been superintended by priestly sects. The bundled-up offerings included pristine handaxes, deformed bronze swords, shields, hammerheads and axeheads; Pryor thinks some of these were manufactured in bulk as facsimiles.

Ridges and Romans

Ancient pre-Roman tracks followed elevated ridges across the lowlands. These were formed by impact of countless Mesolithic and Neolithic human feet, animal hooves and the wheels of cart and chariot. They criss-crossed linear scarplines of well-drained sandstone, chalk and limestone elevated up to three hundred metres above surrounding vales. Their hard, well-drained surfaces made ideal cart tracks and survive today as public paths. Examples are the eponymous Ridgeway from Wiltshire to Goring and the Chilterns and hence, as the Icknield Way, to north-west Norfolk.

During the Claudian invasion, geology played a crucial role in determining marching routes (Fig. 8.3). After landing at modern-day Richborough, Kent the invasion force of four legions would have made their way east along the sandy Paleogene foothills of the North Downs chalk dip-slopes. They were aiming for the first Thames ford crossing (across from Southwark) to take them up to friendly Camulodunum (Colchester). This was the de facto British capital, where King Cunobelinus

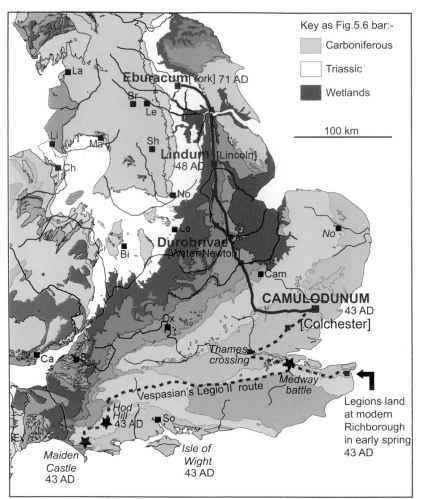

Key as Fig.5.6 bar:-

- Carboniferous
- Triassic
- Wetlands

100 km

Figure 8.3 Geology and the route of the AD 43 invasion (dashed red). Subsequent expansion northwards from Camulodunum (solid red) was along the elevated Middle Jurassic limestone scarplands (mid-blue) that avoided wetlands. Data generalized from British Geological Survey, *Times Atlas of the World* and other sources.

of the Catuvellauni was in power. After a fierce battle against his son Caratacus at the Medway crossing, Vespasian (future emperor) and his XXth legion were detailed to quickly counter potential hostility along their western flanks from the tribal territories (around modern Dorset) of the hostile Durotriges. To avoid the poorly drained and heavily wooded Weald, he would either have embarked by ship from the Medway to Poole or thereabouts and/or taken the ancient North Downs land route from south of the Medway Gap, force-marching his legion westwards along firm chalk scarpland tracks. Tacitus writes that he rapidly took numerous hill-forts, the Isle of Wight and, as archaeological evidence shows, the chief Durotrige stronghold of Maiden Castle. Despite their prominences and ramparts, the British hill forts were no match for Roman siege tactics.

A few years later, expansion of the Roman province northwards to Lindum (Lincoln) made use of the Middle Jurassic limestone scarp that fringed the impassable wetlands of the Lincolnshire fens to the east (Fig. 8.3).

In the 70s, internecine troubles and their aftermath in the Brigantes territory of the southern Pennines were checked by the IXth Legion in a move northwards from Lindum. We can trace their likely route crossing the Humber into friendly Parisi lands, along the subdued Jurassic scarp to the north and hence the chalk foothills of the Wolds. Finally left-hooking across gently rolling Triassic hills, they arrived at the site marked out for the legionary fortress (and later colonia) of Eboracum (York). This was strategically sited above a constricted portion of River Ouse floodplain, partly provided by a Last Glacial terminal moraine. From here the legion could dominate all the Pennine exit drainages, the North Yorkshire hills and all easy land and river routes southwards and northwards. They were also close to the outcrop of that fine Permian building material, Magnesian Limestone, which their engineers and masons soon made use of.

During this rolling colonization many ridge routes were doubtless spotted and enlarged by the keen and

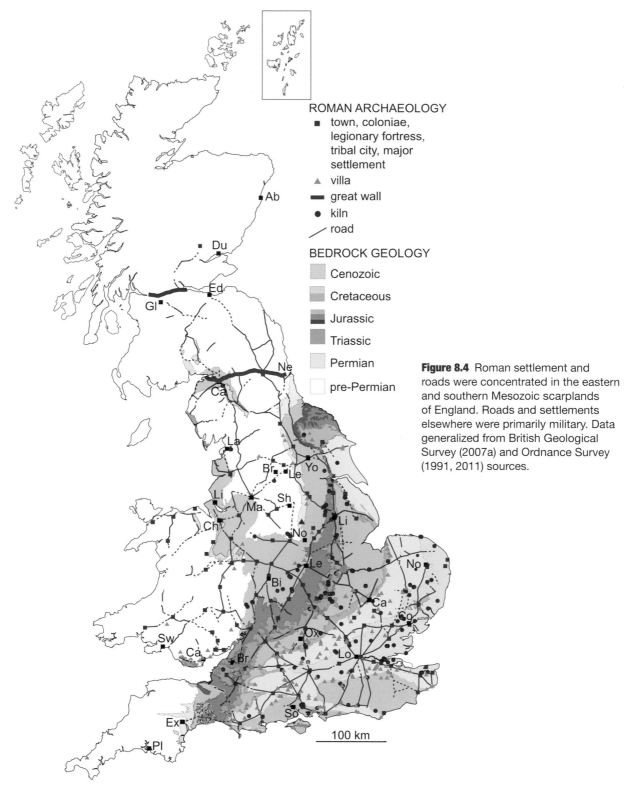

ROMAN ARCHAEOLOGY
- ■ town, coloniae, legionary fortress, tribal city, major settlement
- ▲ villa
- ━ great wall
- ● kiln
- / road

BEDROCK GEOLOGY
- Cenozoic
- Cretaceous
- Jurassic
- Triassic
- Permian
- pre-Permian

Figure 8.4 Roman settlement and roads were concentrated in the eastern and southern Mesozoic scarplands of England. Roads and settlements elsewhere were primarily military. Data generalized from British Geological Survey (2007a) and Ordnance Survey (1991, 2011) sources.

100 km

knowing eyes of Roman military engineers. These elite troops, like the modern Royal Engineers, were permanently attached to the legions. They subsequently laid out compact linear tramping-ways across thousands of kilometres of the Island (Fig. 8.4). The map of early Roman British roads shows linear routes joining London, Colchester, St Albans, Lincoln and York in eastern England. They largely follow more elevated and resistant Jurassic and Cretaceous scarpland rocks. Along these were spaced small military staging posts, marching camps, larger military encampments for seasonal use, and full legionary fortresses. The latter were spaced at sufficient intervals for a whole legion to march in a few days.

Few traces of the once extensive Roman road network remain today. Many became the foundation upon which Saxon, medieval and early modern roads were built. A mooted preserved stretch is in the Millstone Grit foothills of modern day Lancashire/Yorkshire border country, though to some it is an eighteenth-century packhorse route. Here the road's dressed gritstone setts and central gutter curve gently up from the Rochdale basin over the Millstone Grit of Blackstone Edge via Littleborough towards its supposed ultimate destination of Roman Verbeia (Ilkley). The stone road looks strangely but majestically out of place in today's peaty moorlands. Yet regardless of its true age, one's imagination senses here the almost tangible presence of Roman engineers and the busy noises of their work gangs. Surveying kit hangs nearby; ox-drawn wagons wait full of locally hewn stone; labourers tamp setts into place on cambered templates of smooth-raked aggregate.

Villas were part of large country estates, concentrated on the lower slopes of Jurassic limestone and Cretaceous chalk downland landscapes in the Island's southern and eastern rain shadow– sites above flood levels that were sunny, fertile, spring-watered, well-drained and easy to plough. They also had pleasant outlooks and generally lay close to a decent road. The majority, at least in the southwest, appear to date from the mid to late third century.

Marching Routes, Turnpikes and Canals

Military communication has always been a prime necessity, allowing rapid deployments to the scene of particular trouble spots or incursions. So it was that the Saxon leaders who reunited the English kingdom after the Scandinavian incursions of the ninth and tenth centuries made use of much of the old Roman road network. What was then called Watling Street rattled with the sounds of Harold Godwinson's mounted housecarls and thegns as he force-marched his fyrd from London to York and back again across the Mercian scarplands in the summer of 1066. All this forced marching can only have been possible across a seriously well-maintained road system.

Thereafter, the network of Saxon trading routes across the Island was used for centuries. A much later episode of road construction engendered by military needs was begun in both lowland and highland Scotland in the 1720s after the first Jacobite rebellion of 1715. These are 'Wade's roads' named after the General in charge of the project, located on modern Ordnance Survey maps as 'Old Military Road'. They were also constructed to march soldiers rapidly to the west coast ferry ports that led to the troubled island of Ireland. After the second (and last) Jacobite uprising in 1745–46 the government built an extensive system of marching roads and defended forts deep within the Highland homelands of the Catholic clans. The road between the garrison towns of Fort George, Inverness and Fort William joined the northern and southern limits to the strategic reaches of the Great Glen along its linear fault-line.

In the eighteenth and nineteenth centuries the courses of canals and railways were surveyed and then constructed using gunpowder, steam drainage pumps and mechanical excavators to overcome hitherto unconquerable bedrock obstacles. William Smith, our geological pioneer (Chapter 3), was one such canal surveyor in south-west England, but the most celebrated was James Brindley, active in the West Midlands and Lancashire and responsible for the very first, the Bridgewater Canal (see Chapter 28). Today in countless Midland and Pennine valleys, sandstone canal lock walls and contemporary adjacent mills sit squarely in the landscape: human-constructed certainly, but entirely natural in their own way (Fig. 8.5). Over the remainder of the Island a revolution in road transport occurred when the privately financed turnpike system was created in the late eighteenth century (Fig. 8.6).

The often deep cuttings of the developing railway system provided a rich geological bonus: virgin strata were exposed to be recorded in the field notebooks of both professionals from the infant Geological Survey and amateurs alike. In our own era highways appear to randomly criss-cross the Island with impunity. Yet the role of the engineering geologist, like their Roman, Georgian and Victorian forbears, will never cease. All excavations and the siting of tunnels, cuttings and bridges must be fitted into allowable safety margins whose limits are still set by natural rock structure: the possibility of structural collapse due to poor foundations, unstable rock-cut slopes or poor drainage causing landslides are ever-present hazards.

Figure 8.5 Early-19th-century gritstone-built three-storey architecture along the Rochdale Canal, Hebden Bridge, West Yorkshire (53.740042, –2.009897). The stone-mullioned windows let in plenty of light for the hand-loom weavers whose properties these were.

Figure 8.6 The early-19th-century toll stop, Steanor Bottom Bar, Todmorden Turnpike, East Lancashire (53.67502, –2.0845). Payment for the six hilly miles between Todmorden and Littleborough was between 5d and 9d per draught animal, a considerable sum for the time. A typical gritstone edge, the outcrop of the Fletcher Bank Grit, towers above the road to the upper right.

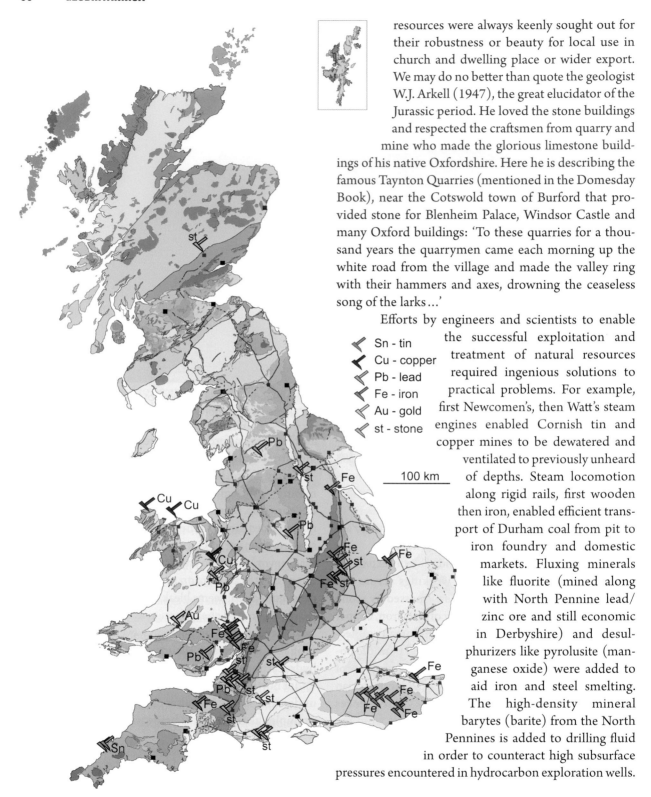

resources were always keenly sought out for their robustness or beauty for local use in church and dwelling place or wider export. We may do no better than quote the geologist W.J. Arkell (1947), the great elucidator of the Jurassic period. He loved the stone buildings and respected the craftsmen from quarry and mine who made the glorious limestone buildings of his native Oxfordshire. Here he is describing the famous Taynton Quarries (mentioned in the Domesday Book), near the Cotswold town of Burford that provided stone for Blenheim Palace, Windsor Castle and many Oxford buildings: 'To these quarries for a thousand years the quarrymen came each morning up the white road from the village and made the valley ring with their hammers and axes, drowning the ceaseless song of the larks…'

Efforts by engineers and scientists to enable the successful exploitation and treatment of natural resources required ingenious solutions to practical problems. For example, first Newcomen's, then Watt's steam engines enabled Cornish tin and copper mines to be dewatered and ventilated to previously unheard of depths. Steam locomotion along rigid rails, first wooden then iron, enabled efficient transport of Durham coal from pit to iron foundry and domestic markets. Fluxing minerals like fluorite (mined along with North Pennine lead/zinc ore and still economic in Derbyshire) and desulphurizers like pyrolusite (manganese oxide) were added to aid iron and steel smelting. The high-density mineral barytes (barite) from the North Pennines is added to drilling fluid in order to counteract high subsurface pressures encountered in hydrocarbon exploration wells.

Sn - tin
Cu - copper
Pb - lead
Fe - iron
Au - gold
st - stone

100 km

Figure 9.3 Proven sites of Roman mining and quarrying activity with major settlements and roads (red) superimposed on a geological map. In an empire rich in mineral resources, lead ores and their contained silver mined from the Carboniferous Limestone and the gold of south-central Wales were undoubtedly the most profitable and sought-after from Britannia province. Geology colour key as Fig. 5.6. Data generalized from British Geological Survey (2007a) and Ordnance Survey (1991, 2011) sources.

Although many abandoned workings for stone and minerals have suffered the ignominy of landfill sites, the survivors still pockmark the countryside. Their ponds, lakes and tree-lined coverts add pictorial and ecological richness to deserts of industrialized farmland or suburban estates. Edward Thomas featured one such in *The Chalk-Pit* (1915):

> Is this the road that climbs above and bends
> Round what was once a chalk-pit: now it is
> By accident an amphitheatre.
> Some ash trees standing ankle-deep in brier
> And bramble act the parts, and neither speak
> Nor stir...

The most spectacular abandoned lowland workings are in the extreme east, the fabled Norfolk Broads, now reed-lined lakes of stunning natural beauty strung out along the floodplains of slow-flowing, meandering rivers. Once they were the sites of noisy medieval peat diggings, quarries for cheap domestic fuel visited round the clock by wagons, just like from modern opencasts. Yet the majority of abandoned areas of mineral workings shown in Figure 9.1 were those of the now extinct coal industry, now largely unseen to us: vast flooded and collapsing networks of subterranean shafts, levels and adits. These form Ted Nield's '*Underlands*' (2013) which he describes as 'Britain's lost landscapes'.

A Legacy of Words

There remains a final legacy from the extraction of natural resources by mining and quarrying. It is a lexicological one: the many hundreds of words used by miners and quarrymen for their working tools, the ores and stones they dug out, and the geological features that in different ways affected their day-to-day working. Some of these have spread into the wider English language. This richness of language has yet to be explored in a systematic way, though in this book we noted certain usages in connection with the development of the geological sciences (Chapter 3). The valuable compendium by W.J. Arkell and S.I. Tomkieff (1953) is now long out of print.

Chapter 10

Building Stone and Aggregates

So that water might be
Elephantine and pinpoint,
What an industry of air,
What transformation of heat.
And water goes off – it skulks
Through the cracks in rocks,
Jemmying them open ...

Norman MacCaig, *Water* in *Collected Poems* (1990)

The Variety of Building Stone

It is a glory for us that we can still see the local stone used for over two millennia of building construction: in bridges, cottages, colleges, houses, shops, markets, factories, mills, legislatures, shrines, ruined monasteries, churches, cathedrals and castles. The very ancient examples that remain, small and large, are simply treasures – walls, gates and forts of Roman ruins or Anglo-Saxon churches, some of the latter perhaps still in use. They daily remind us of stone's ability to provide shelter and space for living, travel, work, play, defence and glorification of the soul.

When surface sedimentary strata are worked for building stone, workable masses are levered out along the bedding planes and intersecting joints that divide and limit the strata. Joints also occur, cutting igneous and metamorphic rocks. They are more-or-less planar fractures (Fig. 10.1), but unlike faults they have had no shearing movement along their length; they just opened out. Some joints are parallel to the ground surface and formed from the release of pressure as rock was brought to the surface by erosion. Others were induced by tectonic forces or as igneous rock solidified and contracted from a molten state. These are often vertical and intersect at high angles. Joints of both kinds enable masses of hard rock to be easily separated by mechanical means.

There is no one single quality of rock that governs its use as building stone, except that durability and strength are essential. This rules out all soft and soluble types, together with any whose internal characteristics encourage rapid weathering by frosts, percolating water and mineralogical alterations of various kinds. Roofing stone must be impermeable in the bulk but, equally important, be readily separated into strong flat slabs, the thinner and lighter the better. So all roofing stones possess fissility, i.e. they are mechanically splittable along more-or-less parallel planes (Fig. 10.2). Late-medieval terms for such rocks, probably originally from Old Norse-English in northern England, are 'sclatston' (pronounced with a very long *a*, as in modern West Riding speech) and 'thackston'. The surname Slater is common today, Theakston less so. Both point to the widespread practice of the trade in medieval and earlier times.

In sedimentary rocks, fissility arises due to internal closely spaced, usually centimetric scale, planes of bedding and lamination. They denote rapid fluctuations or pauses in deposition that produced the original sediment. To the geologist they are 'planar-laminated' sandstones or limestones. To roofing technicians they are stone slates, and to pathlayers they are flagstones. Today all are often lumped together as the generic 'York stone'; highly regarded and increasingly expensive. Their preservation and conservation is a matter of high importance to the look and feel of rural and urban architecture and landscape. Theft of municipal flagstone and wallstone is a growing problem in the rural northern stone districts of the Pennines.

'True' geological slates are metamorphic rocks, but relatively low-grade, produced at a couple of hundred degrees Centigrade at the most. They are wonderfully strong and can be split very thin indeed because their fissility is due to rock cleavage, a splitting fabric on a millimetric scale produced by crystallization of minute platy clay and mica minerals in parallel growth. This occurred during tectonic shortening and metamorphism during mountain-building. They are epitomized by those from the Ordovician of Snowdonia, but also from the Upper Devonian of Delabole and at other horizons and sites in Cornwall, the Lakes and Highland Scotland.

Figure 10.1 Joints and granite tors.
A Hound Tor (414m), east Dartmoor (50.596999, –3.777977). Photo: Shutterstock 208508935©David Brian Williamson. Pillars of granite separated along vertical joints with fine sub-horizontal 'unloading' joints that mimic the bedding of sedimentary rocks.
B Sharpitor (402m), south Dartmoor (50.515211, –4.033100). Vertical joints separate 20–30cm thick courses of granite, suitable for building or walling stone. Sub-horizontal 'unloading' joints feature in the background.

Figure 10.2 Various building sandstones. **A** Carboniferous Elland Flags split and cut for paving, with their characteristically streaky surface decoration. 'Labour in Vain Yard', Norwich. **B** Middle Jurassic ferruginous sandstone blocks, Whitby Harbour, East Yorkshire (54.491090, −0.610163). The small rectangular holes were for insertion of lifting tackle. **C** Stoneyard with Millstone Grit ashlar quoin blocks, stone slates and much else. Hebden Bridge, West Yorkshire (53.743642, −2.010874). **D** Pegholed roofing slates temporarily awaiting re-use, Askrigg, Wensleydale, North Yorkshire.

Building stone for structural features – walls, edging quoins, pediments, pillars, pilasters, arches, mullions, etc. – must not split internally or weather unduly so as to lose their inherent strength. For this reason the very best rocks to cut for ashlar stonework are termed freestone, in the sense that they lack such decoration and so can be cut in any direction. Most igneous but only some sedimentary rocks have this property: geologists call them 'massive' or 'structureless'.

Weathered again

Our everyday experience of weathered sedimentary stone, particularly limestone, can be exhilarating; the Permian dolomitic limestone forming York's walls and its Minster are the colour of clotted cream with a texture of granulated sugar; Oxford colleges' oolitic limestones have hues of honey and mead. Louis MacNeice (1939) light-heartedly captured the variousness of limestone weathering there as he walked through old student haunts:

I roll on
Past walls of broken biscuit, golden gloss,
Porridge or crumbling shortbread or burnt scone,
Puma, mouldy elephant, Persian lamb…

Atmosphere, water and living organisms are responsible for modifying the colour and texture of pristine, freshly exposed rock surfaces (Fig. 10.3). Stone blocks used for interior work change little over the centuries; some used outside may also change little in colour. But stone like the Lincolnshire Limestone that was once a uniform hue can develop light yellow-brown tints, not a uniform coloration, though the original creaminess is still present in parts. Individual spherical oolite grains (like fish roe) and larger fossil shells are clearly visible. They stand proud, paler cream amongst the more yellowy calcitic matrix of crystalline cement. Alternating shellier and more oolite-rich layers are also apparent; these define a crude layering that cuts obliquely across the block as cream-coloured slashes. The overall effect is a pleasingly mellow mixture of diffuse banding.

Viewed from a distance, the whole west front of Lincoln Cathedral above the Gothic frontage arches is a multiplicity of such weathered blocks, each individual interacting with its neighbours. The effect is of subtle variation on a theme, a oneness of architectural form made up of a myriad of individual contributions. This is what stone weathering does to a magnificent building – a bit like the Church itself, we suppose, or a great ash tree whose elongate leaves all dance in a breeze, showing off sunlight at slightly different angles yet uniting in the unmistakable curves and gay flourishes of branches that constitute the tree itself.

Colour variation like that in the Lincolnshire Limestone is due to chemical weathering. The cementing calcite crystals contain a tiny amount (a few score parts per million) of reduced (ferrous) iron within their crystal forms. Exposure to the oxidizing atmosphere over the past millennia causes this iron to change its state by oxidation: the light brown to yellow tinge that so captivates the eye is caused by ferric hydroxide staining. By way of contrast, the constituent ooliths and fossil shells have no such iron impurities and so the calcite weathers 'true' to cream and white. Such are some of the subtleties of long-weathered stone, something that stone masons intent on restoration must know about and allow for.

Contrast this variegated weathering with the creamy richness of Barnack Stone (Fig. 10.3A, B), also of Middle Jurassic age, or the stone-washed whiteness of Upper Jurassic Portland Stone. The latter is the immensely strong limestone used extensively by Wren to rebuild London's churches after the Great Fire, and in many later Neo-Classical urban developments. The calcitic cement of both limestones has little (Barnack) or no (Portland) internal ferrous iron like that at Lincoln.

Physical weathering by percolating water is an out-and-out enemy to decorated building stone. It acts within tiny fractures, along fissures, cracks and bedding planes. Across these there may be great variation in porosity and permeability, especially when the stone contrasts in these qualities with its cementing mortar. Such defects and contrasts are worked on chiefly by frost weathering as interstitial water expands during freezing and contract during thawing. Such 'freeze-thaw' cycles have been repeated tens of thousands of times during the lifetime of a medieval building stone. The result may be spalling, crumbling or splitting.

Finally, the role of epiflora: lichens, mosses and algae. Many rural churches and houses exposed to 'pure' air have irregular surface encrustations of these organisms. Their forms and colours are legion, interacting with the natural primary and secondary weathering colours of the stones. There are the concentric rings of the ghostly, granular cream-grey lichen *Porocyphus*: frayed circles of raggedy gossamer. On the higher walls, tower, arches and roofs are the egg-yolk splats of *Xanthoria parietina*. Reddish-brown species and thin green films of chlorophyll-stained algae coat some shady damp north-facing

Figure 10.3 Various building limestones. **A** Barnack Stone (Barnack 'Rag'), a weathered bioclastic limestone with traces of bedding, exterior of Peterborough Cathedral (52.572483, –0.239210). **B** Unweathered Barnack Stone from Peterborough's nave columns. **C** Yellow-brown- to cream-weathered Lincolnshire Limestone forming the frontage to Lincoln Cathedral (53.234297, –0.536016). **D** Replacement limestone blocks in badly weathered exterior curtain wall around Lincoln's Norman castle (53.234604, –0.540714). **E** St Cecilia's, West Bilney, Norfolk (52.707971, –0.539539) has 'mongrel' stonework added over several hundred years – a stunning mix of creamy Barnack, 'gingerbread' Carstone, deep-brown ferricrete and black flint.

stonework. The overall effect is of a patination so exquisite that it could not be invented by human creativity.

Stone Skeletons

The history of architectural use of stone partly reflects the changing importance of relatively local versus far-sourced materials. Some combinations are obvious, but others remain inexplicable. Stonehenge has the massive trilithons in light-coloured siliceous 'sarsen-stone' from the Wiltshire downlands. These were originally a very pale yellow/white colour after fresh hand-masoning but are now mostly lichen-encrusted and grey-white. The older inner sanctum has the far-sourced 'bluestone', once a light-speckled deep grey-green, now much-weathered. Their source was traced in 1920 by H.H. Thomas, chief petrographer of the Geological Survey, to 'spotted' dolerite (aka microgabbro) intrusions in the Preseli Hills of Dyfed, south-west Wales. Recent geological detective-work (Bevins *et al.* 2012, 2014) points to a particular ridge, Carn Goedog in the eastern Preselis, with minor rhyolite stones from the Fishguard Volcanic group at Port Saeson.

Despite Stonehenge's uniquely distant provenance, most Neolithic mason-builders of henges and tombs used mostly local raw materials toppled or levered from nearby craggy outcrops or dug from shallow quarries and pits. Roman quarrymen and military engineers attached to the permanent legionary forts and associated civilian settlements widely exploited local building materials. They left their graffiti on worked faces at many quarry sites, though nearly two millennia of weathering have reduced most to mere ghosts. The stone – sandstone, limestone, dolerite and granite – was used structurally for building and in all-important subsurface conduits: water supply, hypocausts (underfloor heating) and sewers.

There is good evidence for a widespread trade in stone within Rome's Britannia province. This was chiefly into the new port of London, especially Barnack limestones from Cambridgeshire (Fig. 10.3A, B) and Ancaster limestones from Lincolnshire. Fine freestones were also brought in from the wider empire. Discoveries in Kent and elsewhere reveal some cross-channel imports from Gaul: oolitic limestones from the Boulonnais and fine-grained limestones from the Paris basin. But most spectacular is a metamorphic rock, the estimated 400 tonnes of prized Carrara marble from the Apuan Alps used to face the great imperial triumphal arch at Richborough in the late first century.

Anglo-Saxon, Norman and later masons required compact, flawless freestone for religious buildings; others they used as facing over rougher structural masonry. Later, from the Renaissance onwards, Palladio-inspired architects began to demand ashlar freestone for whole buildings. This caused massive supply problems in many areas. For example, the Cotswold limestone quarries that supplied Oxford's colleges sometimes contracted substandard stone that weathered badly.

The source of stone used in the late Anglo-Saxon church-building boom is of great interest (Fig. 10.4). To cater for demand, abandoned Roman buildings were initially quarried for their stone. Later in the pre-Norman era an indigenous quarrying industry developed rapidly in tandem with this recycling. At the height of Anglo-Saxon culture and economic advancement (caused by an influx of silver from Saxony; Sawyer 2013) an astonishing range of materials was being dug out of bedrock, sometimes transported for scores of miles along ridge-track and river. The main quarrying concentrated in the Mid-Jurassic and Liassic scarplands.

The Normans were large-scale stone importers, chiefly into areas of the east and south-east devoid of local freestone. They favoured their native Caen limestone, a creamy fine-textured or oolitic Middle Jurassic freestone. A massive sea-borne trade ensued across the Channel into the navigable estuaries of south-east and east England for over 400 years. With this stone they wholly built or decorated many of their early cathedrals and castles, e.g. Canterbury and Norwich cathedrals (Fig. 10.5), the Tower of London and Norwich's Royal Castle, using it more sparingly for quoins and windows in less-grand parish churchs in many areas. Some of the fine-grained varieties of this stone used in external work has not weathered well over the past millennium: it has a tendency to eventually flake and crumble.

Thus began the mid-Medieval to Renaissance building boom that was only briefly interrupted by Crusade and Plague. The cross-channel trade in Caen stone trade ended from about the mid-fifteenth century as links to Normandy withered completely during wars with an expanding France. Quarrying eventually ceased at the type locality, and English masons were forced to substitute other, different-textured and different-weathering, limestones. Subsequently the remarkable qualities of native Jurassic limestones were rediscovered, encouraging the widespread opening of freestone quarries in suitable outcrops, especially in the eastern and southern English scarplands. Elsewhere a wide variety of

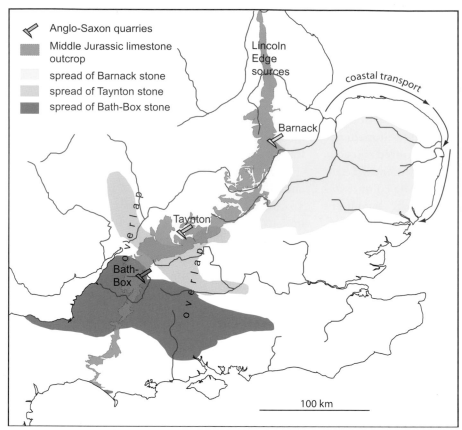

Figure 10.4 The outcrop of Middle Jurassic limestone-bearing strata in central southern England, the chief known Saxon quarries, and the dispersion of stone into Anglo-Saxon churches of the surrounding lowlands. In most areas the stone was shipped by river and/or coastal transports. Data generalized from Jope (1964), British Geological Survey (2007a) and Ordnance Survey (1991, 2011).

Legend:
- Anglo-Saxon quarries
- Middle Jurassic limestone outcrop
- spread of Barnack stone
- spread of Taynton stone
- spread of Bath-Box stone

Map labels: Lincoln Edge sources; coastal transport; Barnack; Taynton; Bath-Box; overlap; 100 km

Figure 10.5 Norman stone, Norman masoned. West view along part of the nave of Norwich Cathedral (1096–1145), (52.631897, +1.301183). Caen limestone makes up the Romanesque nave columns, arches and arcades. The robust column in the right foreground has fine helical moulding. The fan-vaulted roof with its celebrated bosses is 15th-century Perpendicular Gothic work: the stone ribs, slightly darker than the Caen limestone, are of Northants Jurassic limestone.

local stone was used in cathedral construction: purple Cambrian sandstones at St David's, Dyfed, yellow-weathering Coal Measures sandstones at Durham, and austere grey granite at Truro, Cornwall.

Natural roofing materials have been widely utilized across the island (Fig. 10.6). These include the thinnest and choicest limestone stone 'slates' of the Jurassic scarplands: Collyweston, Stonesfield and Cotswold. After quarrying or mining these were traditionally split naturally by exposure to a winter's frost and trimmed by hand using a particular double-chiselled 'slat hammer'. Carboniferous stone slate in West Yorkshire was hand-split and trimmed using sharp wedge-shaped hammers. The coming of the railway age saw the dissemination of

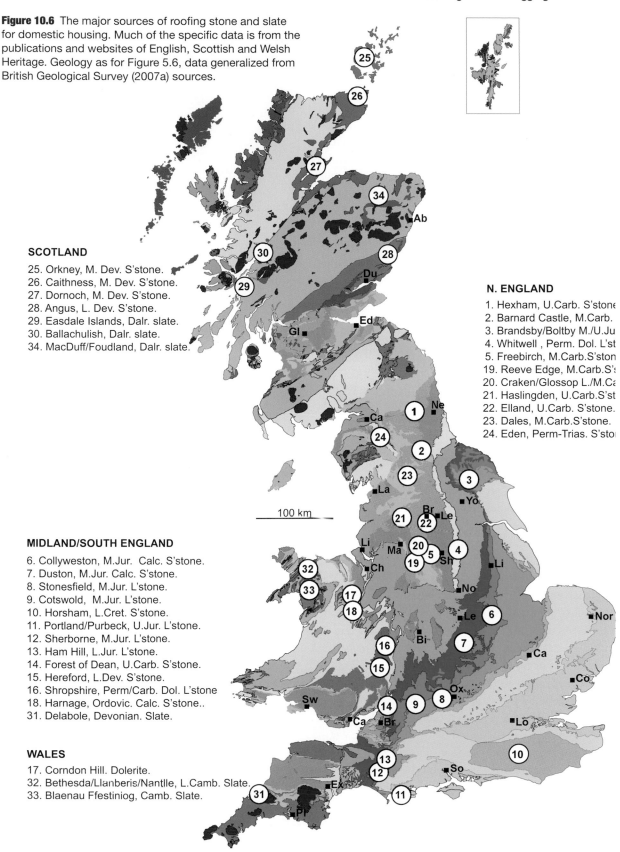

Figure 10.6 The major sources of roofing stone and slate for domestic housing. Much of the specific data is from the publications and websites of English, Scottish and Welsh Heritage. Geology as for Figure 5.6, data generalized from British Geological Survey (2007a) sources.

SCOTLAND

25. Orkney, M. Dev. S'stone.
26. Caithness, M. Dev. S'stone.
27. Dornoch, M. Dev. S'stone.
28. Angus, L. Dev. S'stone.
29. Easdale Islands, Dalr. slate.
30. Ballachulish, Dalr. slate.
34. MacDuff/Foudland, Dalr. slate.

N. ENGLAND

1. Hexham, U.Carb. S'stone
2. Barnard Castle, M.Carb.
3. Brandsby/Boltby M./U.Ju
4. Whitwell , Perm. Dol. L'st
5. Freebirch, M.Carb.S'ston
19. Reeve Edge, M.Carb.S's
20. Craken/Glossop L./M.Ca
21. Haslingden, U.Carb.S'st
22. Elland, U.Carb. S'stone.
23. Dales, M.Carb.S'stone.
24. Eden, Perm-Trias. S'sto

MIDLAND/SOUTH ENGLAND

6. Collyweston, M.Jur. Calc. S'stone.
7. Duston, M.Jur. Calc. S'stone.
8. Stonesfield, M.Jur. L'stone.
9. Cotswold, M.Jur. L'stone.
10. Horsham, L.Cret. S'stone.
11. Portland/Purbeck, U.Jur. L'stone.
12. Sherborne, M.Jur. L'stone.
13. Ham Hill, L.Jur. L'stone.
14. Forest of Dean, U.Carb. S'stone.
15. Hereford, L.Dev. S'stone.
16. Shropshire, Perm/Carb. Dol. L'stone
18. Harnage, Ordovic. Calc. S'stone..
31. Delabole, Devonian. Slate.

WALES

17. Corndon Hill. Dolerite.
32. Bethesda/Llanberis/Nantlle, L.Camb. Slate.
33. Blaenau Ffestiniog, Camb. Slate.

100 km

North Wales roofing slate across much of Britain. Whole districts became grey-capped as the stone provided shelter for mills, factories and homes in rapidly expanding industrial cities from Glasgow to Birmingham. Glinting dark-grey in the northern rain, the stone successfully engages the red brick and yellowish sandstone detailing of much Victorian and Edwardian building.

Today, international flavours are widely available in the stockyards of stone suppliers. Fortunately, planning permission in conservation areas requires local materials to be used in mending, replacement and alterations. Some is actively recycled from derelict buildings, but much still has to come from working quarries. These may often need re-opening from defunct sites, or extending to provide exact or close matches to old stone. Since a roof is often the most obvious external feature of a house, much attention has been paid in recent years to the conservation need for natural roofing materials. Consider the extreme effort by English Heritage to restore the beautiful stone slate roof of Pitchford church, Shropshire (Wood 2003): a temporary quarry was set up locally to work the Lower Palaeozoic Harnage shelly sandstone that originally roofed it.

Bricks, Tiles and Pots

Caesar Augustus was proud that he had changed Rome's public building façades from brick to marble. In the province of far-away Britannia, his nephew Claudius and imperial successors utilized freestone for only the chief public buildings: brick and tile for the rest. The Romans were skilled makers, carefully choosing local clay that would fire a particular hue of red-brown: a product of standardardized and closely regulated kilns. The distinctive thin, rectangular tiles were often laid as successive flat double- or triple-courses upon which freestone blocks or more irregular local stones such as flint nodules and rubble were placed and mortared (Figs 10.7, 13.3). Their wide re-use in later structures, particularly Anglo-Saxon, Norman and early Gothic churches, is well documented.

Roman pottery and tile manufacture was sited in clay-rich vales, often making use of the muddy deposits of former glacial lakes or soft plastic Eocene clays at localities around the western Hampshire Basin and elsewhere. Particularly noteworthy around Poole and Purbeck was the vigorous continuation of an Iron Age tradition of hand-built pottery; 'Black Burnished'. This

Figure 10.7 Wroxeter (Viriconium) legionary fortress baths/ basilica complex (52.674207, −2.644122). At 7m, one of the highest surviving Roman walls in the Island, known locally as 'The Old Work' and constructed in AD 121. Seven reddish double-tile courses separate regularly upwards-diminishing intervals of purplish red-brown hewn sandstone blocks quarried locally from the late-Carboniferous/early Permian Warwickshire Group, probably at Acton Burnell (see Toghill 2006). The scaffolding holes were subsequently rendered but have much enlarged by weathering over the past 2 millennia.

was fired in clamps and widely exported as cheap, robust and functional ware in both Britain and northern Gaul, particularly popular in military establishments (see Mattingly, op. cit. Chapter 16)

After the Roman withdrawal, domestic brick-making was entirely lost until the thirteenth century. As late as the founding of Eton College in the fifteenth century, bricks were expensive imports, mostly from Flanders; two million were estimated to have been used at the school. Then the technology was once more learnt in the Island, probably from Flemish refugees and immigrants. From the seventeenth century onwards their use became standard as medieval stone-built and wood-framed structures became unfashionable during successive rebuilding booms in city and countryside. Exceptions were always the Pennine stone belt, the Jurassic scarplands and the Midland Valley, and elsewhere in Scotland and Wales where near-perfect quarried freestone ruled the roost until the recent concrete revolution.

It was not until the coming of canals and railways that uniform, even-textured and durable mass-produced brick could be widely disseminated cheaply from huge brickpits and industrial kilns like those of the English Midlands and East Lancashire (Chapter 13). Before that, brick clay was usually locally dug and fired, often with poor strength due to inadequate purity, milling and poorly regulated firing. Large medieval kilns were located at coastal sites, for example Humberside, from where the bricks were transported, sometimes as ballast, to large cities like London by barge or coastal vessels. It has been estimated that the cost of transporting such bulky and heavy materials by road was so high that they doubled in price every ten miles.

Uniform and near-surface clays suitable for brick-making occur in rocks of almost every geological period in the Island. It is usually preferable to quarry clay that has little of the minerals gypsum (a hydrated calcium sulphate), pyrite (iron disulphide) and siderite (iron carbonate). These break down during firing, liberating gases that can affect local brick colour by reduction of iron in the clay.

Bits and Pieces – Cement, Concrete, Mortar and Aggregate

A world without cement or concrete seems inconceivable. These relatively cheap materials have made possible increasingly flexible architectural forms over the past 100 years. It is not exactly a new material: the most famous classical-age Greek engineer and architect, Eupalinus, used concrete around 550 BC for the aqueduct to his native Megara. The Romans developed and made extensive use of self-hardening (hydraulic) cements; the world's largest concrete roof still remains the coffered dome of the Parthenon in Rome, now over 2000 years old. Yet, as with brick, between the fall of the western empire and the fourteenth century the techniques of concrete manufacture were lost from the Island. Portland cement and its huge variety of 'hydraulic' derivatives were originally sourced from limestone containing interstitial clay or interbedded with claystone. Many such strata are found throughout the geological successions of the Island, known as 'cementstones' or 'hydraulic' limestones. Most notable are those of the Inverclyde and Border Groups in Midland and Southern Scotland, the Jurassic Lias and hydraulic limestones of England and South Wales.

Mortar is the essential link that bonds brick to brick and stone to stone. It is traditionally a powdered product of burnt lime mixed with quartz sand and slaked with water to make a workable paste. Such mortars provide a firm and durable bond, but one that is not as strong as the brick or stone itself, so it allows some compaction, settling, re-use and ease of repair. As important is its permeability to water, allowing structures to 'breathe' and to dry out quickly after wetting. Modern mortars using Portland cement powder as the bonding agent, though very strong, have low permeability and so act as a barrier to drainage adjacent to porous stone. This concentrates flow along the junction, causing uneven weathering, erosion and alteration – not good outcomes for repair or conservation efforts on older mortared-stone structures.

Many modern gravel and sand pits are worked for building construction aggregate in our larger river floodplains. Some of these are deposits of Quaternary glacial outwash that now define river terraces above the level of the modern channels. In an unexpected bonus, such workings continue to provide archaeological finds that shed light on the lives of our Neolithic to Iron Age ancestors in lowland valley and wetland environments. Most notable at time of writing is the so-called 'British Pompeii', a late Bronze Age hut circle at Must Farm, Whittlesey, preserved during a conflagration by wholesale collapse into a former river course.

Coarse aggregate for road construction needs to be tough, homogenous, relatively finely crystalline, and the individual pieces need to be angular in order to optimize internal friction and minimize compaction. Dolerite, felsite, diorite and tonalite all fit the bill, sourced from large quarries and crushing plants (Fig. 28.2) in many parts of the Island: Mountsorrel, near Leicester (Fig. 4.6), is reputedly the largest quarry in Europe.

'Silex Scintillans' – Flint

The mineral is a form of hydrated chalcedony, an amorphous or microcrystalline form of silica. Its abiding attributes are hardness, chemical stability

Figure 10.8 *The Stonebreaker* (1857–58) by John Brett (1830–1902) ©Walker Art Gallery. Newly quarried flints, like bonbons with their dusty, porous white rinds and black centres, await the stonebreaker's attention. The milestone reads 'London 23 miles'. In the background across the Mole valley is Box Hill, part of the Chalk scarp of the North Downs. The artist was a Pre-Raphaelite and geology afficianado (like Ruskin). He wrote of this plein air work: 'I can only work on it in sunny days – I hope it will look sunny… I am gypsyish sun-tanned all over now.' The luminous foreground colours fade into the distance (atmospheric perspective), the sinuous composition leading the eye around the painting, then off into the distant hills.

to weathering, a lack of internal 'grain' (no bedding, lamination or cleavage) and its brittle fragments possess intensely sharp edges. This unique cutting, building and striking material has had practical uses for hominids over at least 500kyr. It occurs as large decimetric-scale chalky-rinded nodules and tabular masses in the extensive chalk outcrops and coastal cliffs of eastern and southern England (Fig. 10.8).

Flakes may be struck off in various sizes by blows aimed at particular points using a hammer or other hard stones, a technique known as knapping, the oldest tool-making practice. The impact spalls off material along concentric ridges, which increase in circumference outwards from the percussion point. This type of fracture pattern was seen by the ancient Greeks in obsidian, a natural volcanic glass. It resembles the growth lines on mussel-like shells, *konche*,

hence the name for the fracture pattern – conchoidal. Flint makes thinner flakes more easily than obsidian, but to its disadvantage they are slightly more brittle. When flint is struck against materials like iron pyrites ('fool's gold') or steel, sparks are emitted.

Neolithic peoples took particular care in mining high quality flint suitable for the manufacture of working axe-heads and other implements. The most spectacular site to appreciate this nowadays is at Grimes Graves, west Norfolk (Fig. 10.9). Here are the surface traces of back-filled shafts that mined a prized flint-course 15–20cm thick, a tabular layer of black, wax-lustred flint known as the 'floorstone'. Scores of shafts were deliberately dug to this layer 6–12 metres below ground surface; how the layer was originally discovered is a great mystery. Radial networks of horizontal tunnels were dug outwards from

Figure 10.9 Aerial view of the pockmarked site of Grimes Graves (52.476870, 0.676312) in its protected periglacial landscape within Thetford Chase. Photo: ©English Heritage. Each pock (6–8m diameter) is the site of a single backfilled shaft excavation with the rims formed of spoil material left over. Note the very close spacing of neighbouring shafts.

the shafts with antler picks and deer-scapula shovels to exploit the near-horizontal floorstone. The flints and associated chalk debris were probably shovelled into sleds and moved to the shaft bottom, thence either carried up in baskets or perhaps lifted by windlass to the surface. The waste chalk was chucked out radially, raising a rim round each working. Unlike most other stone, the flint needed no drying or 'seasoning' because of its impermeability. Once divided into regular segments (blanks) it would have been either exported as blanks or manufactured *in situ* using stone hammers of various compact, tough rocks found as erratics in local glacial deposits. Radiocarbon dating of abandoned antler-pick tools shows a consistent span of Neolithic activity in the area ranging from 4.9 to 4.3ka.

Given the widespread outcrop of chalk scarplands in southern England, it is not surprising that many other local centres of Neolithic flint mining existed apart from Grimes Graves. We know of over a dozen along the South Downs chalk outcrop alone, with concentrated

activity evident around Cissbury. Some of the Sussex mines date to the early Neolithic, around 6ka, much older than Grimes Graves, but mining techniques seem to have been essentially the same. All the mines indicate impressive degrees of organizational capability in the Neolithic societies that nurtured them. This is because the safe execution of shaft-to-gallery mining calls for organized teams of workers and the application of accurate surveying to maximize the yield of flint. We infer that planned industry existed in the modern sense: overseers, surveyors, miners, surface workers, knappers, hauliers; probably even salespersons promoting their flinty wares far and wide for sale or barter. Discovery of votive offerings in various shafts at Grimes Graves indicate perhaps that the Neolithic miners were conscious of a debt to 'deities', perhaps subterranean ones like Anubis, the Egyptian jackal-like creature. It is interesting to note that Neolithic Seahenge's oak-palisaded ring has been interpreted by Francis Pryor (2003) as originally a shrine or mortuary whose upside-down

PART 5
Mineral GeoBritannica

This focuses on the location and history of extraction of mineral material from the Island's bedrock. Mining created subterranean landscapes down the ages as ore was hewn and shovelled from the dark underworld of adits, shafts, stopes and levels, as well as from opencasts. Wilfred Owen wrote one of his finest poems, *The Miner* (Chapter 15), commemorating a disaster in Staffordshire just months before his own quietly heroic death in 1918. He composed it before a spluttering and smoky fire in his billet-room in Scarborough, placing the dead miners with the lost souls of 'his' soldiers who had died semi-subterranean agonies in foxholes and craters – ruins he himself had suffered. We have now lost the mining profession almost entirely, though we continue to depend on overseas miners and our own mining graduates to provide mineral products from elsewhere. We may forget and disregard the physical realities of subterranean danger that underpin our entire civilization.

Chapter 11

Metals and Mineral Salts

Wealth piles on the mountains.
But where are the people?
We stand by watching the parades
Walking the deserted halls
We who are locked in the pits of gold.

Mazisi Kunene, *The Gold-miners* in *Zulu Poems* (1970)

Concentrated Essences

Many chemical elements useful to humans are rare, their average concentration in the crust only a few tens of parts per million or less. Silicon, calcium, potassium, sodium, iron and aluminium are notable exceptions with abundances of several tens to a few percent. In order to form economic mineral deposits, natural processes of enrichment must take place. The easiest to imagine is the formation of salts by the progressive concentration of seawater in a partly-enclosed sea, lagoon or lake, as in the Dead Sea. Just as successive boilings of kettlefulls of hard water eventually lead to a precipitated coating of calcium carbonate, so elements may concentrate and precipitate from solution as mineral evaporites in hot, arid climates. Common salt (the mineral halite) is the most abundant, but calcium sulphate (anhydrite and gypsum) and potassium chloride (polyhalite) also appear, the latter only when a brine is highly saline. The geological record contains several instances of entire seas that were progressively concentrated in this way – by successive 'toppings-up' from the world ocean. Such thick and extensive deposits are termed 'saline giants': a ready example in Britain and beyond is the huge Zechstein evaporite of Permian age (Chapter 7). It stretches from under Hartlepool to Cracow by way of the southern North Sea basin, the richest potash deposits being under North Yorkshire.

The concentrative power of petrifying springs is most obvious in limestone country. Their waters precipitate calcium carbonate as calcitic tufa or travertine as they emerge to de-gas their excess carbon dioxide on contact with the atmosphere. Their subterranean equivalents circulate warm aqueous fluids across pressure gradients or by convection. They may become heated by magma or even partly originate from that source themselves. Warm, briny fluids in particular are able to scavenge metals from rocks as they pass through them. The circulating hydrothermal waters preferentially pick up certain elements and concentrate them as they pass through cracks and pores in the crust. When such fluids are forced to cool or encounter changes in chemical conditions, mineral matter may precipitate out of solution: veins of mineral are the result. These may have successive layers of different minerals, witness to changing composition and temperature of the mineralizing fluids over time.

The older rocks of the Island are often shot through with mineral veins formed during deep burial and mountain-building episodes. They usually comprise common minerals like quartz or calcite with few other visible forms. But many mineral deposits of tin, copper, lead, zinc and gold also originate in this way, as a result of precipitation from heated waters. Minute specks of fossil fluid are often preserved in the precipitated minerals. Such 'fluid inclusions' can be treated experimentally to determine the temperatures that occurred during precipitation; commonly several tens to a couple of hundred degrees Centigrade.

Pits of Gold

Although metal-mining has left a more localized impact than that of coal (Fig. 11.1), there remain today many communities clustered around ancient mining centres: Truro, Redruth, St Austell and Bodmin in Cornwall and Devon; Matlock and Castleton in Derbyshire; Scunthorpe, Corby and Middlesbrough in Lincolnshire, Northamptonshire and Teeside; Greenhow and Swaledale in North Yorkshire; Millom in Cumbria; Alston and St John's Chapel in Co. Durham; Leadhills and Wanlockhead in the Lanarkshire and Dumfriesshire Southern Uplands. Although no production occurs today, in stone and bone tools to realistically and economically dig such a structure. Modern resurveying estimates that over 5km of workings were developed there, with greater than 40,000 cubic metres of ore extracted in the thousand years between 1700 BC and 700 BC; that's about 120 tonnes per year. It was probably one of the biggest copper workings in northern Europe. Certainly no other locality has Bronze Age galleries so well preserved: the modern visitor can safely walk out the mined levels, courtesy of an inspiring Visitor Centre and Museum.

Shaft-mining techniques were brought back to the Island by the Romans in their exploitation of veins of lead ore and gold. Most sites in Wales, Mendip and Derbyshire were of inclined-vein type. Many were doubtless discovered by the technique of 'hushing', where ponded reservoirs release their water suddenly down hillsides to erode surficial loose sediment and soil, revealing any metalliferous veins present in the bedrock. Once excavated at surface, these were exploited thereafter by adit and stope mining. Near-horizontal drives and levels were excavated (stone-lined and roofed where necessary) outwards towards any inclined veins in the vicinity. Basic Euclidean geometry and accurate surveying enabled calculation of the required orientation, dimensions and lengths of such tunnels.

To extract ore from an inclined vein, Roman-age miners had to stand below (in the 'footwall') hammering and drilling across or upward to the top of the vein (the 'hanging wall') using iron tools (hammers, hand-held drill bits, crowbars) to extract the ore. This was potentially a dangerous and bloody business and it cannot be doubted they would have worn serious protective gear. Fragmented ore was shovelled from the footwall floor into tubs or sledges and the process repeated. A stope or scoop was thus formed out of the vein. A level was then driven parallel to the vein to extend the stope laterally (Fig. 11.4). The extracted ore always came back on sleds to the beginning of the stope, then back along the level roadway to the shaft for transport to the surface.

As veins were exhausted at one level, the shaft was lowered and a new system of deeper levels established. The lower limit to mining was always determined by the ability of the pumps available at the time to drain out groundwater faster than its natural influx. Otherwise drainage had to be by the cutting of special adits from lower elevations that could drain by gravity. These were called 'soughs' in the Derbyshire lead mines of the seventeenth century, first devised there by a Mr Bushell. We know nothing of the Roman skills in such matters, but

Figure 11.4 An early (1890s) flash photograph taken underground of a level in the stope of East Pool tin mine in Cornwall (50.230847, –5.261911). View shows the gap left along a worked-out and steeply dipping tin vein with its massive timber supports laid normal to the vein walls. Photo: http://min-eng.blogspot.ie/2015/05/an-appreciation-of-jc-burrow-pioneering.html

by the early seventeenth century, Cardiganshire mines belonging to Hugh Myddleton were being drained by waterwheel-driven mechanical pumps, perhaps originally a German invention out of Agricola's *De Re Metallica* (Chapter 3).

Iron in the Soul

Although iron is abundant, in the top eight of crustal elements, it usually occurs as mineral oxides, silicates and carbonates that must be smelted. The native (elemental) metal is only available from meteorites, an extremely rare and spatially random source, though fortunately quite distinctive chemically, being rich in nickel and cobalt. Hence iron's probable status as the ultimate precious metal in pre-Iron Age history; witness the recent spectacular discovery of a gold-handled and scabbarded knife in Tutankhamun's tomb (1323 BC). It has an expertly forged blade of meteoritic iron.

Iron-smelting from naturally occurring ores was discovered 300–500 years later, heralding the Iron Age and the birth of the extended Smith family. The starting point were oxides and carbonates of iron reduced at very high temperatures by charcoal. Soon these early smiths hammered out wrought-iron rods for trading or made copies of copper and bronze edged-implements of revolutionary efficiency in agricultural, hunting and military endeavours. Others were used for direct casting or to make metallurgical steel by heating with charcoal for long periods.

During the early development of smelting in the iron-rich Sussex Weald, the trickle of released iron mixed with embers and surrounding soil into malleable knobbly masses called 'blooms'. The ore here was clay ironstone (siderite, an iron carbonate), subsequently smelted in small blast furnaces. These were constructed against natural or artificial ridges to allow a throughflow of air for more efficient combustion. Developments then involved furnaces with a top-down delivery of raw materials and a lower outflow for draining off the metal into bar-shaped moulds before cooling. These became widespread by the seventeenth century, with over seventy known from the Sussex Weald alone. Here is the late-sixteenth-century revelation that the Sussex High Weald was:

> Full of iron mines, all over it; for the casting of which there are furnaces up and down the country, and abundance of wood is yearly spent; many streams are drawn into one channel, and a great deal of meadow ground is turned into ponds and pools for the driving of mills by the flashes, which, beating with hammers upon the iron, fill the neighbourhood round about it, night and day with continual noise…

Production sharply declined after the Civil War in favour of localities like the Forest of Dean and the valleys of Glamorgan and Monmouth, which had more accessible and larger ore and timber reserves. Water power was essential to power the forge hammers so that good valley-damming locations for reservoirs determined the sites of both 'hammer-ponds' and the smeltworks to which the iron ore and charcoal were brought. Widespread charcoal-based smelting for iron is often said to have caused massive deforestation, but it is forgotten that charcoal burners are (and must always have been) canny, far-seeing folk. In both Sussex and the Forest of Dean pollarding and coppicing must have sustained livelihoods over the many generations since

Roman times. Eventually, demand exceeded supply and legislation was placed to restrict use of Sussex timber for this purpose.

A stupendous and far-reaching advance in iron smelting came in the early eighteenth century when coked coal was used as a substitute for charcoal by Abraham Darby at his Coalbrookdale works in Shropshire. Coke manufacture led to the spread and rapid growth of iron smelting in any countryside that lay adjacent to coal-bearing outcrops. Also there, to the delight of the early industrial iron masters, were contiguous deposits of clay ironstone, called 'blackband' ores from their interstratification with coal and black shale. Further technical advances in the iron industry were improved ore selection, smelting and empirical metallurgy, including the role of limestone and dolomite as fluxes and removers of phosphorous impurities. Manganese ores from Harlech and Ireland were added as de-oxidizers and de-sulphurizers.

By the mid-nineteenth century, industrialization of the steel industry enabled successful smelting of high-phosphorous chamositic (iron silicate) sedimentary ores from the eastern Jurassic scarplands of the East Midlands and Cleveland. This led to the growth of entirely new industrial centres of population: mid-Victorian Teesside (Middlesborough), Edwardian north Lincolnshire (Scunthorpe) and 1930s north Northamptonshire (Corby), the latter populated by many Scottish and Irish families seeking new lives in the depths of the Great Depression.

The rich haematite iron ores of south Cumberland were low in phosphorous and began to be exploited for use in Bessemer steel plants from about the mid-nineteenth century, leading to industrial developments at Barrow-in-Furness. Discovery of similar deposits around Bilbao in the Basque country led to imports of up to five million tons per annum by the mid-1880s, but this source had declined by the 1930s.

The British iron-ore extractive industry that enabled continuation of mass steel-making through years of submarine blockade in two world wars is now extinct.

'Great Store of Copper Ores and Rich Leaders'

As discussed above, copper ores were initially mined and smelted as part of the copper 'pyroculture', but copper's greatest use in antiquity was in the making of bronze, a harder alloy with about seven percent of tin. As industrial archaeologist Arthur Raistrick stresses (1973), the transition in organization from medieval to industrial practices in the sphere of exploitation and smelting of non-ferrous metals like copper, tin, lead and zinc took place in the sixteenth century. It began in continental European centres like Saxony and Bohemia, where water power was harnessed to the multiple tasks of ore-lifting, dressing (pulverizing) and smelting, as summarized in Agricola's *De Re Metallica* (Chapter 3). At this time England's mining industry was probably less advanced than that left by Rome. Faced with the unparalleled invasion threat from Philip II's Spain and national shortages of metals for casting of bronze naval cannon (the metal was mostly an expensive import from Hungary), Elizabeth I turned to both European expertise and the expansion of commercial ventures originally set up by her father.

Hence, one Daniel Heckstetter came from Germany in 1563 and visited all the main mineral fields then known in the country, from Cumberland in the north to Cornwall in the south-west. Following his advice, Elizabeth created in 1568 two chartered companies with joint-stock (i.e. mixed private and royal monies) to mine, smelt and refine copper (the Mines Royal) and another (the Mineral and Battery works) to make bronze and brass armaments and for drawing iron wire. With German engineers and supervisors, the Lakeland copper mines of Keswick and Newlands, were soon producing many tons of refined copper weekly. By the end of the sixteenth century, cast iron was more widely used for cannon, though production of copper continued in the Lakes until the Civil War, when the mines and plant were inexplicably destroyed.

Much later, copper became a key material in the Electrical Age instigated by the practical and theoretical discoveries of physicists Faraday and Maxwell. The chief economic occurrences in the Island worked during the Industrial Revolution were now in Devon, Cornwall and Anglesey, the latter by opencast working, the former by deep shaft mining courtesy of efficient drainage by James Watt's formidable steam pumps.

The tin ore necessary to make bronze alloy is restricted in its distribution to Cornwall and West Devon. Indeed, it is scarce elsewhere west of the Orient, save for lesser deposits in southern Brittany and north-west Spain whose geological histories are similar to that of Cornwall (Chapter 7). The name 'Cassiterides' (Κασσίτερος is Greek for tin) or 'Tin Islands' was applied to all these regions in Phoenician and early Greek usage. This has suggested to some that widespread trading links for the metal existed from one or all of these places during the early Bronze Age. Recent research (see Cline, 2014) suggests, however, that imports from Afghanistan into the great Bronze Age civilizations of the Middle East and Aegean may have been predominant. Eventually though, Cornwall and West Devon became the largest producers by far, dominating world production well into the twentieth century. 'Stream tin' was probably the earliest source, concentrated at certain levels in Pleistocene river and estuarine sand and gravel, where it was got by sluicing, dredging and panning (Gerrard, 2000).

Lead and Zinc

The Carboniferous Limestone of upland Britain was host to a remarkable concentration of vein-type mineralization involving the chief lead-bearing mineral, galena, a lead sulphide. After smelting and silver extraction, the Romans used the metal in roofing, water piping and as a lining for coffins and baths; all the baths at Aquae Sulis (Bath), for example, were lined with lead sheet, thirty tons of it being removed and sold during nineteenth-century renovation of the Great Bath. As noted previously (Chapter 8) large-scale mining began in the Mendip Hills of Somerset only a few years after the Claudian invasion. The Romans knew from their experience in Attica that galena gave up a small but significant amount (c.0.4%) of silver as a by-product during smelting. Surviving lead pigs from Mendip (Fig. 11.5) and the English Midlands (the ore probably from Derbyshire) bear inscriptions to the effect that silver extraction was probably the more profitable activity. Thus we read on one: 'G(aius) Iul(ius) Protus Brit(annicum) (Sociorum) Lut(udarensium) ex argentariis', loosely translated by Arthur Raistrick as 'British (lead) from the silver works of the Lutudarum company (under the management of) Gaius Iulius Protus'.

Many galena veins were also rich in zinc ore, as in the Mendip and North Pennine orefields, the latter also rich in the aforementioned non-metallic minerals, barite and fluorite. The Romans greatly expanded the use of zinc in the manufacture of brass, an alloy with copper. Until the mid-eighteenth century calamine was the main ore exploited. Subsequently the sulphide, zinc blende (sphalerite), was extensively mined after the development of a roasting process for its treatment. A major increase in zinc mining followed the introduction of galvanized iron and steel in the mid-nineteenth century.

Figure 11.5 Two Roman lead pigs (around 60cm long, weighing about 70 kilograms) from Charterhouse, Somerset bearing the letters of Imperial ownership: Emperors Vespasian (AD 69–79) and Antoninus (AD 138–161). Photo: by kind permission of Bill Rea ©the estate of Arthur Raistrick.

Gold

Gold amulets and neck-pieces (torcs) seem to have been *the* Iron Age bling. They turn up magnificently fresh in hoards, especially in arable East Anglia, a region far from any known gold mining locality, but where the plough furrow has always run deepest and most commonly. Smaller objects like buttons, brooches and pins also turn up across the country in countless inventories of excavated grave goods from the Neolithic onwards. Accepting that some, at least, of this gold must have been imported, especially from Ireland, much surely would have come from indigenous sources.

Gold mineralization mostly occurs in Palaeozoic and Precambrian rocks. Alluvial gold from Scotland once figured high in production terms, with the Leadhills area outstanding and smaller amounts from further north at Helmsdale. Veins in bedrock of the Dolgellau 'gold belt', North Wales and the Dolaucothi mines of Central Wales have provided by far the greatest production for 2000 years. The veins are predominantly of crystalline quartz, gold tending to be richest where the mineralizing fluids cut Lower Palaeozoic rocks rich in organic carbon, so-called 'black shales'. The gold precipitated from solution along with the far more abundant silica. Between 1860 and 1909 it is recorded that 3500 kilos of gold was mined from the Island, 90 percent from the Dolgellau and mid-Wales gold belts. At the peak of recent gold prices in the aftermath of 'Brexit' this would have been worth around £140 million.

Mineral Salts

As noted previously (Chapter 9), evaporite salts are precipitated from aqueous sources when the climate is arid and hot. So we find them interbedded in Lower Carboniferous, Permian, Triassic and late-Jurassic strata. The Permian magnesium-rich brines penetrated deep below the surface, changing the more permeable parts of originally calcareous sediments (notably reef buildups) into dolomite to become the Magnesian Limestone. Only traces of other more soluble salts remain at outcrop, the bulk having been dissolved over the past few thousands of millennia by percolating fresh groundwaters. The pristine anhydrite, halite and polyhalite is now encountered in wells, boreholes and deep mines under Teesside and North Yorkshire and in the offshore North Sea basin.

Here they have served for 200Myr as an impermeable seal for large reserves of natural gas that leaked from the underlying Coal Measures deep below the sea bed. The potash salts of North Yorkshire are amongst the world's largest, currently mined at Boulby near Saltburn, close to an old ironstone mine, and set to be substantially enhanced in the next few years by a new mine further south along the Yorkshire coast (subject to planning permission). Elsewhere, important deposits of gypsum, alabaster and anhydrite occur in the Permo-Triassic marls of the East Midlands and the Lakeland periphery.

Common salt (the mineral halite) is prominent in the Cheshire Triassic basin. Beginning in Roman times around Droitwich, it was concentrated in lead pans by boiling off natural brine sourced from natural springs and shallow pits. Other important sites were estuarine salt marshes close to supplies of wood or coal. The prime example was the growth of medieval and later saltworks along the north-east coast of England, utilizing the easily mined Northumberland/Durham coal as the fuel source: North and South Shields were major centres.

Chapter 12

Coal, Peat and Oil

Should our coal mines ever be exhausted...we should lose many of the advantages of our high civilisation... That there is a progressive tendency to approach this limit, is certain; but ages may yet pass before it is felt very sensibly; and when it does approach...the transition may not be very violent...

Phillips and Conybeare, *Outlines of the Geology of England and Wales* (1822)

The Old Black Stuff

To most of us over a certain age, mining is synonymous with coal. The Island's extensive outcrops of coal-bearing Upper Carboniferous rocks define the 'exposed' coalfields (Fig. 1.1). Many of these have been exploited in both shallow and deep workings and in opencast pits since Roman or earlier times. At their margins they may be overlain by younger strata through which shafts were sunk to mine the coal down to maximum depths of a kilometre or so. These are the 'concealed' coalfields, recognized and exploited since the early eighteenth century.

Only the Kent, Selby and Belvoir coalfields were totally concealed in that they had no link to contiguous outcrop: their discovery was by geological prediction and borehole testing.

The national demise of coal working over the last forty years gives perhaps the most obvious and widespread example of abandonment of mineral extraction in our recent past, with all its human consequences. It seems astonishing now, but at its peak there were nearly a quarter of a million people at work in a coal-mining industry that raised hundreds of millions of tonnes per year. As we saw in Chapter 1 (Fig. 1.1), the traces of the extinct industry are marked by late-Georgian, Victorian and Edwardian industrial population centres based on the exposed coalfields. Also the hundreds of 'pit villages' in the concealed coalfields that lie under great swathes of rural northern, central and eastern England. The terraced housing of these settlements, either as suburbs around ancient nuclei or as Victorian or Edwardian new-build (Fig. 12.1) bear silent and awkward witness to the death of coal. Though still viable as settlements, the loss of

Figure 12.1 All over the rural areas that lay above concealed coalfields, small 'pit-villages' grew up. These could comprise just a couple or so of brick-built terraces for one particular pit, like this late-19th-century example at Battram, Leicestershire (52.681383, −1.363507).

their reason for existence creates images as emblematic and moving (Nield, 2013) as the faint remains of medieval deserted villages that litter the landscape of eastern lowland and midland England or the ruins of a Highland Clearance settlement at the head of a loch in Assynt.

Peatlands

Since the late Bronze Age, after nearly three millennia of wetland growth, the surfaces of upland western Britain still contain extensive areas of sphagnum mires. These are raised into dome-like carapaces fed by rainwater with thick underlying accumulations of peat. Peats also formed widely and thickly in the wetland margins to lowland river floodplains and estuary margins since sea level highstand 6–7ka. In areas no longer receiving detrital sediment these also developed extensive raised bogs. Though many have been dug out for peat, there remain written records concerning their inflation after heavy rains and subsequent contraction during periods of drought.

Both upland and lowland environments of peat formation were a godsend to later generations, from medieval times to present. Many upland communities, especially in Scotland, still retain ancient rights to cut their own peat from neighbouring bogs (Fig. 12.2). It was also widely used in Pennine lead smelters in combination with locally derived coals (Chapter 26). In most areas all that remains of a once-widespread source of fuel are the shallow overgrown scars of old peat diggings and the stone remains of peat-drying sheds.

Along lowland river floodplains, particularly where adjacent heavily populated fertile lands had been deforested, widespread medieval peat-digging took place. As briefly noted in Chapter 9, this is most evident in East Anglia where the many 'broads' – large, open, shallow-water lakes surrounded by remnants of wetland fen, carr and reedbed – are the flooded legacy of near-industrial scale peat extraction. At the time, probably between the fourteenth and sixteenth centuries, fuel-hungry Norwich, then the second city of England, and its many satellite market towns and villages were the populous centres of a great medieval weaving industry (Chapter 34). Though we have little idea of how peat extraction was governed, transport of the peat to market was almost certainly by shallow-draught boats (wherries) along the riverine arteries of the region. Later peat-cutting activity in areas subject to enclosure was strictly controlled by parish ordnances. For example, in 1815-16 we read of the following admonitions from the Overseer of the Poor for Bressingham, a village close to the wide wetland flanks of the Waveney valley in south Norfolk: 'Whereas it doth appear divers paupers have trespassed upon the poors fen by selling turf to owners and occupiers... we the undersigned do agree to proceed against all such offenders in future...'

Figure 12.2 Fresh-cut turf with a previous cut drying out behind. Photo: Shutterstock 250560337 ©DrimaFilm. Once a familiar site throughout Britain's highland and lowlands, from the bogs of A'Moine in Sutherland to the Norfolk Broads and the Somerset Levels.

The background to this was that Bressingham Fen was part of a common that had been enclosed in 1799: Georgian parliamentary enclosure did not just include farmland. Some acreage was obviously left as the regulated 'poors fen' from which the temporary and long-term unemployed had stolen and sold on their peat. Many such fens are now protected nature reserves – significant and precious surviving wetlands.

The New Black Stuff

Liquid hydrocarbons seep naturally from outcropping sedimentary rocks in many areas, notably around Castleton, Derbyshire; Formby, West Lancashire; and Kimmeridge Bay, Dorset. The most widely distributed and largest production in the nineteenth century was from retorted oil shale quarried and mined from the Oil Shale Group of Lower Carboniferous sedimentary rocks in West Lothian, central Scotland. The industry grew up under James Young at Bathgate, largely to supply kerosene (paraffin) for lighting and heating from the mid-nineteenth century (the world's first commercial oil refinery) to the 1960s. Although reserves are still large, the process is no longer economic (or probably, desirable). All that is left are reddened 'bings': spoil tips of the retorted shale residue, some since quarried for aggregate but nowadays mostly flat-topped, landscaped and protected hills rising significantly from surrounding low ground.

Hydrocarbons are nowadays pumped from their pressurized subterranean reservoirs by deep production wells. The earliest commercial discovery dates from exploration of Carboniferous prospects in the First World War and the drilling in the early 1920s of a successful well at Hardstoft, Derbyshire. Production was never large and the well was re-opened at the beginning of the Second World War. In the annus horribilis of 1939 the Eakring-Duke's Wood prospect in Nottinghamshire was drilled, coming onstream in 1943 (Fig. 12.3) and subsequently at a score or so of other small fields located in the Carboniferous rocks of the East Midlands. In the past twenty years larger fields have been discovered in onshore southern England, chiefly in the Wessex and Weald basins, involving Jurassic and Triassic reservoirs and traps created by the last throes of the Alpine mountain-building. The largest by far is at Wytch Farm, Dorset, which continues to produce oil and natural gas from its well-camouflaged well-heads.

Figure 12.3 Eakring-Duke's Wood oilfield, Nottinghamshire (53.13418, −0.98532). The square-jawed and broad-shouldered *Eakring Roustabout*, a sculpture by Jay O'Meilia of Tulsa, Oklahoma to commemorate the United States oil contractors from there who in 1943 helped drill scores of production wells in this, the largest wartime-producing oilfield.

The geological story of offshore oil and gas exploration and production from the waters of the Island's surrounding seas is too large an endeavour to fit within the current volume. Yet we point to its obvious consequences: the engineering and geological expertise accumulated by the world's first and largest offshore industry, which enabled onshore terminals to process the incoming streams of hydrocarbon, from north Norfolk to Scapa Flow and Sullom Voe via the Humber, Teeside, Buchan and Moray; the rise of Aberdeen as the new-age industrial nexus of the north; an inheritance that enables other offshore industries to develop – wind power, solar, tidal.

PART 6

'...to show to the world what exists in Nature'

How landscape is seen depends not only on its undoubted physicality but also who is doing the looking. Artists, sculptors, architects and writers have enriched our appreciation of landscape down the ages in many different ways. Chapter 13 looks at the collection of made-objects that constitute our architectural and monumental heritage. Chapter 14 explores the sculptural representation of landscape and the vibrant showing-off of other sculptural forms within landscape. Chapter 15 traces the tradition of modern landscape painting from its watery origins in the Netherlands to the sublime mountain landscapes of upland Britain and in various seascapes, skyscapes, moonscapes and riverscapes. Chapter 16 follows the rich vein of landscape that runs through our literature: from times before cogent explanations for natural phenomena to the last two or three generations of poets and novelists who confidently use some landscape-founded geological argument, term or metaphor in their creative work.

Chapter 13

Architecture and Monuments

A brick or roughcast house is a home and quite satisfactory as such, but a stone house, even if it is only a low stone cottage, has something of the monument about it. Who can maintain that the same effect would be produced if the piers of St Paul's were made of nine-by-three-inch bricks, or the towers of Gloucester Cathedral were carried out in pebble-dashed cement rendering?

Eric Benfield (former Purbeck stonemason), *Purbeck Shop* (1940)

As mentioned earlier (Chapter 4) the relationship between people and their material environment has been a close concern for millennia. Since the Industrial Revolution this relationship is most immediately expressed in urban built environments. Eric Benfield's traditional view of buildings and materials (see chapter motto above) is fine and good, but imagine sourcing enough stone for all of today's new builds! Contemporary sociologist Richard Sennet has focused his interest on the development of modern cities, of urban life and the nature of work. What matters is a deep involvement and understanding of materials, whether using concrete, installing solar panels or building a stone wall:

> The work of the hand can inform the mind. The material world speaks back to us by its resistance, its ambiguity, by the way it changes as circumstances change, and the enlightened are those able to enter this dialogue and, by so doing, come to develop an intelligent hand…

A Roof Over the Head

Architecture concerns human-constructed edifices with a function: the main requirement for any building is that it should provide shelter. Anyone who has built a den or a bender knows that the main architectural difficulty involved here is the roof, more precisely its span and support. Buildings have a profound and sometimes lasting place within landscape, the most visible evidence of any culture. Indirectly they express social fabric, beliefs and relationships with nature; they present us with a dialogue enacted between architect, materials, mason and landscape. Some buildings and monuments disregard landscape, others work in harmony with it. Throughout history this relationship has been ambiguous, by turns respectful or exploitative; think of benders, wigwams, cob cottages, follies, pyramids, city walls, castles, cathedrals, henges, stone circles, adobe pueblos, cliff dwellings, skyscrapers.

Before the Industrial Revolution there was a closer relationship between rural peoples and the land. Building and roofing stone, for example, were generally sourced close-by for economic reasons, chiefly the high cost of land transport. Gradually the vernacular architecture of villages and towns became an expression of a region's landscape and geology; and so many remain. Before that, in prehistory back to the Mesolithic peoples, the use of stone was very restricted and most upper building frameworks were of wood.

Architects have always plundered history (think of Victorian Gothic): contemporary buildings can utilize raw materials sourced worldwide and Post-Modern architecture can reference many styles within the one building. Yet so many modern buildings could be anywhere; they have little connection to a particular site's urban or rural landscape. On the other hand, we have all benefited from the use of made materials; the free forms of modern buildings are liberating in their effects on our awareness of both internal and external space. The Scottish Parliament building in Edinburgh, Selfridges in Birmingham, and 'The Deep' in Hull (Fig. 4.11) all utilize their contrasting materials with magnificent flights of the imagination. In addition, modern technology has led to designs that are sensitive to energy conservation, the use of light and general function. Yet having said all that, there must be 'horses for courses', as Eric Benfield reminds us.

Rock Shelters to Monuments

In warmer prehistoric stages Palaeolithic humans found natural shelter, rock caves providing ready-made roofs. Excavated Mesolithic homes were family units occupying interior floor areas of around 20 square metres (about 5m width). They featured an external low palisade of post-timbers and stakes, probably mostly birch, upon which raked roof-poles bore a fixed covering of turf, straw or reed. The earthern floor spaces had hearth fires with a surrounding litter of bone and flint fragments.

This basic structure was used for millennia and was basically unaltered even as the incoming Neolithics and their Bronze Age descendants and immigrants brought over their skills as monument builders (Chapters 4, 10). Thousands of megalith (Greek: 'large rock') monuments are scattered over the Island, from the Orkneys to Land's End. They can be crudely divided into dolmen (cromlech) tombs and linear, circular and henge monuments. Dolmens generally comprise a portal entrance made using two or more upright stones with lintel stones on top (Fig. 13.1). These may lead into a chamber or chambers, hence their classification as passage graves. The size of the chambers and their purpose, whether for single or multiple burials, varied. After construction a mound of rubble and earth covered the stone skeleton of portal and passage. The external platform of the resulting mound (barrow) could be circular, rectangular or oval. Chambered burial mounds were often linked or near to long ceremonial or ritual walkways (cursi) bounded by banks and ditches; the longest is the Greater Cursus (3km) north of the main Stonehenge site. The linear and circular framework monuments were constructed entirely in stone, usually won from local outcrops: Cornish stone circles are granite, Outer Hebrides gneiss and Orkneys flagstone. The chief exception is Stonehenge (Chapters 4, 10; see below).

Megalithic tombs point to a belief that the dead had a continuing existence in a material or spiritual form and needed shelter from the elements. Perhaps even more important, it was no longer just the acquisition of hunted game, shucked oysters or netted fish that concerned the people. Solstice-aligned megaliths (Fig. 13.2) indicate a measured reaction to the timing of seasons that brought the life-nurturing elements of rain and sunshine, enabling crops and livestock to flourish. The sun's rays, the earth beneath, soil and projecting hard rocks, had become more directly relevant than to Mesolithic ancestors. It is useful to see them in the light of the ancient Greek myth of Demeter and Persephone: the daughter released from rocky under-the-earth Hades to herald fertile spring in the sunlit surface landscape.

Figure 13.1 Bryn Celli Ddu, Anglesey (53.207753, −4.236126), a later Neolithic passage tomb built over a previous millenniums henge monument, with evidence for even older (Mesolithic?) postholes. **A** This mid-19th-century engraving (from Daniel 1963) shows the passage stones visible and a scrap of barrow earth (with sapling!) remaining. The passage is aligned with midsummer sunrise that let light into the chamber. **B** Rear view (western) today after modern excavation and reconstruction. Inside the chamber is a unique free-standing, smoothed and tapered schistose pillar.

Figure 13.2 Winter solstice scene at Drumbeg stone circle, Roscarberry, West Cork. Many astronomical sites were designed for this celebration in Neolithic Britain. The setting sun's rays are striking the shorter, flat-topped 'altar' stone to the right (by the dog) and then onto the large backstone at the extreme left (in front of the person).

Orkney was clearly a major centre of Neolithic mason technology. Its inhabitants made full use of abundant local Middle Devonian flagstone (fissile but tough sandstone) in both passage tombs, stone circles and, uniquely, in local vernacular architecture as at Maeshowe (see Fig. 19.4). Stonehenge is, of course, something else. The stone itself is all exotic, imported to the site from far and near. The earliest came the furthest. The central-standing 'bluestones' were erected in the latest Neolithic, a bit before 4.5 millennia ago. They are the chief geological wonder here – intrusive sub-volcanic rocks from the Lower Palaeozoic of Dyfed, astonishing geocultural erratics here in the Mesozoic chalk downlands of Wiltshire. Jacquetta Hawkes suggested that they came from some pre-existing monumental circle, perhaps gifted as an already sacred contribution to the early site. But near is still far when we consider the labour involved in moving the younger, but much larger, sarsen stones some time shortly after 4.5 millennia ago. These huge rocks were hand-masoned to become the outer, once-continuous lintelled circle; the later inner arc of trilithons face into the midsummer sunrise-aligned avenue to the east. They probably all came from the Marlborough Downs some 50km to the south-west: tough silica-cemented sandstone (called silcrete) formed below-surface by warm, humid weathering of early Paleogene sands that once overlay the local Chalk. The labour involved in surface-masoning such rock using hand-held stone axes (themselves probably also made of silcrete) is almost unimaginable.

Hill Forts to Roman Planning

There was usually no great change in the way people lived and built during the Bronze Age except that barrows generally became smaller. The ritual henges continued to be used and elaborated, like the Stonehenge triliths. By way of contrast, the Iron Age was clearly a troubled interval, with bitter inter-tribal rivalries between migrant Gaulish, Belgic and Germanic tribes lasting right up to the Roman conquest. Their architectural legacy was the hill fort. Thousands sprang up, concentrated across the length and breadth of the western and northern 'rocky' parts of the Island. Longhouses and large roundhouses were built within resistant hilltop scarps of sandstone, limestone, basalt, dolerite and granite. Most houses were sited in the shelter of the innermost bounding defensive mounds. Inboard were raised granaries, rock-cut grain storage pits and votive shrines. Many were part-walled or made use of natural outcrops overlain by turf and wooden ramparts.

It would, however, be a mistake to regard the whole of the Island in the Iron Age as just a scattering of hillfort settlements. Practical and cultural influences determined that almost the whole of the central and eastern scarplands of England, then as now rich agricultural land, were dominated by 'open' tribal settlements. In his account of the first invasion, Julius Caesar referred to these as 'oppida', familiar to him from his Gaulish campaigns. Such Iron Age communities pioneered agricultural settlement and deep ploughing with oxen of long field systems on well-drained terrace lands along major southern, eastern and midland rivers and estuaries. Their settlements had low 'preservation potential' for the archaeologist compared to those in the hilly west.

Despite a shaky start and some later scares, Roman Britain was a stable and altogether centrally planned enterprise, based firmly on the Imperial prototype tried out in numerous provinces from Syria to Spain. The Roman cultural package had town architecture at its very heart. New towns were planned from scratch, undefended by walls initially, and designed using a grid based on two main streets intersecting at right angles with central forum, basilica, market place, shop-lined streets, and some little way out of town the amphitheatre. The architecture of major public buildings generally used dressed and mortared freestone with fired red clay tile intervals as spacers to line a rubble infill.

Roman public buildings overcame the age-old roof span problem by the use of sturdy columns supporting rounded barrel arches with timber rafters supporting pantiles. The barrel arch had featured in Etruscan and Hellenic architecture and became the chief roof-support structure in all their larger buildings (Fig. 13.3). Housing for the majority of the urban population was done with brick, tile, wood, wattle and render. The country villa lifestyle spread widely over the southern and eastern island. The villas were like the manor houses of medieval times: luxury homes and estate outbuildings within large-scale managed farmland. They were extended stone and brick

Figure 13.3 The 1st century AD Balkerne Gate, Colchester (Camulodunum): the oldest and finest Roman gateway preserved on the Island (51.889649, −0.894526). Photo: Wikimedia Commons ©Tanya2501. It led traffic to and from Londinium. Both barrel-vaulted arches show inward-tapering voussoirs hacked from the same rectangular tiles that make up the tiled courses between the mortared wall flints.

structures with all the paraphernalia of Roman culture: colonnaded walkways, baths, domestic temples and dining rooms with floors of brilliant mosaics.

Anglo-Saxon to Norman Masons

From about AD 200 many Roman towns with sea-going estuarine connections were enclosed by fortified and gated curtain walls (Fig. 13.4) as protection from Saxon raiders. The Roman imperial economy that had sustained urban life in Britain finally collapsed in AD 410. The concept of a long subsequent 'Dark Age' is a historical exaggeration. Yes, much practical knowledge in the building, utility and extractive industries was lost (including the barrel arch), but the incoming Anglo-Saxons were no barbarians just because they were pagans. They initially made use of the gradually crumbling remains of Roman towns, quarrying them for stone. They also built new, using timber frames with thatched roofs. The grave goods found in the famous pagan ship-burials of coastal Suffolk and Essex witness their high levels of artistic and practical culture. These form the nuclei for the last great series of barrow monuments known in the Island, at Sutton Hoo above the banks and sloping valley sides of the Deben estuary. They were the result of interments for about twenty-five years either side of AD 600.

Christianization began shortly after, with local populations demanding their own neighbourhood churches. Some were built within Roman standing ruins, as at Canterbury. Others were new-builds erected by the efforts of groups of local freeholders who wanted places of worship within convenient distance of their dwellings. During the later Saxon church-building boom (up to 1066) stone was used extensively, often employed in a long-and-short arrangement of freestone quoins at corners. However, the extreme shortage of indigenous freestone in the major centre of population, East Anglia, led to many churches there being of round-tower construction (no quoins needed; Fig. 13.5). Interestingly, all Anglo-Saxon cathedrals seem to have been wood-built; perhaps it was the roof problem we alluded to previously.

After the conquest, the highly skilled Norman masons, many doubtless having had close contact with Mediterranean and Arabic architecture, began building

Figure 13.4 York's Multangular Tower (51.889649, −1.087005). The lower half of the bastion is perhaps Roman Britain's most aesthetically pleasing surviving construction. Photo: Wikimedia Commons©Mkooiman. It features small uniform-masoned blocks of Magnesian Limestone set in perfectly constant courses. The medieval top half uses more or less the same Tadcaster stone, but the blocks are larger and their coursing less uniform and clumsier.

Figure 13.5 Howe parish church, South Norfolk (52.549905, +1.354223). The unknapped flint- and erratic-coursed Anglo-Saxon round tower has exquisite proportions under its pantiled cap. The porch is later, in Decorated style, with a nice knapped flint frieze around the arch.

on a massive scale. All the Saxon timber cathedrals were demolished, built over, or the ecclesiastical site moved elsewhere. In the Domesday survey of 1088 the most populous eastern counties of Linconshire, Norfolk and Suffolk had an astonishing number of churches; the two thousand or so medieval examples that survive are testament to a fervent pan-European movement of Christian faith. In the words of a contemporary French commentator, Europe was 'cladding itself everywhere in a white mantle of churches…': a statement of fact concerning the widespread use of limestone in ecclesiastical architecture. Its spiritual desirability was aided by its undoubtedly favourable physical properties to the quarrier and mason: strength and workability.

The Normans were also great castle builders, rapidly building thirty strategically placed timber structures immediately after the initial conquest. Later they imported their own Caen limestone to build permanent structures (Chapter 10). During the following century castles, abbeys, monasteries, parish churches and cathedrals gradually made use of indigenous stone. Examples are London's White Tower (Kentish Rag, though with

Caen dressings originally); Ely and Peterborough cathedrals (Barnack limestone); and Durham (Coal Measures sandstone from local Wearside quarries). Castle-building in local stone reached its zenith with the ring of mighty fortresses constructed around the heartlands of what was once independent North Wales by Edward I (Longshanks). The most technically perfect surviving fortifications are the Henrician walls of Lower Carboniferous sandstone that envelop the late-medieval centre of much-fought-over Berwick-upon-Tweed.

Norman masons first utilized the building forms and techniques used by the Romans, but also in the related Arabic architecture of Mediterranean buildings with which the far-travelled Normans would have been familiar. The trademark barrel arch was structurally essential where it supported a vaulted nave or ceiling. Because barrel arches and vaults generate large lateral stresses on their supports, Romanesque churches feature large-diameter aisle piers and very thick walls, often leaving room in the lower levels for only small windows. Hence the sight on entering a Norman church or cathedral is often that of shadowy light cast mostly from the higher glazed vaults. Especially noteworthy are the massive piers of Durham cathedral with their astonishing chequer and chevron patterns deeply incised into masoned sandstone building blocks (Fig. 13.6).

Gothic to Classicism

Durham's very early rib vaulting and more elaborate later developments like fan vaulting came to dominate the pointed arches and windows of English Gothic church architecture. The three phases are Early English (Salisbury, Lincoln, Wells cathedrals); Decorated (Exeter); and Perpendicular (much of Canterbury). York and Winchester have mixes of all three. Each used progressively more advanced and delicate vaulting with the external walls featuring extensive stained window glass. In the Perpendicular style (1330–1540) the glass was set within long, thin vertical mullions, usually of pale limestone. These let natural light into the church interiors through exquisite hand-coloured glass panels.

By the sixteenth century, although Gothic was still dominant, the Renaissance idiom had begun to appear, initially as ornament to buildings that were medieval in conception, as at Hampton Court. At this time brick began its steep ascent as the major building material of the Island, notably in Oxbridge colleges (Jesus and Queens' in Cambridge) and larger country houses

Figure 13.6
Romanesque nave piers, arches and arcade in Durham Cathedral (1093–1135) (54.773233, −1.576392). The emphatic incised zig-zags cut across ashlar courses of locally quarried Coal Measures sandstone.

(Tattershall, Oxburgh, Herstmonceux). By Tudor times the southeastern counties followed the lead given by eastern coastal counties from Essex to Lincolnshire in brick construction of larger buildings. By the mid-seventeenth century there was general use of locally fired brick in more substantial houses outside of the Jurassic stone belt. It aided the fashion for Dutch architecture throughout England, especially in East Anglia, where the Dutch gable is a feature of many buildings. Here and elsewhere most vernacular cottages and yeomen's houses were still of timber frame construction well into the eighteenth century. Brick found favour for outbuildings like stables, for there was a widespread view that stonework was less permeable (often true) and encouraged horses to sweat in damp weather.

Most late-medieval urban houses were thatched. The steady loss of whole streets and districts to fire, culminating in the Great Fire of London, led not only to the eventual growth of Insurance Companies, but also to widespread civic ordnances that stipulated all

rebuilding and roofing was to be in either brick, tile, slate or stone. The stage was set for Neo-Classically derived eighteenth-century Georgian brick townhouses to dominate the expanding urban landscapes of the Island. This was the glory-time of locally fired red brick, mostly hand-moulded with increasing attention to composition and standard firing temperatures. The finished product has weathered to hues of russet and port wine, setting off their thickish mortar and accompanying stone quoins and window mullions. Such brick gradually became universally available; cheap, reliable and strong. Brick clay was especially valued from the Triassic Marls of northern and midland England and from a myriad of local pits dug by hand in softer Cenozoic and Quaternary clays.

A residual taste for the 'pure' classicism implicit in white freestone led great brick-built houses like Holkham Hall to be externally rendered to imitate stone (a fashion still visible in parts of pretentious Victorian London). There was also the widespread use of Portland stone, that strong, compact Jurassic oolitic limestone (Chapter 4), by Wren, Jones, Vanbrugh and Hawksmoor in the late-seventeenth-century to early-eighteenth-century rebuilding of London and elsewhere. Stone continued to be fashionable in Georgian new-builds like Regency Bath (mostly brick with stone facing) and it had rarely given way to brick in the urban and rural stone districts of the west and north. A fine example is provided by eighteenth- to nineteenth-century developments in Midland Scotland: Glasgow and the west utilized the brown-maroon Permian sandstone of Mauchline, Ayrshire while Edinburgh's New Town and the east used the yellowish/pale brown weathering Lower Carboniferous sandstones of the Lothians.

Industrial and Civic Pride

Late Georgian and early Victorian industrial landscapes featured mechanized mines, canal buildings (Fig. 13.7), bridges, mills, factories, railway sheds (Fig. 13.8), smelters and foundries, mainly built of local stone or, less commonly, local brick. They include surviving engine-houses (Devon, Cornwall, Durham) and mills (Calderdale, Bradford, New Lanark) that are now listed buildings; even the engine-houses speak of Neo-Classical elegance in their sparse-proportioned architecture.

Mid-Victorian civic pride was inflamed after parliamentary reform. Together with the Corporations Act of 1835, it led eventually to a boom in public buildings in the newly enfranchized northern and midland industrial towns and cities. Notable previous mid-Georgian and early Victorian emblems of mercantile and industrial pride were the town halls at Liverpool (Neo-Classical) in local Triassic Sandstone, and Birmingham (Fig. 13.9 – astonishingly pure Classical), brick-built and faced by Anglesey Carboniferous Limestone.

Later, civic halls sprang up all over Lancashire and Yorkshire, built according to typically mid-Victorian jackdaw-like eclecticism; styles labelled Gothic Revival (Manchester, Rochdale); late-Neo-Classical (Huddersfield); Neo-Classical/Baroque (Leeds); and Italianate Gothic (Bradford, Halifax; Fig. 13.10). Exterior building material was almost entirely sandstone or pebbly sandstone quarried from the Pennine Carboniferous, usually various members of the Middle Carboniferous Millstone Grit. After post-industrial cleaning the stone has successfully weathered to exquisite pale yellowy-brown hues (Manchester, Halifax,

Figure 13.7 Millstone Grit barrel-arched road bridge spanning the Rochdale canal and towpath just west of central Todmorden, West Yorkshire (53.720245, −2.103246). A toll-booth window can be seen immediately through the arch

Figure 13.8 Glass, iron girders, faux-Romanesque limestone arches, port-holed arcades and central cast iron columns in York's unique curved railway shed: designed to take the line sharply away from the historic centre (53.957980, −1.093190).

Bradford, Huddersfield). The coarser sandstones (Rough Rock) used in plinth and groundwork at Leeds attract modern urban pollutants unduly and always seem brutal rather than warm by comparison.

A small disaster for aficionados of gently weathered vernacular stone and brick landscapes occurred in the 1850s. The unleashed forces of industrialization not only began to coat buildings with lashings of soot, but they arrived at the gates of the brick-manufacturers themselves. Clays were soon mechanically worked all the way from bedrock to finished brick. They were dug, pulverized, mixed, extruded, dried, and baked in metal moulds within Hoffman kilns (invented in 1858) that could run continuously under effective thermostats with adjustable feeds of coke or coal. Deep, thick claystones of low initial plasticity were now deep-quarried in gigantic opencasts. These were epitomized by the Upper Jurassic Oxford Clay outcrops east of the Middle Jurassic

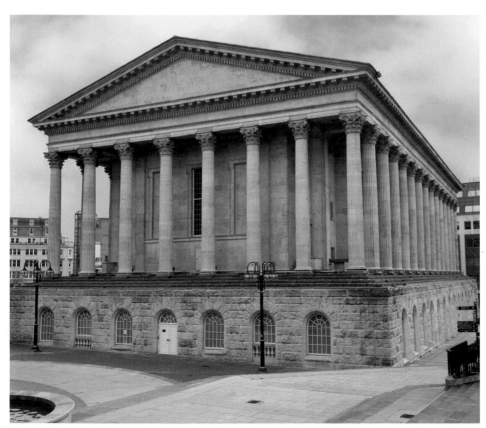

Figure 13.9 Birmingham Town Hall (52.479447, −1.903649). Photo: Wikimedia Commons©Very Quiet. Pure classical lines at the dawn of the Victorian era.

Figure 13.10 The Barrys' 1860s Halifax Town Hall (53.724452, −1.860681); perhaps the most elegant and lovable of all the mid-19th-century civic halls in the Pennine Carboniferous stone belt. Photo: Shutterstock 211719283 ©Alastair Wallace. It was built with 24,000 tons of local Carboniferous Ringby quarry sandstone (the Upper Carboniferous Elland Flags on Swales Moor). Stone-cleaning has emphasized that choice of stone is a major contribution to a building's character.

stonebelt from Bedfordshire to Yorkshire, especially around Bedford itself and the area south of Peterborough, where production around Fletton reached over 150,000 bricks per day by 1889. Similar clays were widely exploited in the Upper Carboniferous of the north and midlands, notably from the 1880s at Altham near Accrington, Lancashire.

The result was the spread by rail and canal of a blanket of often raw-red, smooth-surfaced, iron-hard brick whose perfect uniformity of shape, size, edge, composition and colour allowed for absolutely no subtlety and no possibility of amelioration by patination due to slow atmospheric weathering. Such was their perfection of shape that even the spread of intervening mortar became just a millimetric smear, hardly breaking the monotony of bricked surface. Clad with grey Welsh slate, the brick industrial and terraced domestic architecture of large areas of midland and northern England (outside of the traditional Pennine stone belts) became, to many brick-haters, glaring eyesores until they were soon mercifully covered in soot. Even worse for some critics was the new fashion for glazed red brick: in the words of one grumpy architectural writer: 'Their proper habitat is the public lavatory, and perhaps the dairy…'. Since the advent of stone-cleaning, the modern eye sees the bright colour contrast engendered by such brick with ashlar sandstone in a more merciful light. Luckily, fashion changed, and by the mid-1930s brickworks were offering more varieties of less emphatic (and more weatherable) variegated red- and pink-tinted wares for the massive rebuilding of workers' housing in the council estates of inter-war and post-Second World War suburbia.

Arts and Crafts and the Seeds of Modernism

Building urban housing for an ever-expanding population has been a concern since people began moving in large numbers to manufacturing towns and to London during the Industrial Revolution. Today, just over 80 percent of the British population live in urban centres, the area under asphalt equivalent to that of Leicestershire. Town planning has had a profound impact upon the landscape and is often a controversial process. Initial practical reaction against the uniformity and grim excesses of industrial landscape came from William Morris's nineteenth-century Arts and Crafts movement. He established a decorative arts business with Dante Gabriel Rossetti and Edward Burne-Jones. Centred around Morris's Red House at Bexley Heath (Fig. 33.5), this became the 'Arts and Crafts Movement', promoting traditional craftsmanship, the artisan, the physical act of making (as in pre-Industrial Revolution) and an intimate knowledge of the physical properties of materials (as in traditional skills). Responding to the perceived evils of an urban and capitalist society, it was informed by a genuine wish for economic and social reform, addressing the nature of work and the moral and social health of the nation. Anti-industrial, the movement looked back to medieval and folk art by utilizing motifs from the natural world, its flora and fauna. Its influence, particularly on the decorative arts, was felt well into the twentieth century. Morris's own 'Red House' (Fig. 33.4) was designed to embody these beliefs.

Charles Rennie Mackintosh's 1897 Glasgow Art School carries on the impetus of the Arts and Crafts movement. Yet his foray was a transitional one, for the building crafted in brown Ayrshire sandstone and the exterior, with its strikingly large Georgian-style window spaces, appears unique. Internally it is a triumph of functionalism: analytical, thoughtful, stripped of ornament, and designed mostly for the convenience and enjoyment of the internal consumer, in this case students and faculty.

Town Planning

William Morris's aesthetic was underpinned by his socialist philosophy, and similar intentions lay behind the enthusiasm for later nineteenth-century utopian communities and projects. The relationship between people and the urban environment was addressed by Ebenezer Howard in *Garden Cities of Tomorrow* (1898) in which he pioneered the concept of town planning in harmony with nature. The garden city movement grew popular, with towns like Letchworth and Welwyn planned with open spaces around clusters of civic buildings. Many of the aforementioned pre- and post-Second World War council estates owed much to the garden city concept. Their rows and avenues of brick-built semis featured public spaces (some with modernist sculpture), trees, grass, large gardens and enclaves of local shops, public amenities and schools.

The town planner, Peter Hall, an admirer of Howard and the garden city concept, was interested not just in cities but in regions. He helped invent the concept of 'the city region'. He comments: 'We don't build enough and that what we do build is too often ugly, alienating and poorly served by infrastructure.' He chooses Freiberg, Germany as an outstanding example of how to do it:

> They have created good jobs, built superb housing in fine natural settings and generated rich urban lives. But not only that: simultaneously these cities have become models of sustainable urban life, minimizing energy needs, recycling waste and reducing emissions…

Architecture can be used to provide a total environment; topographical features can be preserved or played off against architectural form and its social and individual requirements. This can be done with realistic provision for industry and transport – in fact, much like living organisms on a gigantic scale.

Modernism

Truth to materials, design closely aligned to function, and the impetus to begin afresh led to sometimes stark but certainly much-simplified modernist design. Le Corbusier's statement: 'A house is a machine for living in' was made almost a hundred years ago; it has unfortunately been interpreted too frequently by architects as a fundamental contradiction between town and country, and therefore landscape: something he would have deplored.

Modernism aimed from the outset to free buildings from previous architectural forms by having sparse, clean design and taking advantage of the new possibilities brought by materials such as concrete, steel and glass. It gave us the prefabricated house and most of the developments in technology, design and materials that we take for granted in our buildings today. It also inspired 1960s brick- and concrete-built communal living: huge blocks of flats in industrial cities built according to an understandable urge to improve the lot of slum-dwellers. Yet many are now demolished; in their size, construct and failure to engage landscape they also failed to provide human relief from what came to be regarded as urban prisons. Brutalist architecture was perhaps the most confident expression of modernism, and also its most extreme expression of 'truth to materials': reinforced concrete, steel and glass. It has been much criticised in the past, but now buildings such as the Barbican Centre and the Trellick and Belfron towers are much admired, especially by their occupants.

Postmodern Times

Rowan Moore, the *Observer* newspaper's architecture critic, wrote in December 2015 that Postmodernist architecture was: 'anarchic and anti-establishment... a reaction against the impersonal, abstract and rationalist qualities of modernist architecture (and) which instead favoured decoration and references to historical styles.' It can be playful and individualistic. A fine example is John Outram's pumping station on the Isle of Dogs, an eye-catching 'temple'. Postmodernist buildings are very 'in-your-face'; they want to dominate and be seen but often do not relate to the landscapes they inhabit.

From the 1980s there was a decline in public building and with the rise of neoliberal economics an urban boom developed for privately owned iconic buildings. In and around the City of London a new 'macho-modernism' of non-linear Gherkin-type erections arose to dominate the skyline: the trapezoidal fragment that is the 'Shard'; the 'Gherkin' itself; the buttress-like 'Cheesegrater' and others. The rise of these buildings, though synonymous with the rise (and fall) of a dubious money culture, does however give some cause for some optimism. As mentioned earlier (Chapter 4) the ongoing revolution in ultra-lightweight construction materials looks like giving us more interesting non-linear shapes to fit into the landscape in future times: witness the elongated and curved enteromorphic shapes promulgated by Zaha Hadid and others. It is often forgotten that natural landscape itself is rich in curved and irregular shapes – in the often chaotic non-linearity of rock outcrops, in scarplands, valley sides, concavo-convex hillslopes, bournes, combes and dells. Some of the finest modern buildings like the Kielder Observatory in Northumberland (2008) marry the function of the building with materials, technology and the landscape. Also, but for the sensitive application of such approaches to conservation efforts, many fine old buildings would no longer exist.

So, an optimistic perspective couples the advances made in technological design and the use of materials with the kind of thinking that underpins current advances in sustainable living. For example, the Centre for Alternative Technology at Machynlleth in Powys, mid-Wales promotes building and farming practices that work in partnership with the environment. They provide education and courses relating to organic farming, renewable energy, sustainable building (using low-impact building materials, including thatch and adobe) and much more. It is somehow pleasing that adobe, one of the oldest known solid-building constructs (a sun-dried mix of pulverized mud/mudstone/marl and straw) could be making a major comeback in twenty-first century architecture. It could find a suitable place in our Island landscapes, as its Devon predecessor and equivalent, cob, did in previous centuries.

Chapter 14

Sculpture

Pebbles and rocks show nature's way of treating stone – smooth-worn pebbles reveal the contours inherent in stones, contours determined by variation in the structural cohesion of stone.

The critic and writer Herbert Read commenting on Henry Moore's sculpture (1934)

Materials

Works of sculpture depict something happening within multidimensional space. They are non-functional and made of almost anything: stone, mineral, clay, wood, glass, metal, plastic, light, fabric, skin, bone, ice, or any combination thereof. They are produced by carving, engraving, casting or assembling but could also be light projected onto something, or even heard. Some, like Richard Long's footprints through grass, *A Line Made By Walking* or Katrina Palmer's *Loss Adjusters* will leave behind no material trace. Only stone, mineral and clay come directly from the landscape, taken from outcrops or quarries whose complex forms and textures themselves define 'natural sculpture' or 'found objects'. Other sculptures of glass or metal are sourced from natural ores and minerals in a more indirect way: by smelting, fusing, refining, etc.

Permanent sculpture intended for the outdoors needs to be durable, as are many rocks, and metal alloys like bronze. Contemporary sculptors assess additional environmental factors when choosing materials, for just as in stone buildings, outdoor sculpture can be sensitive to weather, changing light and season (Fig. 14.1) and

Figure 14.1 Two different views in contrasting weather and light of Bronze Bowl with Lace (2013–14) by Ursula von Rydingsyard (b. 1942) at the Yorkshire Sculpture Park. The 6m-high bronze is cast from assembled cedar wood bricks. To the left it manages to seem both organic and definite, stratified with growth bands like a huge fossil coral. To the right it appears gracefully arboreal, an integral part of a misty and diffuse winter-wooded landscape.

both natural and human pollution. Stone is commonly chosen as a sculptural material by reason of its variety (see Moore and Larkin 2006 for many specific examples). The best stone for sculptural carving is the same as that used for fine architectural work: freestone quarried from unweathered strata, dried out *in situ*, strong, homogenous in texture, with a lack of fissility and so on.

Texture and colour are everything in the appearance of sculpted stone, as in the colour, spacing and density of annual rings in different types of wood. This is most obvious in the variegated banding seen in metamorphic schists, gneisses, and in some marbles. Amongst igneous rocks the lighter overall colour of granitic types come from their more abundant quartz, white mica and alkali feldspar crystals. Two-mica granites reveal their 'salt-and-pepper' textures. The darker intermediate to basic rocks, diorite (Fig. 14.2), gabbro, dolerite and basalt, are richer in minerals containing iron, like olivine, pyroxene, amphibole and black mica. Rapidly

cooled basic lavas like basalt give a uniform deep, blue-black sheen, especially when polished. Intermeshed coarse lighter crystals (a texture known as porphyritic, pronounced *por-fear-ri-tic*) may add a pronounced dapple of light and colour, the feldspars sometimes pinker and very obvious. These are like blotches, freckles or moles on skin.

There are also a multiplicity of textures and colours in sedimentary rock. Light-coloured limestones and calcareous sandstones often find favour with sculptors, with texture provided by variable proportions of ooliths, pellets and fossil fragments. Carbon-rich microcrystalline varieties are jet black, often used in the seventeenth to eighteenth centuries for vault slabs in churches and cathedrals and in contemporary sculpted objects, such as those surrounding Sir Thomas Browne's famous contemplative statue in central Norwich. Generally speaking, the sculptor must pay more attention to the likely future colour of weathered limestone than to those rocks of igneous origin. As we previously discussed for Lincoln cathedral (Chapter 10), limestones with iron-rich crystalline cementing material tend to weather to yellow-brown hues over time. Oolitic ironstones (like Henry Moore's favourite, Hornton Stone; see Fig. 14.8) weather from their pristine dark-green-black to a rich variety of ochreous hues.

Place and Time

Although sculptural objects always express artistic intention, impact and meaning change according to placement. Siting is thus important, enabling multi-directional viewing with different aspect and lighting; this makes outdoor sculpture more interesting as it resonates with landscape, establishing a relationship with it in a way that few paintings can. This applies to the memorial of an unknown soldier, an abstract figure in parkland, or a light sculpture projected into a building. Publicly accessible sculpture

Figure 14.2 The Egyptian goddess Sekhmet from the Temple of Thebes (*c.*1350 BC) in the foyer of Liverpool Museum. She is sculpted in dark diorite whose blotches of lighter feldspar and more uniform aggregates of dark pyroxene give a freckling effect to the polished torso. The sculpted (seated) lower body was brutally broken by Second World War bomb damage that destroyed the Museum. It reveals the raw stone's rough crystal fabric and intimates the immense labour entailed in its original polishing.

parks themselves are a concept of the last thirty years, though they nod to private outdoor displays by the eighteenth-century wealthy in the grounds of their great houses.

Sculptures commissioned or intended for a particular place are site specific – such as the carefully placed stone-carved holy statuary in both internal and external niches within the fabric of medieval churches and cathedrals. Even those niches now empty due to destruction or neglect give witness that a former sacred object was once placed there. A contemporary example is Anthony Gormley's *Another Place* (Fig. 14.3), whose life-sized figures stand ramrod straight and face out to sea in all weathers on a gently shelving shoreface. Nothing comparable could be achieved indoors. Yet Flor Kent's *Kindertransport* bronze (Fig. 14.4) of refugee Jewish children waiting by their suitcase on a plinth in Liverpool Street station concourse also uses its outside/inside location in a telling way.

Some public sculptures are close to the places where individuals did great deeds: like Winston Churchill in Parliament Square, London. Others are placed to stress or reinforce the hierarchical power and wealth of an elite or to commemorate individuals deemed important in science, culture or commerce. Many have become the focus of great civic pride, both local (e.g. Liverpool's *Liver Birds*), national (Landseer's Trafalgar Square lions) and international (Queen Victoria statues from London to Sydney).

Many cultures contain stories about people who are turned to stone or mineral; we have mentioned a fictional instance already from A.S. Byatt (Chapter 2). The stone carved into tombstones portraying knights and ladies emphasizes stone's durability as distinct from human mortality. However, as Thomas Hardy's poem *During Wind and Rain* (1917) reminds us, carved words on stone always dissolve away, no matter how slowly it may seem to us:

Ah, no; the years, the years;
Down their carved names the rain-drop ploughs.

Throughout history, sculpted and undressed stone has been used to commemorate the dead, mark a grave or construct a mausoleum. Cemeteries dot today's landscapes and occupy land adjacent to every village, town

Figure 14.3 *Another Place* (2007) by Anthony Gormley (b. 1950). (54.487690, −3.053500). One of many identical cast-iron figures (based on a cast taken from the artist's body) in the installation at Blundellsands, Merseyside.

Figure 14.4 *Kindertransport* (2003) by Flor Kent (b. 1961). Photo: Wikimedia Commons ©Myanotheraccount. A bronze memorial at Liverpool Street station, East London. It was to here, via the Harwich ferry, that Austrian and German refugee children arrived in 1938. It stands as a reminder of Britain's generosity towards refugees in the past.

and city. In them we see tombstones whose names and dates have been near-obliterated by natural weathering. Yet the toppled and buried carved stones of Roman Britain's wayside memorials that lined the main roads exiting towns and cities have often been astonishingly well preserved over two millennia. Their inscriptions are the richest source of our knowledge of legionary soldiers, administrators and ordinary citizens – their kinship groupings, social make-up (Romano-British intermarriage), status (freed slaves) and homelands (Batavian and Tungrian cavalrymen).

Modern-age stone grave-markers and civic war memorials for the fallen eloquently express the prevailing culture when they were made. Some were first set up on plinths in public places after the shock and carnage of the Boer War. After 1918 hundreds of thousands of Portland stone grave markers and cast-metal war memorials were erected throughout Britain, France and Belgium. Their simple elegance still has the power to move.

Sculpture can also express the dark and fantastical; witness the heads and zoomorphic creatures entwined in Celtic and Anglo-Saxon ornaments, medieval gargoyles, eighteenth-century grottoes and late-twentieth-century garden gnomes. They may mark a search for perfection, order and the ideal – humankind exerting its superior will and intelligence over the wild forces of nature. Sculptures may also be landmarks, a concept perhaps first realized by the Ancients (a stone-carved, bronze-faced Helios or the 'Colossus of Rhodes'). A notable modern example is Anthony Gormley's steel *Angel of the North* (Fig. 26.5) near Gateshead. We will never know whether Stonehenge was intended as a landmark sculpture. Notwithstanding its astronomical use, like all megalithic ritual monuments it seems to have such a quality. To the Neolithic peoples it might have had the awesome brutality of a 1960s tower block; few would call its lichen-encrusted grey stonework or the Lego-like arrangement of the remaining trilithons beautiful. In 2014, in keeping with its possible original cognate role as a death memorial, footage of First World War soldiers was projected onto it to commemorate those who were trained nearby on Salisbury Plain. A similar connection was made by Paul Cummins's 2014 installation, *Blood Swept Lands and Seas of Red*: 888,246 ceramic poppies in the moat of the Tower of London, each marking a British or Colonial fatality in the First World War.

Outdoor garden sculpture also provides visual focus for contemplation; formal gardens and public parks combine with a sculptural heritage. Home gardens may contain fountains and water features, sculptures ranging from original pieces to, say, a concrete Venus de Milo (assorted sizes available in your local garden centre) and, of course, a garden gnome or two. Like an official stamp of ownership, they express a confirmation that the local landscape is ordered, sorted – even the humble gnome creature perhaps an expression of the dark forces of nature tamed.

Early Heritage

The earliest sculptures in Britain are either abstract or geometrical. The most remarkable are late-Neolithic to Bronze Age carved stone 'balls with knobs on', found mostly in Aberdeenshire. The majority are sandstone, greenstone and quartzite. Those of soft serpentinite have fine and detailed carving, as in the exquisite 'Towie Stone' from Glaschul Hill, dating from between 3200 and 2500 BC. Carefully chosen for fine carving using obsidian or flint tools, it has faces raised into button-like protuberances featuring decorations of rings and spirals. The piece is obviously special, a highly sophisticated and probably precious relic.

Most numerous are thousands of outcrop-carved Bronze Age 'cup and ring' designs, mostly in the north of England and Scotland (Fig. 14.5). The earliest were carved onto stone outcrops in elevated positions, landmark sites or possibly burial places. Later examples were carved onto standing stones, as on five of the megaliths at Stonehenge. Iron Age petroglyphs are also abstract and geometric, the symmetry of Bronze Age cup and ring marks giving way to a more free and ornamental style that flows with swirling ornament and symbolic pattern; zoomorphic detail is sometimes included amongst the decoration.

Faces of Rome

After millennia of abstract geometrical carving and monumental placements, the Roman figurative sculptural heritage arrived with a bang. This was no big shock to client tribes in the south, for they had long adopted coinage on whose faces were to be found the characteristically frank (ugly) images of the early Caesars. The Romans were strongly influenced by Classical and Hellenic figurative stone carving in limestone and marble. They too sculpted in these materials and also in clay, bronze and other metals, introducing statues and statuettes (for domestic use) of Roman gods in order to spread their beliefs

and mix them with native deities throughout their empire. Most interesting are a group of third century AD bronze statues of Roman gods and goddesses found in a pot at Devizes, Wiltshire. The sculptor was probably a local following native British traditions; the figure of the god Mars, who usually holds weapons, grasps two snakes with horned heads. Snakes and dragons were important symbols in Iron Age culture, also common in the early Christian era, for example the decoration of early manuscripts.

Pictish Slabs, High Crosses and Hogbacks

Much Anglo-Saxon art has been lost, plundered by Vikings or destroyed during the Reformation. They introduced a sophisticated artistic culture into England that looked back to Iron Age art rather than to Roman heritage. It is full of decoration, intricate interlacing, fluidity, subtle symbolism, and references to the natural world. The silver-gilt square-headed brooch found at Chessel Down on the Isle of Wight, for example, contains twenty-four different heads and beasts tucked into its ornament. Whale bone and ivory carving were also popular.

Most has been learned about Anglo-Saxon art and culture from the early-seventh-century Sutton Hoo ship-burial in Suffolk, thought to be the grave of Raedwald, King of East Anglia, who established Christian rulership in England. Though the roof of the mound covering the ship and its chamber had long since caved in, preserved in the earth beneath were the undisturbed remains of what was essentially an enormous and lavish 89ft funereal installation. The body, the ship and the chamber were all decorated with stupendously fine objects. The metal has survived, but sufficient remains of textiles and other perishable artefacts complete a picture of great wealth and sophistication.

Anglo-Saxon stone carving survives chiefly in ecclesiastical buildings and generally in the Byzantine style, symbolic rather than representational, an art form introduced when Rome was Christianized. The Hedda Stone (*c*.AD 870) from Peterborough Cathedral is pre-eminent (Fig. 14.6). It stylistically represents Abbot Hedda and his monks, who were murdered in a Viking raid. It is a beautiful object, carved from delicious cream-white oolitic Barnack Stone, the figures showing movement and gestures. Holes bored somewhat crudely and positioned in its sides suggest that at some time it was carried on rod-like supports, perhaps in procession as a much-venerated relic.

It was also during the Anglo-Saxon period that many of the beautifully carved sandstone High Crosses of northern England and Scotland were made (Figs. 23.4, 24.3). The purpose of these was probably manifold: to mark a holy place; the grave of an important person; a meeting point from which to celebrate mass; or to educate, since

Figure 14.6 The Hedda Stone, 100x75x30cm, Peterborough Cathedral. Photo: Wikipedia Commons©NotFromUtrecht. A cast is in the 'Cast Court' of the Victoria and Albert Museum, London. This large (>300kg) Anglo-Saxon piece, ark-like in form, has floral and figurative carvings on the sides and the steeply pitched roof. On the pictured face, six monks from the mid-eighth century are depicted in various attire and accoutrements. We assume they include Abbot Hedda. Their abbey, Medeshamstede (forerunner of Peterborough), was destroyed and they were murdered.

many are decorated with panels and stories depicting scriptural scenes. They also contain intricate strapwork interlacing, thought to have developed initially from interlaced gold strips in metalwork. The Ruthwell High Cross in Dumfriesshire (Chapter 4; Fig. 23.4) is unique in having a runic poem chiselled around its margins.

Pictish carvings deserve special note, for the Picts possessed a unique 'vocabulary' of symbols. Converted from pantheistic worship to Christianity by the late-sixth century, their early stonework reflects this duality. Mostly of sandstone, the earliest were carved on sandstone boulders and outcrops. Later, dressed slabs (Fig. 21.4) feature both native symbols of mirrors, zig-zags and snakes, together with Christian cross motifs. They finally evolved elaborately shaped and carved cross slabs, one dated to the ninth century by an inscription to the Dalriadan King Caustantin.

The Vikings destroyed much, but they also left their mark in stone, usually sandstone. Several Viking burial markers contain Scandinavian decoration and reference both Norse mythology and Christian stories. There are numerous finds, chiefly in North Yorkshire and Midland Scotland, of 'hogback' burial stones or memorial markers (Fig. 14.7). The hogback shape, like an upturned boat with a ridge on the top, perhaps represents Norse-style buildings. The 'roofs' usually feature a cover of carved shingles. On the main body and ends they may be decorated with Saxon-style plaitwork and interlacing. From

Figure 14.7 The impressive group of tenth century Viking 'hogback' stones (Stephen Driscoll tells us they are carved from local Carboniferous sandstone) at Govan Old Parish Church, Clydeside (55.863508, −4.313406). Photo: Wikimedia Commons©Deadmanjones. After marauding Dublin Vikings sacked Dumbarton in AD 870, Govan became the ecclesiastical centre of the Christianized kingdom of Strathclyde.

Swaledale, North Yorkshire, one stone has beasts resembling bears clutching both ends. There are no indigenous Scandinavian parallels to these forms, so it is thought that since the stones are generally found in graveyards, Christianized Viking settlers evolved their own sculptural forms.

Zig-Zags to Pointed Arches

In the Norman culture, stone carving was important for decoration, veneration and education. Skilled stone masons were much in demand the length of the land to feed a demand for sculpture, chiefly architectural: doorways, windows, chancel arch and capitals were enriched with relief carvings, often of chevrons, as notably in Durham cathedral (Fig. 13.6), but also a more difficult form, the helix (a 3D spiral) also at Durham and on the piers around the former nave altar at Norwich (Fig. 10.5).

The rich Anglo-Saxon cultural heritage did not disappear after 1066. Initially the movement towards figurative sculpture that developed on the continent was slow to spread to Britain. It is evident in examples like the mid-twelfth-century chancel arch carved in oolitic limestone of St Peters, Tickencote, Rutland. Here, six successive zones of carvings contain the grotesque heads of birds, beasts and demonic humans peeping out from intricate interlacing and the familiar zigzags. Generally the often-grim message was carved in the tympanum over the main doorway. It is rarely so explicit as that at St Swithin's, Quenington where Christ touches a figure in hell with his cross to reclaim him or her. As Denise Levertov (1923–1997) writes in her Ikon (1989):

Down through the tomb's inward arch

He has shouldered out into Limbo

To gather them, dazed, from dreamless slumber...

Churches and cathedrals built in Gothic style (1150–1600) soar elegantly upwards, their glass and sculpture part of the overall ascendancy. An example is King's College chapel, Cambridge whose fan-vaulting drips with ornament, 'looking more like lace than Barnack limestone...'. The stone, we should add, was quarried from desecrated Fenland monasteries and abbeys after the Dissolution, courtesy of the college patron, Henry VIII. Pre-Reformation Gothic churches had façades stuffed with life-sized statues of saints and bishops, ornate blind arcades and sculpted decoration (e.g. Salisbury, Lincoln, Wells). Most of the congregation would have been illiterate, so the sculptures were intended to inspire awe and to educate. Much of such statuary from the period was deliberately defaced or destroyed, especially after Edward VI's 'Putting Away of Books and Images' Act of 1549, which ordered the destruction of all ecclesiastical images. For example, of the seventy-nine original fourteenth-century figures on the Great West Front of Salisbury Cathedral only seven survived. The remaining seventy-two we see today are mid-Victorian or later.

Little public sculpture at this time was independent of building, though private demand for ivory-carved reliquaries was high amongst the wealthy. A particular fashion from the fourteenth to the early sixteenth centuries was for alabaster statuary showing richly detailed religious figurative work, sometimes coloured in polychrome. The mineral came from Triassic gypsum deposits in the English Midlands and the statues were exported far and wide in Europe. After passage of the 1549 Act, many were sold on in France and elsewhere, the industry ceasing until re-activated in the more tolerant nineteenth century.

Baroque to Neo-Classical via Grotesque Chaos

The influence of the Italian Renaissance was slow to spread to England because of the Reformation and the rise of Puritanism. Baroque (c.1600–c.1700) is a theatrical and energetic sculptural form. In the parks of French grand houses, statuary and nature were often manipulated to create complex works of art. In England Baroque was more restrained. The finest piece of sculpture of the period is the bust of Christopher Wren by Edward Pierce. He catches a lively youthful idealism and genius in his informal image of Wren.

Gentlemen on the Grand Tour (from the 1720s onwards) saw not only the work of Renaissance artists, but also classically inspired villas and gardens built during and after the Renaissance. Upon return to England they tried to emulate both the buildings and the formal gardens, sometimes shipping back to England antique statuary to erect both inside their houses and in picturesque outdoor settings. The inclusion of statuary, ornamental fountains and follies in private and later public settings partly defined the playful/decadent Rococo style that also included interior design, furniture and chinaware. 'Grotesque Chaos' expressed the fondness for decorated grottoes. Alexander Pope established his own at his Twickenham villa. In the words of the Twickenham Museum guidebook:

In 1739, inspired by a visit to the Hotwell Spa at the base of the Avon Gorge he developed an interest in geology and mining and, over the next four years transformed his grotto, decorating it with ores, spars, mundic, stalactites, crystals, Bristol and Cornish diamonds, marbles, alabaster, snakestones and spongestone. Much of this material was sent from Cornwall by Dr William Borlase. Other material came from all over the country and abroad, given by friends, acquaintances and those who wished to be known to him.

Neo-Classical was a stern style, latterly expressing an enthusiasm for the 'ideals' of the French Revolution and a protest against the excesses of Rococo. It took inspiration from antique Greek/Roman sculpture fired by high-minded intentions such as heroism and patriotism, especially if it was in any way Napoleonic. The British passion for all things classical was intensified by the discoveries of Pompeii and Herculaneum and by Greek treasures brought back to England. Sculptor John Flaxman saw the Elgin Marbles when they first arrived; he worked for Josiah Wedgwood (Chapter 28) designing jasperware and basaltware, many of his designs inspired by Sir William Hamilton's collection of Greek vases.

The late-eighteenth-century aristocracy may have had a passion for all things classical but Rococo, also called the French or Modern style, had much appeal to the burgeoning middle class. Wedgwood cleverly tapped into the taste for the Neo-Classical, but Rococo-style (highly ornate and gilded) porcelain ware (shepherd swains, Arcadian lovers, etc.) was all the rage, notably from Falconet's factory at Sèvre, copied by many Stoke-on-Trent potters.

Victoriana

Victorian sculpture generally had a Neo-Classical narrative, preferably personifying some moral exemplar and/or Hellenic ideal. Though later Victorian society was superficially prudish and buttoned-up, the bulk of stone carving of the period depicted smooth, naked white flesh, sentimentality, a preoccupation with death, and the material trappings of wealth. A love of this kind of sculpted ornament found full expression in funerary sculpture. Today the 'Magnificent Seven' cemeteries in London are all tourist attractions. In Highgate, Abney Park, Brompton and the rest, sculpted gravestones vie with one another in the splendour of death. Contrast these with the Victorian pauper gravestones in the walled former General Cemetery at St George's Fields in Leeds. Now landscaped, its walkways are paved with scores of large headstones cut from local flagstone and carved

with names that celebrate multiple grave-occupants, up to a dozen in each.

Reacting against what was perceived to be a paucity of non-classical contemporary sculpture, the later nineteenth-century Arts and Crafts movement, its chief proponent William Morris, looked to simple forms, traditional craftsmanship, and folk motifs and decoration inspired by the natural world. Ruskin's writings inspired the movement, and so did the general passion for geology. The movement had a profound influence on twentieth-century sculptors, particularly the modernists.

Direct Carving Indoors and Out

Natural and native stones of distinct textures and colours now took centre stage, inheriting a long but punctuated tradition of sculpting in marble and its pure-white clones (see Moore and Larkin, 2006). The work of British modernist sculptors during the first few decades of the twentieth century shares numerous traits: a wish to be true to the quality of materials, an emphasis on respect for the stone itself, and craftsmanship. Henry Moore, Britain's most famous modernist sculptor, believed that Eric Gill, Jacob Epstein, Frank Dobson, Henri Gaudier Brzeka and Barbara Hepworth all contributed to 'the practice and recognition of direct carving…'. Amongst Moore's first direct carvings was his *Mother and Child* (1932) (Fig. 14.8). He and the other carvers drew inspiration from the world tradition of stone sculpture as housed and exhibited in the British Museum, especially from Central America and the Cyclades, but also the Neolithic use of stone in megalithic monuments. Jacob Epstein said of Moore's own sculpture that: 'It contains the austere logic of ancient sculpture.'

All the modernists would have agreed with Eric Gill that 'to imitate the characteristic beauty of flesh and blood in a material so different as stone is absurd…' So both Henry Moore and Barbara Hepworth reduced shapes to their most elemental and both embraced abstraction. Moore, like Picasso and Marcel Duchamp before him, collected pebbles and bones and utilized found objects in sculpture. He spoke of walking along a beach and seeing stones that fitted with his current 'form interest…', also noting the bumps and hollows of weathering. Moore intended his later huge black and patinated bronzes to be placed in landscape; their curved sitting and leaning postures can be viewed in a context

Figure 14.8 *Mother and Child* (1932) by Henry Moore (1898–1986). Green Hornton Stone. 99.5x53.5x38cm. Acc No LH12. Sainsbury Centre, University of East Anglia. Photo: ©Henry Moore Foundation Archive. An immensely strong, watchful and protective mother puts a massive shoulder up at an unseen approacher, protecting her child. Green Hornton Stone is a calcareous ironstone from the Middle Lias of North Oxfordshire. Its greenish colour comes from the iron silicate mineral, chamosite. Weathering along joints has caused oxidation to brown iron hydroxides, clearly seen (but not in this view) like a quite natural skin blemish on the figure's back.

of hillslope, valley bottom, or skyline. The piercings pioneered by him and by Barbara Hepworth enable the viewer to see a landscape from within, as it were. Many remain in disparate places: on Galloway hillsides, in the sheep-grazed valley of the Yorkshire Sculpture Park, and in public spaces and university campuses across the land. Barbara Hepworth's sculpture garden at St Ives, Cornwall was created with a similar intention (Fig. 32.5); her pierced stone figures include the spectacularly beautiful polished Delabole Slate grouping *Two Figures (Menhirs)* in the Tate Gallery, London.

Post-Modern Outdoors

Post-Modernist sculptors reject 'traditional aesthetic and material concerns', and from the 1960s onwards began to explore new ways of making: exploring video, the environment, light, installation, ice, kinetic sculpture and so on. In conceptual art the idea takes precedence. John Frankland's *Boulder* (Fig. 4.3) is one, uncrafted but giving a nod to megalithic forms. Anthony Gormley (Figs 14.3, 26.5) explores the inner (imaginative) and outer spaces inhabited by humans.

Another take on landscape and sculpture is the use of *in situ* stone. At Tout Quarry on the Isle of Portland in Dorset, visiting sculptors have carved their own works onto the weathered old working faces of the Portland limestone remaining in the quarry. Landscape sculptor Andy Goldsworthy restored forty-five stone-built sheep folds in Cumbria, sometimes enclosing huge boulders within them, and in doing so making them seem precious. His *Striding Arches* in the Southern Uplands are featured later in this book (Fig. 23.5). Richard Long's work is close to the land but is often deliberately ephemeral, using natural materials like wood, earth, and snow. He explores humankind's relationship with landscape on his walking journeys. Critic Charlotte Higgins notes: 'On his journeys, he has arranged stones by roads, made circles from boulders, aligned pebbles in riverbeds, traced furrows in sand.' Since his sculptures are ephemeral, they are installed, then either disassembled or left to disintegrate naturally. Much current conceptual 'sculpture' is of this nature.

The boundaries between traditional sculptural and other artistic disciplines have blurred. Many contemporary artists use everything at their disposal. Ilana Halperin (see Chapter 4), for example, uses sculpture but also performance, printmaking, film, writing and more. Her work is typically global, outward-looking in its concerns: 'a means to connect our immediate lived experience with the harder to comprehend aeons of geological time…'.

Chapter 15

Painting

He was one of the great appointments to show to the world what exists in nature but which was not known till his time.

John Constable writing about Richard Wilson

'Landschap Kunst'

Artists help us to see the world anew, perhaps most obviously in landscape art. The word 'landscape' comes from the Middle Dutch '*landschap*' and it was in the newly independent sixteenth- and seventeenth-century Netherlands, as artists turned away from Catholic religious themes, that landscape art (our subtitle above) had its first golden age. It was from there that it spread to Britain in the eighteenth century. For many today it typifies English art as a whole, Turner and Constable in particular, expressing a love of the pastoral, certainly, but more generally of physicality, how seasonal weather and ever-changing light alters the forms of land, coast and sea. British landscape art often places an emphasis on such ephemeral events; watercolour often seems to be the medium best suited to capture these. But as we saw in Paul Nash's *Pillar and Moon* (Fig. 4.2) an artist's intention might also be to explore not just the physical world but also the human condition.

In Western art generally, topographical landscape barely existed until the fifteenth century. Early examples are usually fragmentary and incidental – a direct consequence of prevailing cultural influences. Classical Greek civilization was informed by humanism, expressed directly through depictions of the human figure. The Romans admired nature and decorated their villas with illusionistic landscapes. Early Christian painting and manuscript illumination portray landscape through decorative designs and symbols. These are usually highly formalized heavenly, historical or idealized settings for idealized figures where landscape is all foreground, a unifying space. Here the intention is often didactic, with an emphasis on the world yet to come rather than this one: why focus on life in this world when it is so brutal

and short; Paradise will be far more beautiful than anything the landscapes of this world have to offer? Nature was often perceived as dangerous, wild and untamed, seen as evidence of divine might and power.

The Renaissance perspective changed this so that the physical space around people and objects became important. Leonardo da Vinci was fascinated by nature and how we perceive it; his notebooks record botany, geology, and meteorology. At the height of the Renaissance, artists were familiar with linear perspective, using light and shade to render a sense of form. Leonardo also noted atmospheric perspective: how a distant landscape appears misty and has a bluish cast. Such burgeoning interest in this world, and the wish to describe recognizable experiences, gained momentum during succeeding centuries. Its progress was accelerated by the Renaissance rediscovery of Classical and Arabic science and the subsequent scientific advances of the Western secular Enlightenment.

A Note on Pigments

Sources for pigment in painting may be animal, vegetable, metallic, chemo-synthetic or mineral. Here we briefly consider only the latter, as used by pre-nineteenth-century artists in particular. Colour was traditionally obtained from powdered minerals by mixing them with a binding medium like water, spit, fat, oil or egg white. The earliest pigments were mostly earthy and ochreous iron oxides and hydroxides, used for funereal body coating (Chapter 4) or in cave wall pictures. These colours – reds, yellows, ochre – were supplemented or blended by Upper Palaeolithic artists with browns from manganese oxides (pyrolusite) and whites from calcite and chalk. All these were readily available in British strata or mineral deposits. During the Renaissance, trade with the Orient introduced the vivid blues and ultramarine from powdered lapis lazuli, with the bright yellow of orpiment (arsenic sulphide) and red from cinnabar, a mercury sulphide, sourced from southern Spain. All were expensive, particularly the former from its one source in

NE Afghanistan. Blue pigment was also widely available from powdered azurite, a hydrous copper carbonate, though it partly dehydrates with age to green malachite, a more stable form of copper carbonate also used for pigment. Chemical processing of organic and metallic ingredients began in the eighteenth century with the discovery of Prussian Blue. In the nineteenth century synthetic pigments like French Ultramarine and Cobalt Blue and a host of others revolutionized pigment manufacture, whilst the invention of collapsible lead tubes in the 1840s enabled artists to paint oils directly out in the landscape (*en plein air*).

Beginnings of British Landscape

In most early- to mid-eighteenth-century British painting, landscape was included, if at all, as an anonymous setting or prop for portraits of wealthy landowners. At this time there was also a fashion for idealized pastoral landscapes recalling a classical Arcadian ideal (nymphs, swains, shepherds, olive trees, grottoes: the lot). Classically inspired English country houses had their surrounding parklands landscaped by such as Capability Brown in keeping with this Pastoral or Picturesque ideal, made fashionable by French artists Nicolas Poussin and Claude Lorrain.

Thomas Gainsborough's Mr and Mrs Andrews (c.1750) (Fig. 15.1) goes much further. The proud sitters (particularly Mr A.; Mrs A. looks primly unimpressed, though she may have a squint) are placed firmly on one side. Their landscape is on the other: observed, recorded and celebrated by the artist with as much care and skill as the landowners have obviously tended their lands; the scene can be recognized today, though the oak is somewhat larger. Gainsborough subsequently became a fashionable portraitist and moved to Bath. His landscapes included in later portraits are, by comparison, stylized and formulaic. He is reported to have complained that his wealthy portrait clientele precluded pursuance of his love for landscape.

Mr and Mrs Andrews shows a love of pastoral landscape during the Agricultural Revolution. Soon, wilder vistas would find fuller expression through Romanticism, but Neo-Classical landscape painting

Figure 15.1 *Mr and Mrs Andrews* (c.1750) by Thomas Gainsborough (1727–1788). Oil on canvas. 69.8x119.4cm. Acc. No. NG630 © National Gallery, London/National Portrait Gallery. Real landscape; morning sun shines through cumulus cloud, the posed figures make way for corn stubble (drilled), neat stooks and folded sheep pasture along the sloping Stour valley. A distant scarp looks steeper and wilder, the Stour here has cut through poor land, Palaeocene sands and gravels on Chalk, as it wends its way past Dedham to the estuarine coastline at the Essex/Suffolk border.

continued to be popular well into the nineteenth century. In *Anecdotes on Painting* (1761), Horace Walpole wrote: 'In a country so profusely beautiful with the amenities of nature it is extraordinary that we have produced so few good painters of landscape.'

Walpole may not have been aware of the work (Fig. 15.2) of the mid-eighteenth-century Welsh artist, Richard Wilson. He is regarded as one of the primary creators of British Romanticism and has been called the father of British landscape painting. He was born at Penegoes, near Machynlleth on the River Dovey a few miles south of the Cader Idris range. His art education was paid for by a cousin who had become immensely rich due to the discovery of lead ores on his land. Wilson closely observed topography and was sensitive to light and weather conditions; he painted outdoors and engaged emotionally with the scene.

The yearning to experience raw and untamed nature gained momentum at the same time as the Industrial Revolution was gobbling up the countryside and drawing millions towards work in industrial towns and cities. The paintings of Joseph Wright capture the excitement of both landscape, the Industrial Revolution (Fig. 27.4) and the Scientific Enlightenment via Birmingham's Lunar Society (Chapter 28). Though his paintings may be considered Baroque in the use of tenebrism (dramatic highlights), Neo-Classical in the factual detail and general didactic intent, he makes dramatic use of light in his studies of sea caves and, most spectacularly, the erupting Mount Vesuvius. His English landscapes are more subtle, featuring both late-afternoon sunshine and luminescent moonlight lingering on the Matlock's limestone crags (Fig. 15.3).

Figure 15.2 *Llyn-y-Cau, Cader Idris* (Exhibited 1774) by Richard Wilson (1713–1782). Oil paint on canvas. 51.1x73cm. Acc. No. 5596 ©Tate, London 2015. Wilson changed elements of such sublime landscapes in the interest of composition. The cwm's backwall is especially accentuated. The tiny figures, the cow, the ardent telescopist, the huge morainic boulders in the foreground all accentuate the study. We know today that an Ice Age valley glacier outflowed here; Wilson's contemporary audience didn't, but they recognized sublime when they saw it.

Figure 15.3 *Matlock Tor by Moonlight* (1777–1780) by Joseph Wright of Derby (1734–1797). Oil on canvas. 63.5x76.2cm. Acc. No. B1976 -70-177 ©Yale Centre for British Art. Here the stepped profile of the limestone tor above the Derwent gorge is revealed by bedding planes with vertical jointing. The joints and related faults often host lead ore in the area; Wright's wife, Ann, was the daughter of a local lead miner.

East Anglian Idylls

The Norwich School of landscape painting (the only school attributed to a named British city) was effectively founded in 1803 as the Norwich Society of Artists by John Crome and Robert Ladbrooke, natives of the city. The contrast between the Norwich School's landscape work and Wilson's and Wright's has much to do with the radically different landscape of East Anglia's gently undulating and glaciated lowlands, slow meandering rivers, soft coastal cliffs and saturated marshes. Their low horizons and the emphasis on horizontals often serve to illuminate drama in the sky above (Fig. 15.4).

Crome, Ladbrooke and the Norwich School generally were much influenced by Dutch landscape painting; there are strong geological and cultural similarities between the Netherlands and East Anglia. Crome saw many examples in the collection of his first patron, Thomas Harvey, which capture the weather and light effects over land and sea. In addition, his landscapes (Fig. 15.5), like Constable after him, contain telling details of minutely observed trees, flowers, working men, women, children and horses.

Notwithstanding Crome, John Sell Cotman is now regarded as the original genius of the School, a skilled engraver and innovator in watercolour of seascapes and landscapes (Fig. 15.6). Unlike Crome, who confined much of his painting to Norfolk, Cotman painted widely across Britain and on the continent. Like Turner (who encouraged his work) he was to travel to Snowdonia to paint and to pay homage to Richard Wilson.

Figure 15.4 *Beeston Regis from the Roman Camp* (1842) by Robert Ladbrooke (1768–1842) Oil on panel. 26.7x44.5cm. Acc. No. NT 1401289 ©National Trust, Felbrigg Hall, Norfolk. North Norfolk's vast coastal sky unfolds its voluptuous clouds over the arenaceous, heathy landscape of glacial outwash and hummocky terminal moraine that is the Cromer–Holt Ridge.

Figure 15.5 *Mousehold Heath* (c.1818–20) by John Crome (1768–1821). Oil paint on canvas.109.9x181.0cm. Acc. No. 0689 ©Tate, London 2015. An early summer landscape of unenclosed heathland commons. The diagonals made by the foreground and the sky frame the sunlit slopes in the centre, along which a rough track meanders. The thistled foreground ridges have calcareous (chalky) substrates; Great Mullein (yet to flower) is evident left of centre. A countryman in a smock and with a long pole points towards the lower ground, perhaps telling of Kett's Rebellion and his followers camping here for six summer weeks in 1549.

Figure 15.6 *Hell Cauldron* (1803–5) by John Sell Cotman (1782–1842). Pen, ink and watercolour. ©Leeds Museums & Galleries (Leeds Art Gallery) U.K./Bridgeman Images. One of an early sequence Cotman made along the River Greta in North Yorkshire, a right bank tributary to the Tees. In them he sees a fluvial landscape of excavated and eroded rock: cascades, waterfalls, crags and narrow gorges cut through near-horizontal late-Lower Carboniferous sandstones. He said that nature guided his choice of colour as he made '…close copies of that fickle Dame.'

The Romantics

Romantic art, like Picturesque gardens, shows a shift in values: natural landscape was not perceived as significant on its own account, but for the way it triggered feelings, memories and ideas in the person looking at it. Much early-nineteenth-century art sought to represent the 'grandeur of nature' and produce in us 'feelings of awe, fear and horror': the Sublime. Romantic artists saw in nature a revelation of God's will. William Blake, Joseph Turner, John Constable, and later Samuel Palmer are the chief British artists from this period. They contemplated nature with devotion: 'the sound of water escaping from mill-runs, willows, old rotten planks, and brickwork, I love such things...', wrote Constable. Their influences were 'back-to-nature' movements, Dutch landscape painting, Jean-Jacques Rousseau and the Romantic Lake poets, notably Wordsworth.

In Constable's paintings, country people, their dogs and plough horses are working in apparent harmony with the landscape; they are full of subtle movement (Fig. 15.7). Almost all were executed in his London studio from working sketches. The Stour Valley paintings show the twelve square miles around his birthplace, East Bergholt; but they don't merely record his homeplace with love, they celebrate nature. His brushwork, full of life and sensitivity, records the play of light and the different moods of the sky and the weather. He wrote: 'No two days are alike, nor even two hours, neither were there ever two leaves alike since the creation of the world...'. He achieved his distinctive landscapes by dispensing with tradition. His use of optical mixing, where the eye of the onlooker mixes adjacent strokes of paint, say red/yellow and green/blue to make brown/green, profoundly influenced later artists, particularly the Impressionists, as did his sensitive,

Figure 15.7 *The Stour Valley and Dedham Church* (*c*.1815) by John Constable (1776–1837). Oil on canvas. 55.6x77.8cm Acc. No. 48.266, Warren Collection/William Wilkins Warren Fund. Image ©2016, Museum of Fine Arts, Boston.The flat-bottomed Stour valley with its meandering channel in a wetland floodplain of freshwater marsh and water meadow. A river terrace is clearly shown centre left. The labourers in the foreground on the arable lands of the sloping valley side are shovelling manure into a cart with two fine draught horses patiently waiting. The dog's stance leads the eye downhill to a horse ploughman at work with his Suffolk Punches.

flickering brushwork recording billowing clouds or the rustle of leaves. His early paintings show a golden age, an endless summer of peace and plenty; he was painting Suffolk during the agricultural boom years leading up to the peace of 1815 and the more uncertain years afterwards. These early works also reflect his inner landscape: he was newly married and happy. In many paintings the drama is present in the quality of light and the sky. His most popular paintings The Haywain (1821) and Flatford Mill (1817) are two examples. There is a canalized river in the latter. Such a landscape would have seemed alien to a seventeenth-century Suffolk farmer when the staple was wool, not corn; the mill was originally built for fulling. In the eighteenth century wool production moved north, and Suffolk farmers turned to grain to satisfy the demand from a rapidly growing London. Locks were built on the Stour for moving grain, with cuts and wharves built beside each mill.

Joseph Turner painted nature at its most violent and sublime, travelling the length and breadth of England and Wales to do so. He admired Wilson and travelled to Cader Idris to follow in his footsteps. At the same time he recorded the drama of the Industrial Revolution's technological advances, notably the new steam age. All of his paintings show the drama of a Nature bathed in sublime light and atmospheric effects. Many are of the shore (Fig. 15.8) or out at sea. He grew up and lived by the Thames, and records a seafaring nation very much at the mercy of the elements. He generally shows Man as pitifully small when pitted against the forces of nature, especially the sea. He took pride in factual accuracy:

Figure 15.8 *Keelmen Heaving in the Coals by Moonlight* (1835) by Joseph Mallord William Turner (1775–1851). Oil on canvas. 92.3x122.8cm. Acc. No. 1942.9.86 Widener collection ©National Gallery of Art Washington D.C. The moonlit and torchlit scene looks eastwards down the estuary of the River Tyne. Shallow-draught vessels of the keelmen are transfering coal, mined from adits cut into outcropping seams along the river's steep banks, into sea-going sailing vessels. This was 'sea-coal' destined for the capital, a trade (together with salt from the Shields saltings) that even at this time had gone on for 500 years.

forms often dissolve in light, smoke, mist and steam, the images at times approaching abstraction.

The search for a spiritual meaning in nature is a recurring theme in British art. Samuel Palmer was much influenced by Blake's imaginative and spiritual view of the world. William Blake found the beauty of nature overwhelming but didn't take up landscape painting. Palmer was wholly committed to landscape. Though forming a precise record of nature, his work is visionary and devotional. Generations of landscape painters have been influenced by him, most notably Graham Sutherland (see below).

During the nineteenth century, topographical prints gained importance. They satisfied the growing demand for 'a view' in those who could not afford to buy an original painting. Turner and Constable were both fascinated by landscape engraving and new methods of printmaking. Both published books of prints, working with printmakers to develop ways to show the subtleties of colour through white light and dark shade.

Pre-Raphaelites Arts and Crafts and the Impressionists

During the later nineteenth century, several new movements developed in Britain, though Neo-Classicism and Romanticism continued to be popular. These new movements, perhaps reflecting an increasingly complex society, rejected traditional artistic conventions, but in very different ways. The Pre-Raphaelites looked back to medieval culture and were part of the rage for all things Gothic. They used strong colours, abundant detail, and tried to imitate nature (with a lot of geology) as closely as possible, including detailed naturalism and symbolic subject matter. They can be seen as a reaction to the picturesque and the sublime, but their naturalism and focus on detail points to an analytical perception of nature, as seen in John Brett's The Stonebreaker (Fig. 10.8) and John Ruskin's The Pass at Killiecrankie (Fig. 4.1). Their paintings are often full of symbolism. Holman Hunt's flock of strayed sheep, for example (Fig. 15.9), are at play or resting amongst the wild flowers. They seem oblivious

Figure 15.9 *Our English Coasts, (Strayed Sheep)* (Exhibited 1852) by William Holman Hunt (1827–1910). Oil paint on canvas. 43.2x58.4cm. Acc. No. 5665 ©Tate, London 2015. The exquisite plump sheep are gambolling and resting on a craggy sandstone outcrop along the Sussex cliff line amongst delicately painted wild flowers (Hemp Agrimony, Red Campion). Compositionally the painting is full of tension and diagonals: a painting with a message.

First with the Whales, last with the eagle skies—
Drown'd wast thou till an Earth quake made thee
    steep—
Another cannot wake thy giant Size!
This is truly landscape of imagination and intellect.

The Pastoral Vision

Pastoral writers generally put their poems, novels and travelogues in 'improved' and productive landscapes in which there is little sign of rocky outcrop. John Clare saw fertile lowlands through eyes that experienced the devastation wrought by enclosure and greed. Robert Burns raged his magnificent heart for the universal entitlement of the common man or woman. The opinionated and irascible William Cobbett is uninterested in the actuality of landscape other than its fertility. He particularly hates infertile heaths, yet sees through their lack of fertility to the substrate beneath and to the base motivations of enclosure, most notably in his account of Thames terrace heaths in his *Rural Rides*:

> A much more ugly country than that between Egham and Kensington would with great difficulty be found in England. Flat as a pancake, and, until you come to Hammersmith, the soil is a nasty stony dirt upon a bed of gravel. Hounslow Heath, which is only a little worse than the general run, is a sample of all that is bad in soil and villainous in look. Yet this is now enclosed, and what they call 'cultivated'. Here is a fresh robbery of villages, hamlets, and farm and labourer's buildings and abodes!'

Victorian to 'TheGeorgians'

In an age when geology and physical landscape began to come together, we see some vivid appreciations of the interrelationship. For example, in Anthony Trollope's *Barchester Towers* there is a delicious description of Ullathorne Court's plant-weathered stone:

> This wall was built of cut stone, rudely cut indeed, and now much worn, but of a beautiful rich tawny yellow colour, the effect of that stonecrop of minute growth, which it had taken three centuries to produce… No colourist that ever yet worked from a palette has been able to come up to this rich colouring of years crowding themselves upon years…

Whereas when Nathaniel Hawthorne visited the Lakes in 1855 he wearily concluded that:

> On the rudest surface of English earth, there is seen the effect of centuries of civilisation, so that you do not quite get at naked nature anywhere. And then every point of beauty is so well known, and has been described so much, that one must needs look through other peoples's eyes, and feel as if he were seeing a picture rather than a reality…

The individual's unique ability to appreciate landscape (which Hawthorne denies) requires especially careful nurturing. According to modernist Herbert Read: 'Memory is a flower which only opens fully in the kingdom of Heaven, where the eye is eternally innocent.' Or, as memorably circumscribed around an eye-shaped aperture through a wooden sculpture in Bergh Apton, South Norfolk: 'You only see a thing for the first time once.' So we can all see anything new for ourselves; something the Pre-Raphaelites realized in their exact and vivid-hued landscapes.

Thomas Hardy's heaths, downlands and monuments often define landscapes of suffering and fortitude (his Egdon Heath in *Return of the Native* especially), yet they also have delightful insights. Such are the chalklands of Flincomb-Ash and Chalk-Newton where Tess and Marian are working in *Tess of the d'Urbervilles*:

> The swede field in which she and her companion were set hacking was a stretch of a hundred odd acres, in one patch, on the highest ground of the farm, rising above stony lanchets or lynchets – the outcrop of siliceous veins in the chalk formation, composed of myriads of loose white flints in bulbous, cusped and phallic shapes…'

By the beginning of the second decade of the twentieth century, poets referred to as 'The Georgians' featured in influential anthologies of that name. Their style, notably in the poems of Walter de la Mare and Rupert Brooke, became fashionable and immensely popular: easy pantheism, gentle pastoral idylls and fairy scenes. Brooke, though, could be sharp, as in *The Chilterns*, shaking off a doomed love affair:

Thank God, that's done! and I'll take the road,
Quit of my youth and you,
The Roman road to Wendover
By Tring and Lilley Hoo,
As a free man may do…

Best known of all, that image riffing on England's native substrate in *1914 V. The Soldier*:

There shall be
In that rich earth a richer dust concealed;
A dust whom England bore, shaped, made aware,
Gave, once, her flowers to love, her ways to roam,
A body of England's, breathing English air,
Washed by the rivers, blest by suns of home…

The Warriors

In contrast to the Georgians, realism and plain-speaking became engendered in the poetry of men on active service who volunteered or were conscripted in the first terrible world war of the twentieth century. The trenched landscapes of northern Europe were no place for pastoral idylls, and sublimity was not high on a list of attractions. For Sassoon, Graves, Thomas, Owen, Blunden, Rosenberg and others there was directness: a variety of rhythm, rhyme, assonance, half-rhyme; a colloquial familiarity. Rather as the Metaphysicals had plundered the dictionary for metaphor and simile that arose out of Renaissance ideals and global exploration of new terrains, so these warrior-poets dug their own rugged verses from the trench-deep, bloody earth.

So, in early 1918 a young infantry officer on training leave with his battalion of the Manchester Regiment in Scarborough sits in front of a smoky coal fire in the elevated turret room of his hotel billet. That day he had news of a terrible mining disaster in the West Midlands, not far from his home town of Shrewsbury. The horrendous memory of the trenches, the solid ordinary endurance of the men in 'his' platoon and the hundreds of colliery casualties co-mingle in his mind with the hissing coal in the grate. Quickly, within the hour, he had the draft of *The Miners*, whose last verses are quoted below. It was published shortly afterwards by a national daily: Wilfred Owen's first published poem:

But the coals were murmuring of their mine,
And moans down there
Of boys that slept wry sleep, and men
Writhing for air.
And I saw white bones in the cinder-shard,
Bones without number;
For many hearts with coal are charred
And few remember.
And I thought of some who worked dark pits
Of war, and died
Digging the rock where Death refutes
Peace his indeed…

As a young man Owen was fascinated by geology and archaeology. In a letter to his mother shortly after the poem was published he wrote, 'a poem on the Colliery Disaster; but I get mixed up with the War at the end. It is short, but oh! Sour!'

From such poems Owen claims his place with Housman, the 'later Hardy' and Edward Thomas as among the founders of modern English poetry. Thomas had lived amongst the chalk downlands of southern England, writing on the countryside, and on other novelists and poets who concerned themselves with the countryside, and by extensive reviewing. He made an outpouring of poetry from late 1914 until his death at Arras in 1917. It drew on his deep reserves of familiarity with English and Welsh landscapes and of the common idioms of speech, often with internal monologue or conversation. His downlands appear time and again, their long-abandoned marl-pits now coverts full of flowers and trees, and the clay-with-flints that often overlie the bare chalk in bournes and hollows. In *Wind and Mist* (1915) he wrote:

The flint was the one crop that never failed.
The clay first broke my heart, and then my back;
And the back heals not…

And:

Yes. Sixty miles of South Downs at one glance.
Sometimes a man feels proud of them, as if
He had just created them with one mighty
thought…

The Shieling (1916) was perhaps written with the upland Wales of his ancestors in mind, though it was actually the name of a friend's house in the southern Lakes:

It stands alone
Up in a land of stone
All worn like ancient stairs,
A land of rocks and trees
Nourished on wind and stone…

The Moderns

To the moderns, rational analysis of landscape, together with the legacy of the Romantics and the insights of pyschology, was grist to their mill. The early poetry of D.H. Lawrence was included in Georgian anthologies but he soon developed his own unique voice. Here he is writing in *The Rainbow* (1915) of a visit to Lincoln Cathedral:

They had passed through the gate, and the great west front was before them, with all its breadth and ornament. It was a false front, he said, looking at the golden stone and the twin towers and loving them just the same. In a little ecstasy he found himself in the porch, on the brink of the unrevealed. He looked up to the lovely unfolding of the stone. He was to pass within to the perfect womb….

Herbert Read was a notable art critic; his eye for the detailed sense of place that stone gave to the landscape in his native North Yorkshire never left him. When describing what landscape meant to him he writes, '… love of landscape must feed on intimacy as well as on magnificence …'. For this reason he loved moor- and fell-country, casting aside the sublime:

> Mountains I have no love for; they are accidents of nature, masses thrown up in volcanic agony. But moors and fells are moulded by gentle forces, by rain water and wind, and are human in their contours and proportions, inducing affection rather than awe …

Poets of the twentieth century carried on such notions of informed, subtle and resonant approaches to landscape: how geological processes and evolution, the concept of deep time can help move a poem's theme along. Consider first Hugh MacDiarmid, already quoted by us (Chapters 1, 2). His birthplace was Langholm, a Dumfriesshire market and weaving town nestled snugly beside the exit of the River Esk from the Lower Palaeozoic massif of the Southern Uplands. His mother was caretaker of the town library; the family lived below it and had the library keys. So young Christopher Grieve (as he was born) had unlimited access to books and became a prodigious reader. *Whita Hill* (1936) shows a typically contrarian attitude towards military memorials set up in Victorian Scotland:

> Few ken to whom this muckle monument stands,
> Some general or admiral I've nae doot,
> On the hill-top whaur weather lang syne
> Has blotted its inscribed palaver oot …

The 'muckle monument' in question stands by craggy outcrops of early Carboniferous sandstone that overlook Langholm, its blocks sourced in an adjacent quarry. The monument is clearly visible from nearby England, the chief source of MacDiarmid's fervent and irritable nationalism. More lyrically he writes in *Tarras Water*

(1934) of shallow turbulence in the salmon-running, left-bank tributary to the Esk:

> Pride of play in a flourish of eddies,
> Bravura of blowballs and silver digressions,
> Ringing and glittering she swirls and steadies,
> And moulds each ripple with secret suppressions…

In 1965, Basil Bunting, then a sub-editor on the Newcastle *Daily Journal*, published the long five-part poem *Briggflatts* after several years spent on its composition and ruthless condensation. Prior to this was a lifetime of wandering and friendships with poets Ezra Pound and Louis Zukovsky. The poem tells of a personal odyssey of love, adventure and exploration mixed with historical and mythical incidents seen through English, Welsh and Norse eyes. The poem can be read (Bunting himself had a still-strong Northumberland accent in the 1960s) as an extended metaphor on the timelessness of rock and stone in relation to human life and both personal and historical memory. Rocks are named (gabbro, limestone, sandstone, slate), climbed, chiselled, hammered, polished and made love on; they act as weather-prone reminders of past love and loss. The poems' essential musicality, muscularity and the riffs on these various themes are immediately obvious and perfectly modulated in the very first stanzas:

> Brag, sweet tenor bull,
> Descant on Rawthey's madrigal,
> Each pebble its part
> For the fell's late spring.
> Dance tiptoe, bull,
> Black against may…
> …A mason times his mallet
> To a lark's twitter,
> Listening while the marble rests,
> Laying his rule
> At a letter's edge,
> Fingertips checking,
> Till the stone spells a name
> Naming none,
> A man abolished…

The bull's brays are accompanied by river-music as the Rawthey runs in spate, swollen by snowmelt from the surrounding fells just emerging from winter. The unsteady turbulent river-roar includes the sounds of pebbles dragged and pounded over its rocky bed. Meanwhile, nearby, another rhythmic sound-association adds to the bull/river madrigal theme; a stone mason chisels a name into his limestone slab whilst a lark sings above him. The chisel makes its own imitation trill as each dull tympanic blow from the wooden mallet creates its own twitter on the steel's ricocheting edge.

As he makes abundantly clear, the youthful W.H. Auden wanted nothing more than to make a career as a mining engineer. On long hikes with his two brothers (the oldest became chief geologist of India) he acquired a thorough knowledge of the North Pennines orefield and its mining centres like Alston, St John's Chapel and Rookhope in Weardale. Here he is writing *Letter to Lord Byron* while journeying with Louis MacNeice in Iceland in 1936:

> And from my sixth until my sixteenth year
> I thought myself a mining engineer
> The mine I always pictured was for lead
> Though copper mines might *faute de mieux*, be
>   sound.
> Today I like a weight upon my bed;
> I always travel by the Underground …

The spell that limestone cast upon the poet is celebrated in his own confessedly favourite poem, *In Praise of Limestone*. It is a paean to the geological landscapes and underworlds of limestone, its theme a universal metaphor for loving understanding between human beings. The longish poem is a sort of fantasy that also involves his family:

> If it form the one landscape that we the inconstant
>   ones
> Are consistently homesick for, this is chiefly
> Because it dissolves in water. Mark these rounded
>   slopes
> With their surface fragrance of thyme and beneath
> A secret system of caves and conduits; hear these
>   springs
> That spurt out everywhere with a chuckle …

He is not only celebrating North Pennine landscapes but is setting up limestone as both metaphor for, and celebration of, his own strong individuality. He imagines a dissolvable stone landscape of tolerance; it memorably ends:

> when I try to imagine a faultless love
> Or the life to come, what I hear is the murmur
> Of underground streams, what I see is a limestone
>   landscape.

Auden's friend from Oxford days and fellow Icelandic traveller, Louis MacNeice (1907–1963), shared with him a love of landscape and elemental rock. He was born and spent his early childhood among the landscapes of Ulster and Connacht, though his schooldays were largely in England after his mother's early death when he was just seven. In his long and popular autobiographical poem, *Autumn Journal* (1939) he describes his move from a school in Dorset to one in Wiltshire:

> Freedom in walks by twos and threes on Sunday,
> We dug fossils from the yellow rock
> Or drank the Dorset distance.
> … And we found it was time to be leaving
> To be changing school, sandstone changed for
>   chalk
> And ammonites for the flinty husks of sponges,
> Another lingo to talk
> And jerseys in other colours.

He travelled widely in the 1930s, including to the Hebrides, where he was disillusioned by the dependence-culture, as he saw it, of Scottish Gaeldom. Here in four lines are his condensation of cultural and geological realities in *The Hebrides*:

> On these islands
> The west wind drops its messages of indolence,
> No one hurries, the Gulf Stream warms the gnarled
> Rampart of gneiss …

In 1963 MacNeice contracted the pneumonia that would kill him, walking back in heavy rain after time spent shivering in the limestone potholes and caves of Settle. Auden wrote a moving elegy to the memory of his old friend, memorably entitled *The Cave of Making*.

To read Norman MacCaig's poetry is to be cast by metaphorical spells that come from ferocious intelligence and deep curiosity. In his own words: 'I have two guzzling eyes for landscapes, seascapes, shorescapes …'. Dozens of his 700-odd selected poems attest to this greed. Several explore the physicality of the Torridonian mountains of Assynt. Suilven and its numerous springs are unforgettably portrayed in *Sandstone Mountain*:

> The bare rock hill turned out to be
> A rocky sponge – it leaked all round
> A maze of trickles …
> … The hill streamed ribbons everywhere,
> Perpetual conjuror, from his sandstone gown …

And, notably at one with the sentiments of Greek poet C.P. Cafavy's masterpiece, *Ithaca* (MacCaig studied and taught classics) is *Landscape and I*:

> Landscape and I get on together well.
> Though I'm the talkative one, still he can tell
> His symptoms of being to me, the way a shell
> Murmurs of oceans …
> So then I'll woo the mountain till I know
> The meaning of the meaning, no less. Oh,
> There's a Schiehallion anywhere you go.
> The thing is, climb it.

Norman Nicholson's poetry explores, often with sensitive physical imagery, a semi-industrial, once coal- and iron-rich seaward borderland fixed to the Lake District's central core. His life was spent in Millom, where he witnessed the discovery of the haematite iron ore that transformed the town from the 1920s until its exhaustion some decades later (see motto to Chapter 9). His poetry draws upon geological history and processes set against the human historical past of his antecedents, from Norse settlers clearing native forest to the miners of coal and iron ore. Often his imagery is geologically direct, as in *Beck* (1975) in which, with a Huttonian emphasis, he examines the effects of erosion on outcrop, hillslope and river valley:

> Slate and sandstone
> Flake and deliquesce,
> And in a grey
> Alluvial sweat
> Ingleborough and Helvellyn
> Waste daily away…

In *Wall* (1975) he imagines a fellside's undulating drystone walls ever-subsiding as the ground under them erodes and settles down – the landscape reducing by shrinkage, the walls moving with it:

> The wall walks the fell—
> Grey millipede on slow
> Stone hooves;
> Its slack back hollowed
> At gulleys and grooves,
> Old boulders
> Too big to be rolled away.
> Fallen fragments
> Of the high crags
> Crawl in the walk of the wall…

George Mackay Brown's Orkney stones are solid and cold but only temporarily lifeless. They are brought back to life by either human passions, the cycle of seasons or incoming starlight from a jubilant and unashamed God. This poet of place, Scotto-Scandinavian Orkney, had the brown-red Middle Devonian sandstone flagstones of the Stromness Group all around him for most of his life, yet not a single tree. In a cognate image to that presented in Bunting's *Briggflatts*, he tersely tells in *Seal Island Anthology* of the carving into song of a gravestone and its subsequent weathering that will render the whole thing mute:

> Suddenly a stone chirped
> Bella's goodness,
> Faithfulness,
> Fruitfulness,
> The numbers
> Of Bella's beginning and end.
> It sang like a harp, the stone!

In the sequence *Daffodil Time* he has this image of spring-awakened stone:

> The stone that wore darkness like the minister's
>   coat
> All winter,
> Where the crow furled, where
> Snow lingered longest—
> Look, now, sunrise
> Is tilting its jar of light over the world
> And now, this noon,
> A random splash has hit the winter stone.

Alasdair Maclean was a poet who lived a mostly solitary life on his inherited family croft on Ardnamurchan, 'a bleak, inhuman, fearful landscape…'. The poem *Stone* sets out the stark dilemma of the Highland people who live amongst glacier-scraped rock and thin acid soils in the many places like Ardnamurchan:

> A long peninsula of solid rock,
> Upholstered every year in threadbare green.
> Stones everywhere, ambiguous and burgeoning.
> …God was short of earth when He made
>   Ardnamurchan.

A contrasting sort of relationship with stone and landscape features in many of Ted Hughes's poems. He grew up in the mill town of Mytholmroyd in the Upper Calder valley of West Yorkshire. The town lies under an immense north-facing outcrop of the local Millstone Grit, and it is not difficult to see its permanent mineral shadow in Hughes's poetry. His possession of 'alive' nature in his poetry; bird, beast and fish, is often complete, whereas his physical world is cold, brutal and unforgiving. There is little sense of redemption in landscape and its mineral architecture. In the remarkable project *Remains of Elmet* with photographer Fay Godwin (see also Chapter 4), humans and landscape are utterly interchangeable; weathering is death, rainwater is sunlight. We cannot escape the fixity of inchoate matter; stone is content with its lot, is indifferent, unable to see, feel, develop or change. It is from the stars, like starlight (his 'star-blaze'). Yet it is empty, stranded here, like us. Over the great emptiness of moorland flickers light and shade, from water and rock. They are:

> a stage for the performance of heaven.
> Any audience is incidental.

The abandonment of moorland cultivation is seen in

crumbling walls, shipwrecks of stone that once provided strong rigging, gathering it together. The contact of humans with rock is where the sole of a foot is:

> pressed to the world-rock, flat
> Warm
> with its human map
> Tough-skinned, for this meeting
> Comfortable
> The first acquaintance of the rock-surface
> Since it was star-blaze…

Or, immaculate in its detachment after the death of a new-born lamb, he tells his daughter that the creature was lucky to have ended its birth struggles on a warm, blue-skied day:

> And the sky line of hills, after millions of hard
>   years,
> Sitting soft.

Glyn Hughes set much of his artwork, novels and poems in the South Pennines: the countryside of his autobiographical *Millstone Grit*. For Glyn this was a country that rises (literally) above economic and personal deprivation and difficulty – a subject for lyric celebration. In these tablelands with their high moors, craggy edges and deep glaciated dales, the strata step successively upwards to over 350m above sea level. He returns to the scene in *A Year in the Bull Box*, his last poetry collection. *Going There On The Long Causeway* is an elegy from this gentle, thoughtful and wryly humorous man, written in the face of the reality of his own death. It begins:

> There are cottages up there with boulders before
>   their doors.
> Stone is the beauty of here, its forms, its shine,
> Its glitter of silica in rain, or laid across fields
> And some of the stones are as large as churches;
> The Pennines' unavoidable substance.
> "Here lies this stone that waits the Resurrection"
> As one in Elland churchyard has it…

He goes on to wish that stone should never become just a tomb-covering of winter grief. It must always be a perch-pedestal on the tops for spring curlews, their staccato cries signifying renewal and resurrection of the spirit. That was how he, at any rate, would take the final:

> …sky-trip clean across the valleys of Padiham and
>   Burnley.

PART 7

GeoRegions – Cameos of Landscape, Cultures and History

We all live in some landscape or other, whether it is by a rocky, cliffed seashore, a port, an estuary margin, in a great city, a village deep within arable-farmed downland or a much-loved national park. Landscape is many things to many people and exists at all sorts of scales: dead-flat drained former wetlands, a local valley side, a cluster of crags, a grassy escarpment, a hilly outlier rising from a level vale, or a mountain range. Here we summarize the seventeen major tracts of land whose geology, topography, culture and history enable them comfortably to be termed GeoRegions.

Chapter 17

Introduction to GeoRegions

Jigsaw Pieces

Earlier the sometimes conflicting emotions engendered by landscape were discussed. Fair questions for the reader might be: 'So what's in landscape for me; where is my place in landscape?' To answer these questions it is best to return to our analogy in Part 4: that of the Island's geological history assembled like a jigsaw puzzle. The assembled puzzles presented in Figure 17.1

have the GeoRegions outlined on them: the 65 million people living on the spaces defined by these maps were placed by mainly economic factors that have affected our societies over the generations.

Now, here's the rub: while viewing our own infinitesimally small footprint in this or that landscape, the past also needs to be represented – a succession of palimpsests down all the years. This was the approach taken by

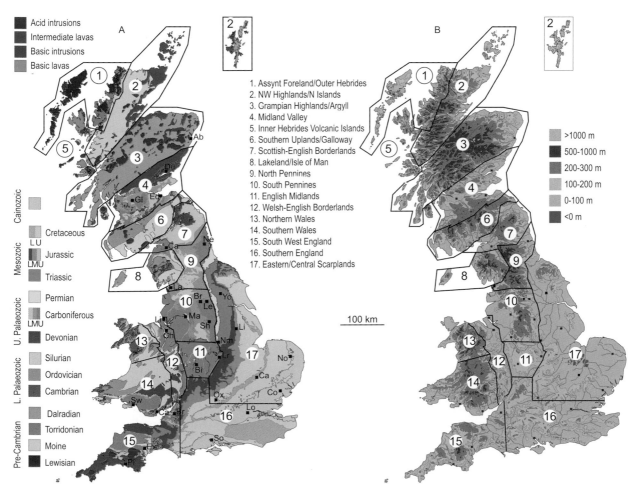

1. Assynt Foreland/Outer Hebrides
2. NW Highlands/N Islands
3. Grampian Highlands/Argyll
4. Midland Valley
5. Inner Hebrides Volcanic Islands
6. Southern Uplands/Galloway
7. Scottish-English Borderlands
8. Lakeland/Isle of Man
9. North Pennines
10. South Pennines
11. English Midlands
12. Welsh-English Borderlands
13. Northern Wales
14. Southern Wales
15. South West England
16. Southern England
17. Eastern/Central Scarplands

100 km

Figure 17.1 Maps to show the approximate distribution of the British GeoRegions. **A** Geological. **B** Topographical. Data generalized from British Geological Survey (2007b) and Ordnance Survey (1991, 2011) sources.

Jacquetta Hawkes, mentioned in our 'Forewords': the individual taking temporary possession of their own footprint on the geological substrate beneath; using family connections, history and geology to reconstruct past time. This book has already provided a basis for the history and geology; the reader must determine their own heritage.

Returning to the jigsaw analogy and moving away from the individual pieces, both space and time need to be considered. A chosen location might be a birthplace or a current or former home. The place has physical attributes like elevation, gradient, topography, geology. It may border water, a lake or the sea; experience particular weather; be sited where it is because of a road or rail junction, a source of water, an old market or a defensible site. It has a history: peoples, names, immigrants arriving, emmigrants departing, battles fought, social protest, industry and so on. Then it has a position in the wider area: neighbours, a regional context, a region, part of an ancient kingdom, perhaps (e.g. Mercia, Strathclyde, Ystrad). Even further back in time comes the greatest change: there are no humans any more. The land is ice-covered or tundra, a warm sea recedes, mountains rise, volcanoes erupt and earthquakes shatter. The Island has disappeared into geological time, as in the ancient geographies portrayed in Chapter 7.

Divvying Up

What follow are 'headlines' for Island landscapes, followed by chapter-length cameos of particular GeoRegions illustrating particular events that have given rich and distinctive heritages. These are written as brisk, affectionate portraits of places featuring a mix of basic geology, topography, history, architecture, industry and culture as appropriate. They are, of necessity, the authors' own subjective accounts, though it is hoped that readers who live in them can recognize at least something of their essence and character. Material from previous chapters is freely made use of in the accounts.

How exactly are these GeoRegions defined? They are rarely mono-geological; they can contain mixtures of rocks of different types and age, which in turn create a myriad of local landscapes. Nevertheless, these unite in characterizing the whole region in question. Particular GeoRegions call for different approaches. For example, in mountainous and generally sparsely populated areas like the NW Highlands, Grampians, Lakeland and Snowdonia, geology dominates everything. Others, like the mixed landscapes of hills, fells, deep valleys and coastal plains in the Midland Valley, North and South Pennines and South Wales, are where great cities grew rapidly 200 years ago on or adjacent to coal- and iron-rich substrates. Such GeoRegions have an often dramatic juxtaposition of rural and industrial lanscape within them. In formerly glaciated lowland regions like the English Midlands, often thick glacial successions usually hide the pre-Quaternary bedrock. They also contain tracts of slightly more elevated former coalfields. More exhilarating uplands like Charnwood come courtesy of highly resistant late-Precambrian volcanic rocks and their igneous intrusions.

Headlines

In what follows we offer brisk 'headline' summaries of the GeoRegions depicted in Figure 17.1. The following chapters carry these on in more detail.

Assynt Foreland and Outer Hebrides

An Atlantic-battered mainland fringed by myriad islands: glaciated coastal platforms, coastal dunes and machair meadows, isolated mountain survivors between scraped-out sea lochs. Ancient geology is here: coastal plinths of Archean-era gneiss below late-Proterozoic red-brown sandstone mountains, with an icing of Palaeozoic limestone and quartzite draping off eastwards. Lands for hardy survivors: crofters and fishermen; popular with climbers, artists and photographers.

North West Highlands and Northern Islands

A mountainous plateau descending east to the fertile coastal lowlands of Ross and Sutherland: Caithness has immense spongy bogs ('Flow Country') within glacier-scoured rock basins. Eastwards the Proterozoic Moine rocks are bevelled by Old Red Sandstone, thrust westwards over Assynt, and cut to the south by the slash of the Great Glen. Offshore across Dornoch are the scattered silhouettes of Orcadian whale-backed islands, fossil fish-bearing flagstones sourcing their Neolithic monumental and domestic architecture. Distant Shetland's peninsula-enclosed bays: Sullom Voe the fountainhead of hydrocarbon.

Grampian Highlands and Argyll

A gathering of high, rocky Dalradian metamorphic places: Cairngorm in the glaciated granitic core, an eastern hinterland for the Picts in their tribal valleys above the Highland Border Fault. To the west, accessible sea-broached Argyll with its elongate islands and long peninsulas, the source of Gaeldom via Ulster. A fertile eastern coastal fringe, historic Moray, Buchan and Angus battered along North Sea-bounding cliffs; Aberdeen the chief locus of Europe's hydrocarbon industry.

Midland Valley

Always the populous central core to Scotland's landmass: a lively and rapidly changing landscape of mostly Upper Palaeozoic rocks; fertile coal-bearing lowlands beset by innumerable crags and fells of volcanic lavas, sills and dykes. Also with scattered and resistant hilly Lower Palaeozoic inliers (Pentlands, Cairn Table, Tinto). The lower ground from Fife and Lothian to Clydeside is an isthmus between the Forth and Clyde. It joins the vibrant modern cities of Edinburgh and Glasgow along an ancient line of kingdoms from Anglian Lothian to British Strathclyde. Crucible of both the Scottish Enlightenment and northernmost outpost of the coal-, iron- and steel-based Industrial Revolution.

Formerly Volcanic Islands of the Inner Hebrides

Poet Sorley Maclean's 'stòitteachd bheanntan', an 'up-thrust of mountain peaks', their Paleogene intrusive centres and stepped basalt lava landscapes rising high above the sea lochs and firths that guard the approaches to the largest islands of Arran, Skye and Mull. Formerly populous crofting communities in sheltered glens and bay-heads united by boat and ferry, long-governed by the Norse: brought low by the drawn-out depopulation that were the Clearances before late rescue by Parliamentary Act in Gladstone's penultimate administration.

Southern Uplands and Galloway

A hilly, elongate east–west massif of mostly Lower Palaeozoic greywacke rock that forces any traveller along just a few valley routes between the Midland Valley or England. A dissected and glaciated plateau of rolling, partly forested hills with the major valleys and settlements within softer Permo-Triassic strata downfaulted into N–S trending outliers (e.g. Dumfries, Stranraer). Galloway in the south-west has craggy mountains founded on a suite of resistant granitic intrusions.

Scottish–English Borderlands

These agriculturally productive lowlands form an arc of Upper Palaeozoic country from the Solway Firth to the Northumberland coast. Mixed in are the Cheviots, a large mountainous inlier of mainly Devonian volcanic rocks and great tracts of formerly wild sandstone fell country, now extensively afforested. Full of ruins (Roman Wall, castles, peel houses, abbeys) these are lands of legends: cross-border brigandage, faery queens, ferocious border battles and skirmishes.

Lakeland, its Surrounds and the Isle of Man

The mountainous Lower Palaeozoic core of resistant sedimentary, volcanic and plutonic rocks has spectacular glaciated headwaters and radiating valleys. Courtesy of its chief native, Wordsworth, and honorary natives Southey and Coleridge (the 'Lakeland Poets'), its landscapes nurture the Island's most intense relationship with the beautiful and sublime. Real day-to-day lowlands surround it, pastoral agriculture and industry co-existing on a substrate rich in coal and metal ores. To the west the fells and peaks of Man are visible, rising up as an island outlier of the northern Lakes.

North Pennines

Northern English limestone fell and karst country at its brilliant white best, cut by the headwaters of numerous east-draining rivers from sourcing headwaters all along Cross Fell and the Howgills, traversing through glaciated dales to the North Sea. Mighty pioneering industry grew up around the Durham coalfield and on Tyneside, where the Roman and medieval nucleus of Newcastle-upon-Tyne stands as an elegant reminder of late-Georgian/early-Victorian urban planning in Carboniferous sandstone.

South Pennines

Hard-worked streams and rivers, a pioneering lock-studded trans-Pennine canal system, a substrate of coal- and iron-bearing Carboniferous around a core of craggy Millstone Grit moorlands. Add the gritty independent people, much abused by successive invaders and exploiters, who not only turned to pastoral agriculture but to yarn, cloth, steel, mining and engineering: the greatest and earliest integrated industrial achievement in the world. Today the weekend fells and dales echo with the tramp of hikers' feet, the tinkle of climbers' harness and careering mountain bikes.

English Midlands

Fertile lowlands spawned the proud, rich and once-dominant kingdom of Mercia. Charnwood Forest's crags reveal a fiery volcanic past as Avalonia rifted from its Gonwanaland nucleus. Coalfield inliers, a smothering of Triassic marls and deposits of Quaternary glacier-lakes complete the geological picture. All is stitched together by the first canal network that carried coal, brick, potter's clay and finished wares: the industrial and populous west linked silently and gracefully on smooth highways of water with the exporting ports of early empire.

Welsh–English Borderlands

Roman fortresses, Deva (Chester) and Caerleon (near Newport) first defined the line that later became an enduring east–west cultural divide between the Seisnig and Cymru peoples: from Cheshire and Merseyside's Wirral in the north to Tintern and Chepstow on Severnside in the south. Topographic diversity comes courtesy of varied geology: faulted resistant inliers (Clwydian Range, Wrekin, Long Mynd, Malvern Hills) and rolling scarplands like Wenlock Edge standing proud above fertile, brown-soiled pastoral lowlands.

Northern Wales

Historic Gwynedd: both the rugged mountain fastness that is the Snowdonia part of the Cambrian Mountains and, across the narrow Menai Straits, the fertile lowlands of Anglesey, historically the 'breadbasket' of northern Wales. Home-grown glaciers from Snowdonia and incoming ice down the Irish Sea from the Lakes and Scotland scoured the usually softer Precambrian rocks of Anglesey. Valley glaciers shaped and sharpened the resistant volcanic rocks of Snowdonia into the scree-lined cwms and U-shaped valleys we see today.

Southern Wales

Bronze Age Dyfed may have gifted its Mynned Preseli bluestones to the British tribes constructing the Europe-wide ceremonial landscape around Stonehenge. In a later age the gales of industrialization blew across the hill-farmed and valley-richness of the Lower Palaeozoic Cambrian Mountains and into the valleys cutting the Glamorgan plateau of the South Wales syncline, first as shallow coal workings along the north crop, and then by deep-mining across the length and breadth of the area. Transported down to the coast, the coal enabled a string of industrial towns and cities to grow up or to expand, the coal trickling round the rest of Wales and southern Britain through a myriad of sheltered anchorages.

South West England

England's long peninsula, its spine a succession of resistant granitic cupolas rising up from deeper foundations as today's unglaciated hilly moorlands. Remote, mineral-rich (probably the Cassiterides named by Herotodus), its mighty cliffs blasted by Atlantic and Channel gales. A long sea-going heritage of trade and naval power from the many ports in sheltered estuarine rias along its southern coastline. A mild though damp climate combined with physical beauty and maritime sublimity made it a premier tourist and artistic destination (the 'Cornish Riviera') after construction of rail links and the near-exhaustion of its mining industries.

Southern England

A folded stitchwork of unglaciated calcareous open downland ridges and white sea-cliffs enclose wide, river-drained clayey vales and bays with subsidiary sandstone and cherty scarps. These rolling, often wooded and hedged arable landscapes flourish under a mostly benign climate (warm and dryish). The rich inner core of a diverse region surrounds the metropolis's remarkable diversity of peoples and talents. The premier city of Europe sprawls across its widest indentation, the Thames estuary, providing access to the Channel and North Sea. An outward-looking mercantile port, originally to the Roman world of the Mediterranean, then to north-west Europe, and finally the wider world. Trading has always defined it: neither Boudicca's conflagration nor Goering's bombers stopped it for long.

Eastern and Central Scarplands

The mostly Mesozoic scarplands (limestone, sandstone, ironstone and chalk) and major river-drained clay-founded vales were extensively glaciated at least four times, eroding bedrock and depositing the fertile sediment partly responsible for subsequent agricultural prosperity, witnessed by many cathedrals and thousands of medieval settlements and churches. Major estuaries facing rich markets across the southern North Sea led to a lucrative trade in wool, later in woven cloth. In modern times great industries grew up from the mining of Jurassic sedimentary iron ore, with coal railed in from the concealed coalfields of the north: survival in two world wars came from ore dug out of Teesside, Lincolnshire and Northamptonshire during times of submarine blockade.

Chapter 18

Assynt Foreland and Outer Hebrides

Glaciers, grinding West, gouged out
these valleys, rasping the brown sandstone,
and left, on the hard rock below – the
ruffled foreland –
this frieze of mountains, filed
on the blue air – Stac Polly,
Cul Beag, Cul Mor, Suilven,
Canisp – a frieze and
a litany.

Norman MacCaig, *A Man in Assynt* (1969)

S I 'n tirsgiamhach tir a 'machair,
Tir nan dithean miogach daithe,
An tir laireach aigeach mhartach,
Tir an aigh gu brath nach gaisear.
('Tis a beautiful land, the land of the machair,
The land of the smiling coloured flowers,
The land of mares and stallions and kine,
The land of good fortune which never shall be
    blighted.')

John MacCodrum, *Smeorach Chlann Domhnaill* (*c.*1750), transl. J.A.
Love (2009)

This is the Island's ancient foreland, its many rocky outcrops exposing early-Precambrian foundations (Fig. 18.1). Deep geophysical soundings indicate such rock also forms the crustal basement to much of Scotland, with nods to similarly venerable rocks in south Greenland and Central Labrador: all once part of an ancient pre-Atlantic continent. The rock successions of single mountains in Assynt can span >2.5Ga, more than anywhere, their building-blocks definite and obvious in the many steep places left after thorough-going glaciations. The oldest parts are light/dark-banded Lewisian gneiss that crops out along the sea-lashed mainland of Sutherland and Wester Ross, from Cape Wrath down to the Kyle of Lochalsh. The NeoProterozoic red-brown pebbly sandstones of the Torridonian lie upon them unconformably (Fig. 18.2) and upon these in turn, also unconformably, are the Cambro-Ordovician quartzites and limestones of the Durness group.

Westwards are the Outer Hebrides, 150km of sheltering islands: Lewis/Harris, the Uists, Benbecula and Barra: poet Louis MacNeice's 'ramparts of gneiss' that breast the full westward Atlantic fetch. The fringing convex archipelago owes its origins to uplift by an offshore fault that bounds it eastwards along the Minches. Its craggy granitic hills in south and west Lewis and Harris and the more smoothish peaks of the Uists are up to several hundred metres elevation. They rise from the subdued ice-scraped gneissic landscape as a myriad of small hills and intervening wetland bogs and lakes, in Gaelic *cnoc agus lochan* ('hill and lake') country. By way of contrast, along windward coastlines and on raised beaches are fertile acres of long-crofted and carefully managed machair with their *dithean miogach daithe* in springtime ('marsh and spotted orchids, Irish lady's tresses'). John Love writes memorably that these calcareous, sandy-soiled meadowlands lie 'directly upon an unforgiving platform of ancient, acid rocks, attractively streaked grey and black...'.

Pioneer Highland geologist John McCulloch, visiting by boat in the early nineteenth century, viewed the Lewisian gneisses at his landing places as a pediment upon which the stately red-brown Torridonian mountains sat, part of poet Norman MacCaig's 'ruffled foreland'. They are exposed in great sea cliffs such as at Applecross or Cape Wrath, and define a discontinuous north to south straggle of isolated peaks, memorably 'filed in the blue air'; stepped-profiled, loaf-shaped ridges between over a dozen deeply glacier-scoured lochs. The Torridonian succession records the gradual burial of a hilly landscape, overwhelmed by riverine detritus sourced from the west. Its nuanced ochreous colours are caused by the thin iron oxide (haematite) coatings on its sand grains. These were precipitated from ancient groundwaters in a semi-arid climate and oxygenated atmosphere a billion years ago.

In low sunshine, Torridonian peaks like Liathach, Quinag (Fig. 18.3) and Bheinne Foinne show an icing-like coating of pale grey Cambro-Ordovician

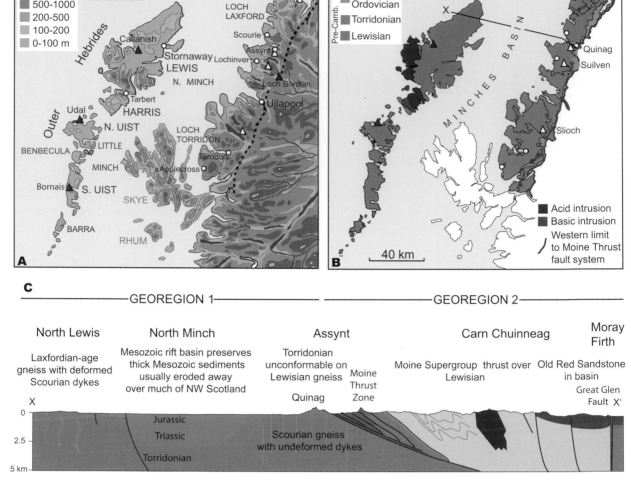

C

—————————GEOREGION 1—————————   —————————GEOREGION 2—————————

North Lewis	North Minch	Assynt	Carn Chuinneag	Moray Firth
Laxfordian-age gneiss with deformed Scourian dykes	Mesozoic rift basin preserves thick Mesozoic sediments usually eroded away over much of NW Scotland	Torridonian unconformable on Lewisian gneiss	Moine Supergroup thrust over Lewisian	Old Red Sandstone in basin

Moine Thrust Zone

Quinag

Great Glen Fault X'

X

0

2.5

5 km

Jurassic

Triassic

Torridonian

Scourian gneiss with undeformed dykes

Figure 18.1 Maps and geological section for the Assynt Foreland and Outer Hebrides. Data generalized from British Geological Survey (2007b), Ordnance Survey (1991, 2011) and *The Times Atlas of the World* (1987) sources.

Figure 18.2 Ben Slioch (981m) and Loch Maree looking north. Photo: ©Vaughan Melzer; camera at 57.630975, –5.349483. Lower levels, including the buttress to the right immediately above the loch, are Lewisian gneiss. A valley eroded into the gneiss was infilled by Torridonian river-derived sediment that makes up the remainder of the mountain. The left-hand valley slope and unconformity is well-defined by the pinkish lenticle of earliest Torridonian strata that descends and thickens towards the buttress.

Figure 18.3 The Torridonian mountain of Quinag (707m) looking SW from Kylescu. Photo: ©Vaughan Melzer; camera at 58.258068, −5.025624. The sandstones, with their gently dipping stratification and gullied cracks along prominent vertical jointing, overlie a plinth of light-weathering Lewisian gneiss seen along the middle right, and also in scattered outcrops in the foreground.

strata dipping gently off their eastern slopes (from the sea the Norse saw Quinag as a 'grey pillar'). Their mapped outcrop wriggles south-west from Cape Wrath to southern Skye. The unconformity is the result of a virtuoso display of coastal erosion in Cambrian times by the newly arrived Iapetus ocean. The kilometre-thick succession of sedimentary rock deposited under shallow-marine conditions is best exposed east of Cape Wrath, by Durness and Loch Eriboll. The resistant quartzites form bare ridges and scarps along the whole length of outcrop, much in evidence along the trace of the Moine Thrust, the eastern boundary to the region (Chapter 19). The limestones add fertility to often poor and acidic lowland soils weathered from glacial deposits, raised beaches and riverine terraces.

Life in the region has always been tough, and the population relatively sparse, on account of the harsh physical and climatic environment. Yet its maritime aspect allowed free communication by boat in fair weather. Mesolithic colonizers 'island-hopped' after the Younger Dryas cold spell into the Outer Hebrides and Assynt's small but sheltered anchorages. Their coastal habitation sites were rock shelters with periwinkle/limpet shell middens and scattered stone-flaked tool remnants and carved bone pieces. The shelter at Sand on the Applecross peninsula is best-dated, occupied around 9.5ka. Rock tools indicative of wider trading contacts include distinctive glassy pitchstone from one or more of Eigg, Rhum and Skye, the once-volcanic Inner Hebrides (Chapter 22).

After sea level stabilized around 7ka the absence of significant riverine sediment influx enabled the clear shallows of coastal waters, especially the windward island coasts of Uist, to produce copious quantities of organic calcium carbonate. Wave and tide pulverized thousands of generations of mollusc and algal remains. Beaches and storm washovers became spreads of abraded shell sand, the habitual onshore winds blowing the finer fractions far inland from a more-or-less continuous apron of sheltering, marram-fixed dunes. By the early Bronze Age, sedimented upon the ancient gneisses and pre-existing glacial soils, the calcareous sediment had deposited sufficient volume to define the fringing machair coastal plain behind the active dunes. Once manured with storm-uprooted kelp and animal dung, it developed great fertility. This fragile environment (John Love writes playfully, 'Oh dear what can the Machair be?') is today vulnerable to Atlantic storms and sea-level rise. Yet it was robust enough to escape the effects caused by moister and cooler climates of the late Bronze and early Iron Ages and mid-medieval intervals of storminess. It has been the key to sustainable Hebridean crofting agriculture for nearly four thousand years: a mixed pastoral/cropping economy that, combined with inshore fishing, weaving and tourism, continues today.

A flourishing late-Neolithic civilization in the region is eloquently witnessed by the astonishing Callanish monument on the western shores of central Lewis (Fig. 18.4) together with other more modest monuments. It is old in human terms, but made beyond ancient by the nearly 4Gyr of mineral that forms its stones. The whole thing, on a splendid coastal site, is designed within a cruciform arrangement of smaller outer stones aligned to three points of the compass. The circle of taller stones in the middle has a monster slab nearly 5m

Figure 18.4 Callanish Stone Circle, Isle of Lewis. Photo: Shutterstock 289654022 ©Mark Heighes. 58.195222, −6.744168. View shows the unrivalled curving elegance of the tall central megalith and its attendant sentinel-like slabs of banded gneiss.

high, curved like a skinny supermodel, at the very centre. Twin stone rows define a long, northerly-aligned and inwardly-tapering processional avenue. Seen from the air it resembles nothing more than a perfectly-executed telescopic sight. The central stone is the target: the whole edifice a Cartesian grid for the mapping of just about any celestial event; solar, stellar, lunar, eclipse and comet. Such features make Callanish more than just a nod to the extravagant contemporary monuments of Orkney and Wick. It points to an organized island agricultural society based on the machair, a stable and well-governed one in which practical, spiritual and phenomenological events were recorded and perhaps celebrated for their predictable recurrence.

The sophistication of the Hebridean monument builders is matched on the mainland by the Loch Borralan chambered cairns of Assynt. These are notable for their prominent positioning, with distant views of Torridonian mountains (Fig. 18.5) from a site that may have lain between fertile and marginal land. They show off the sophisticated and deliberate use of contrasting local bedrock, including slabs of silver-white Cambrian quartzite, iridescent Moine schist, and pink-red

weathering Borrolan syenite. The slabs were worked by harder and tougher quartzite hammer-tools, discarded examples of which were discovered during excavation.

By the Bronze Age the machair lands were being widely cultivated; plough marks are recorded from archaeological excavations. On the machair of the Udal peninsula, North Uist, a settlement of stone houses buried during severe storms in the late seventeenth century proved to have been inhabited since Neolithic times. The oldest deposits consist of a line of stones with a large upright and a shaft with quartz pebbles covered by a whale vertebra.

Iron Age to Early Medieval culture saw construction of the fortified stone roundhouses known as brochs; these continued as habitations during the migration period from Ulster. Christianity came to Assynt (and to much of Argyll) in the seventh century when the Ulster monk Màel Rubai founded an ecclesiastical centre on the Applecross peninsula. His name occurs in twenty-one churches across the region and in Argyll. After doubtless numerous raids the Norse took over the western Hebridean archipelago as settlers by the mid-ninth century. They farmed, with pork, venison and herring

Figure 18.5 *Evening light On Cul Mor, Assynt* (2014) by Keith Salmen (b. 1959). Acrylic and pastel. 30x30cm ©Keith Salmen. The artist describes the scene: 'As the day went on however, the cloud thickened and it became quite gloomy and threatening. We were well down the hill by the time the cloud started to drop onto the summits but I remember it being a very strange kind of light and scene. This painting then is actually based on a viewpoint high up on the hill, but trying to imagine what it would have looked like if we'd still been up there as the cloud started to descend.'

frequently on the menu. A fascinating archaeological find from the midden of a longhouse at Bornais, South Uist is a tile fragment of gorgeous green porphyry, a metamorphosed andesite lava from southern Greece, the much-prized 'Krokeatis Lithos' ('Lapis Lacedaemonius' to the Romans). Greek geological friends inform the authors that this cultural erratic was originally quarried from near Sparta. Archaeologist Niall Sharples says the porphyry was popular in Rome and has been found at a dozen other sites (usually ecclesiastical) across Scotland and Ireland. He thinks its popularity may in some way be associated with the spread of Christianity. At any rate it bears silent witness to the far-flung trading and raiding networks of the maritime Norse culture. Together with Arran, Caithness and Orkney, the region came under the suzerainty of the Norwegian throne via the Earldom of Orkney, till reclaimed by an emergent Scotland in 1266; an easy date, for the English at least, to remember.

Chapter 19

North West Highlands and Northern Islands

Today, the young men, a score
Levered from Vestrafiold
The tallest stone, a star-raker.
Ale-skins were dry
Before the arrival of the stone-dresser.

George Mackay Brown, *The Twentieth Stone* (2005)

There were fine strong men in the Duke's time.
He drove them to the shore, he drove them
To Canada. He gave no friendly thought to them
As he turned his coach and four
On the sweet green sward
By the Place for Pulling up Boats
Where no boats are.

Norman MacCaig, *Two Thieves* (1983) in *Collected Poems* (1990)

This is north-west Scotland east of Assynt (Fig. 19.1): a glacially dissected plateau of corried peaks, high peatlands, blanket bog, moorland, loch and glen. It slopes away eastwards to the fertile lowlands bordering the Dornoch and Moray Firths grounded in Old Red Sandstone. The main mass comprises deformed and metamorphosed Moine bedrock (mostly schists) with scattered Caledonian granitic intrusions. The whole lot is bounded by two iconic faults: to the west the irregular trace of the gently-dipping Moine Thrust and to the south east the vertical linear gash of the Great Glen Fault.

The general topography has the main mountain peaks at around 1000m elevation down the main western watershed: Ben Hope, Ben More Assynt, Ben Dearg, Ben Attow; all corrie-centred sources of valley glaciers. Modern drainage follows the glacially enlarged glens, the longer and larger (Oykel, Farrar, Affric) streaming to the east, often with moraine-dammed lochs along their courses. Over half the region's bedrock is hidden by post-Neolithic peat formed in internally drained wetlands choked with blanket bog. This is the 'Flow Country' of Caithness and Sutherland, said to be the largest area of wetland peat in Europe, much

now preserved from greed and exploitation as denoted SSSIs (see Chapter 1).

As befits a gently tilted feature, the trace of the Moine Thrust (Chapter 7) wanders along below the elevated topography of its hanging wall for 150km from the shores of Loch Eriboll to southern Skye (Fig. 19.1). To the west is Assynt (Chapter 18), to the east the high western margin to the Moine Schists; the latter a weakened junction of contrastingly resistant rock, long the site of differential erosion (Fig. 19.2). The region's southern boundary divides the NW Highlands from the Grampians along glacier-scoured Gleann Mor na h-Alba, the Great Glen of Scotland. It marks a fault trace (Chapter 7) whose linearity betrays its origins as a vertical fracture along which leftwards horizontal sliding took place, in other words a tear or transverse fault. In Middle Devonian times it would have creaked, groaned and crashed with rock avalanches as terrific ruptures (perhaps of Richter scale 7–8) periodically rent it end to end for a few million years.

The region's more populous pastoral lowlands border the North Sea along the sandy promontories of the Moray and Dornoch Firths, fringed by mighty cliffs from Thurso to Wick in Caithness. These and the dozen or so low whale-backed islands of the Orkney archipelago across the Pentland Firth are founded upon Middle Old Red Sandstone. The linearity of the coastline from Dornoch to Wick is due to the presence of yet another bounding fault: a Mesozoic relict that is the largely offshore Helmsdale Fault. It has tiny but significant remnant outcrops of Jurassic and Triassic rocks scattered along its down-dropped eastern hanging wall.

Shetland is another matter; its varied scenery, blanket bog and interspersed hill country (no real mountains) and north–south aligned faulted coastlines are like a miniaturized Highland sampler. Distinctive Moine, Dalradian and Old Red Sandstone outcrops are separated by the Walls, Melby and Nesting Faults, the former being regarded as a major tear fault: the northern extension of the Great Glen Fault. The largest Caledonian ophiolite

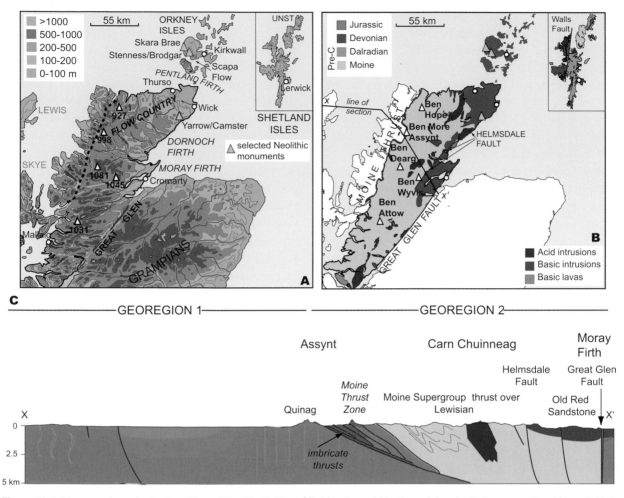

Figure 19.1 Maps and geological section of the North West Highlands and Northern Islands. Data generalized from British Geological Survey (2007b), Ordnance Survey (1991, 2011) and *The Times Atlas of the World* (1987) sources.

Figure 19.2 The Glencoul Thrust, part of the Moine Thrust complex. Photo: ©Vaughan Melzer, camera at 58.241307, –5.006310. Across the loch, the lower bracken-covered slopes have scattered outcrops of Lewisian gneiss. The well-bedded sheet-like crags that slope gently to the right down to loch level are Cambro-Ordovician quartzites: they overlie the gneiss unconformably. The entire upper sequence of irregular and discontinuous crags up to the skyline is again Lewisian gneiss, this time pushed over the quartzites along the Glencoul Thrust.

in Britain, mostly gabbro and peridotite sourced from the Iapetus mantle, occurs on Unst.

Pollen analysis reveals that peat accumulation in the region began in several upland sites of limited outward drainage in the late-Mesolithic, around 8ka. By Neolithic times, Orkney and Caithness (together with the area around Callanish on Lewis; Chapter 18) were clearly major centres of civilization. At that time pollen analysis indicates native valley-floor pine and oak woodlands had been cleared for agriculture on rich brown forest soils in catchments to 300m elevation. Rapid fluctuations in the proportions of hazel/birch and grassland/open country pollen in the lowlands indicate periodic clearance and re-establishment of grazing pastoralism with some cropping. The reduced vegetation cover caused the accelerated soil erosion recorded in lake sediment cores.

The fertile and prosperous lowland areas became the northern nuclei of the Neolithic maritime-linked culture of elaborate ancestor burial and ceremonial, Europe-wide in extent. The area around Wick has distinctive chambered cairns; Camster-type, linear and rounded with horned terminations, presumably ceremonial places. They were carefully constructed on elevated ground using local Old Red Sandstone flagstones capped above by fine corbelled roofing flags, then a jumble of more random grey-weathering boulders. Seen today (Fig. 19.3), with angular swathes of sitka forest and windfarms as backdrops, they seem strikingly flexible

and modern to the eye, undulating in their peaty surrounds as if a prize exhibit by architect Zahir Hadid. Archaeologist Amelia Pannett suggests that the siting of the cairns, close to the interfluve and source of the River Wick, was, in some way, a watery symbolic connection.

On Orkney the inhabitants made full use of local Middle Devonian flagstone in passage tombs such as Maeshowe on the Mainland and in local vernacular architecture, something unique in Europe. Maeshowe's passages and chambers are built with carefully crafted stone slabs. The alignment of the grave is such that the rear wall of the central chamber is illuminated at the winter solstice. A Neolithic 'low road' connects the Maeshowe barrow with the magnificently preserved stone-built village of Skara Brae, the megaliths of Stenness and the Ring of Brodgar nearby. The stone circles have proved to contain slabs worked from known quarry sites, perhaps selected for individual properties of colour and shape, even one half-worked and abandoned: the Vestrafiold quarry, on a hill north of Skara Brae, features in George Mackay Brown's poem quoted above. Maeshowe village itself was buried and preserved by an ancient sandblow and eroded once more by waves and wind in the nineteenth century. It is a breathtaking show-off of the Neolithic mason's art, all the stone carefully laid drystone-style, the roundhouse lower walls of dressed sandstone and fine interiors featuring lintels, box beds, shelves, cupboards and central hearths. It must have been cosy as Atlantic gales howled outside

Figure 19.3 Neolithic linear chambered burial cairn at Camster (the Grey Cairns), near Lybster, Caithness (58.379092, –3.264229). Photo: Shutterstock 3080859©Paula Fisher. The surrounding raised court may be clearly seen. The 60m long, 4.5m high structure continues to the left of this view. Two entrances lead down narrow passages to the internal chambers.

(Fig. 19.4). A pair of houses excavated on the small island of Papa Westray are similarly impressive, the low walls of trimmed flags carefully laid brick-style in stretcher-bond, even the floor areas paved.

The onset of a wetter climate in Iron Age times saw rising water tables, enhanced peat bog productivity, and abandonment of arable intake lands above 200m elevation. Numerous Iron Age brochs in the Caithness coastal region testify both to the continued availability of Caithness flagstones to build them, and the uncertainties of the societal change that made them necessary. Their Pictish descendants who defied the Romans were later Christianized, then Norse-governed and colonized (as the Orkneyinga Saga reminds us) down to late-medieval times, with a boat burial known from Orkney. The modern era brought first the economic misery of the clearances, when the sheltered, habitable places and shielings of the middling and eastern glens were mercilessly depopulated. Many took to picking kelp or were settled in tiny crofts in new fishing villages. Wick prospered and grew on the herring shoals, but for most it was the ultimate emigration: over the oceans to the Dominions, especially Canada. The First World War in Orkney brought enemy submarines to harry the naval ships at anchorages in Scapa Flow. This place later witnessed the greatest loss of shipping tonnage in history: the scuttling of the German High Seas Fleet in 1917. In the Second World War the airfields and raids over the islands' greensward are seen in the atmospheric work of Sandy Wylie's Searchlights in Scapa Flow (Fig. 19.5). Late-twentieth-century prosperity for many came with the universal bonanza of manual, technical, engineering, scientific and academic work provided by the North Sea oil and gas industry and its windows to a wider world.

We conclude by imagining that we are back in time, to 1820 and at Navity, a couple of miles south of Cromarty on the Moray Firth. A broad-shouldered, flame-haired young man of seventeen has come down to the rocky

Figure 19.4 Internal decoration of one of the eight drystone roundhouse foundations in the Neolithic hamlet at Skara Brae (59.048748, −3.341709). Photo: Shutterstock 199270853 ©duchy. Note the coastal setting within the usually sheltered Bay of Skaill. The hamlet sits square on the outcrop of the Sandwick Fish Bed, whose masoned sandstone flagstones were used in construction.

Figure 19.5 *Search Lights (Air Raid in Scapa Flow)* by Sandy Wylie ©Orkney Islands Council 39.7x76cm. Acc. No. PCF6. The hogback silhouettes of the Orkneys (*Orc* is a Pictish root for boar or pig) is starkly realized in this startling picture.

shore from his first morning's work to munch his lunchtime piece. His heavy working clothes, dusty apron and belted hammer show he works his trade in a nearby quarry. He finishes his piece and, being inquisitive, notices a particular rounded rock at his feet. He picks it up, props it on a larger rock for support and whacks it firmly with his hammer. Twenty years later he recollects in *The Old Red Sandstone; or New Walks In An Old Field* that the split nodule: 'contained inside a beautifully finished piece of sculpture, – one of the volutes…Of all Nature's riddles, these seemed to me to be at once the most interesting and the most difficult to expound.'

The young man who found the ammonite eroded from Jurassic strata at Navity was Hugh Miller, a native of Cromarty. His later studies of Old Red Sandstone fish fossils gained the attention of Louis Agassiz (discoverer of the Ice Ages) in the 1840s. Miller came entirely on his own terms into the late stages of the Scottish cultural renaissance. He saw geology and fossils as part of the Earth's heritage, and its message a rational one to the human race; a 'natural theology'. His later Edinburgh newspaper editorials included respect for different religious beliefs and early condemnation of the ongoing Highland 'Clearances'. A fine man.

Chapter 20

Grampian Highlands and Argyll

there are no more nations beyond us; nothing is there but waves and rocks, and the Romans, more deadly still than these for in them there is an arrogance which no submission or good behaviour can escape. Pillagers of the world, they have exhausted the land by their indiscriminate plunder... they create a desolation and call it peace.

Calgacus' reputed speech to his Pictish army (according to Cornelius Tacitus) before the battle of Mons Graupius: *The Agricola*, Chapter 30.

The whole immense head of the Mountain is composed of large loose stones – thousands of acres – Before we

had got half way up we passed large patches of snow and near the top there is a chasm ... other huge crags arising round it give the appearance to Nevis of a shattered heart or Core in itself – These Chasms are 1500 feet in depth and are the most tremendous places I have ever seen ...

John Keats writing to brother Tom about his ascent of Ben Nevis on 2 August 1818.

This region of high places lies between the Great Glen and Highland Boundary faults (Fig. 20.1). It is the deeply eroded metamorphic core to the Caledonian mountain

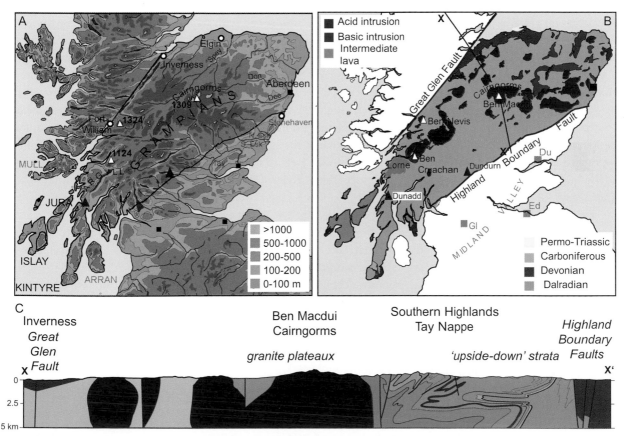

Figure 20.1 Maps and geological section of Grampian Highlands and Argyll. Data generalized from British Geological Survey (2007b), Ordnance Survey (1991, 2011) and *The Times Atlas of the World* (1987) sources.

Figure 20.2 Panorama of the largely granitic Nevis Massif from just north of the Great Glen Fault. Photo: ©Vaughan Melzer. The field of view is about 11km, taken at 56.882113, −5.052663, 9km away at 89m elevation. Ben Nevis is the beak-shaped hill centre-right and comprises trachytic lavas. Two great glaciated U-shaped valleys come out towards the viewer. The western, Allt A'Mhuilinn, between the peaks of Nevis and Càrn Mòr Dearg, is the most spectacular. The col between them and a valley wall to the south are just visible on the most distant skyline: a mighty arête.

belt. Here are most of Scotland's 282 Munro peaks (>3000ft/914m elevation), many of the mightiest made of Caledonian 'Newer Granites': Silurian-age intrusions into the late-Precambrian Dalradian Supergroup. Scores of them, ranging in area from many hundreds of square kilometres down to field-sized masses, dominate a swathe of country from Ben Cruachan in Argyll to Aberdeen. Subsidiary ranges such as Rannoch–Etive–Glen Coe and Rathnagar show off numerous igneous intrusive bodies, as does Ben Nevis itself, high over the southern Great Glen by Fort William (Fig. 20.2). At 1344m, the Island's chief of peaks is sculpted mostly out of granitic rocks, but its curiously beaked-shaped summit comprises tough trachytic/andesitic lavas whose 'Chasms' and rockfalls on its east-facing slopes so impressed John Keats. Such rocks also form the ridges that mark the 'Three Sisters of Glen Coe' and the stacked-up lava flows that form the 'trap' terrain of the Lorne Plateau.

The main high core to the Grampians is the subarctic massif defined by outcrop of the Cairngorm pluton: a windswept, glaciated and boulder-strewn plateau of light-weathering granite, supremely resistant to erosion. Much is above 800 metres, and away from its deeply carved periphery of corrie-sourced glaciated valleys its slopes can be regular and smooth, though boulder-covered from frost-weathering and flanked by vertiginous unvegetated scree (Fig. 20.3). Southerners can imagine the Cairngorms as a glaciated Dartmoor, the ice having taken scoops out of its rim. Here in Cairngorm the modern drainages run out on three sides: the Spey to the Old Red Sandstone lowlands bordering the Moray Firth between Inverness and Elgin; the Dee and Don to Aberdeen with its Rubislaw granite and quarry in the

middle of town; the Esks and Tay cutting the Highland Boundary Fault line to cross Strathmore into the northern Midland Valley, and hence to the North Sea.

The main mass of the more lowly Grampian and Argyll landscapes are founded upon Dalradian metamorphic schists, quartzites, subordinate limestones and basaltic volcanic rocks. Quartzite landscapes are bare or thin-soiled with white crags, as provided by the Eilde, Glencoe, Appin and Jura Quartzites, all agriculturally sterile. The relatively high fertility of certain of Argyll's coastal lowlands and islands (like Islay and Lismore) is due to the sweet-weathering Tayvallich and Lorne basic volcanics, burnt lime from scattered Dalradian limestones, and manure provided by beasts and kelp down the ages. The name 'Dalradian' was coined by nineteenth-century geologist Archibald Geikie from 'Dàl Riata' (Dalriada), the fifth- to eighth-century Gaelic kingdom that spanned Greater Argyll, from Kintyre to Skye. Together with Antrim across the water, perhaps the Gaelic sourceland, it defined a unified maritime culture.

The main mass of the region, from Tayside in an arc northwards to the Moray Firth, remained Pictish. The main tribal units were spread thickly in the lower reaches of the deep Highland glens and particularly along the most hospitable Strathmore, Banff, Buchan and Moray/Dornoch coastal lowlands, founded on Old Red Sandstone. It must be supposed that these Pictish tribes were aboriginal, distinct in language, possibly a variant of Brythonic Welsh, but probably not that different in overall culture from the lowland tribes of Britain. By AD 80 Governor Agricola's legions had reached the outfall of the Grampian glens along the topographic barrier caused by the Highland Border Fault. The Pictish tribes

Figure 20.3 Shelter Stone Crags and the Avon Falls (centre distance) 1.5km NNE of Ben Macdui: the heart of glaciated Cairngorms granite country (57.094666, −3.641922). Photo: Shutterstock 155970062 ©Phil.Tinkler.

were pushed back into a remaining Grampian hinterland, where Calgacus made his last stand. The exact site of the mountainside battle is unknown; many claim his pre-battle 'speech' quoted above was of Tacitus' own invention.

The Picts survived into the post-Roman era as sometimes uneasy neighbours with newly colonized Gaelic Dalriada, the Strathclyde Britons and Northumbrian Anglo-Saxons. Long prior to this, their own distinctive culture grew up in the Grampian margins, reaching west to Dun Duirn, a border fortress with Dalriada in Strathearn. It seems to have been a settled and peaceful time to begin with, gradually Christianized from the west with stone-built dwellings, stone carving (Chapter 14) and writing in Ogham script.

Argyll was first to become Christian: Columba was active here, and a monastic tradition radiated outwards from Iona. We see the Dalriadan culture in its landscape setting best from the vantage point of already-ancient sacred and ceremonial sites around Kilmartin Glen in mid-Argyll. Here, between Crinan and Lochgilphead, at the beginning of the Kintyre peninsula, are stretches of fertile lowland located at the faulted termination to the Dalradian Tayvallich lavas. Easily accessible to the sea from both east and west, there is a proliferation of Neolithic, Bronze Age and Iron Age monuments, carvings and burial sites: a proven tally of 350 in a ritual landscape of some 5000 years duration, perhaps the longest span in the whole Island. The Gaels established their royal palace at Dunadd on a protuberance of Tayvallich volcanics within the meandering floodplain of the River Add (Fig. 20.4).

What a physical landscape they chose! The waters of the Sound of Jura are foreground for the extravagant play of setting sun on the Dalradian quartzite hills of the Paps of Jura away to the south-west. To the north-west over Jura's northern tip and Scarba are the Isle of Mull's peaks founded on Cenozoic volcanic rocks (Chapter 22). There, at that island's western end, was the

Figure 20.4 Natural outcops around Dunadd citadel, 50m or so above the River Add floodplain, near Kilmartin (56.086010, −5.478423). Photo: Shutterstock 1920327 ©Bill McKelvie. The Dalradian volcanic outcrop in the foreground bears the royal 'footprint of fealty' (size 8 if you're interested), thought by some to be part of the Dalriadan coronation procedure. There is also (but not visible here) ogham lettering (?C7th AD) and a Pictish boar carving.

Gaels' spiritual Christian homeland; many-crossed Iona, tucked safely (so they thought) to windward. Dunadd's trade was via their Ulster kinsfolk and traders from wider Europe. There was fine glassware, pottery from Gaul and jewellery fittings, especially the characteristic Hiberno-Pictish penannular bronze brooches used to fix the ubiquitous cloak. As well as its prolific and widely renowned scriptorium, Iona's monastery culture also nourished craftwork; archaeological finds of smelting and moulding around the remains of firepits point to the activities of itinerant artisans working various metals, chiefly bronze. Some brooches and moulds found at Dunadd use Anglo-Saxon motifs, and the citadel may have hosted more permanent craft workshops. These helped to develop the 'Insular Art Style' whose main glory was the *Book of Kells*, begun on Iona around 800, some 200 years after Columba's death. The move of that and other items to the safety of Kells in Leinster after

the first Norse raids of the late eighth century signalled the eventual destruction in battle of both Gaelic and Pictish nations. Not until after the victories of Kenneth MacAlpine in the mid-ninth century and the reigns of his son and grandson did the unified kingdom of Alba (proto-Scotland) gradually emerge.

Modern times, of course, have seen many changes – the Clearances, sheep grazing, landscape tourism, ski resorts in the Cairngorms. Most of all the rise of Aberdeen, once a fishing port that also shipped its tough grey granite from Rubislaw quarry to every corner of the globe. The offshore hydrocarbon industry centred there and along the eastern coastal plain has been a beacon of prosperity for forty-five years. At the time of writing offshore production is in steady, probably irreversible, decline but, like Houston, the region and its university will hold onto its hard-learnt hydrocarbon engineering and geology skills. The tenor of the wider, wild

Figure 20.5 *Cliffs and Sea* (1958) by Joan Kathleen Harding Eardley (1921–1963). 11.3x31.2cm. The Gracefield Arts Centre, Dumfries. Acc. No. DGGAC438R ©Estate of Joan Eardley & DACS 2016. All rights reserved.

north-east coast before such modern transformations is caught unmistakably and in the raw by Joan Eardley's vigorous and uncompromising artwork (Fig. 20.5).

Yet, by Dunadd's shore the waves still ruffle the skerried foreshore of the harbour at Crinan. The imaginative geologist examining turbidites in the Crinan Grits can catch the sound of dipping oars as an incoming merchant's lighter brings treasures from Ulster and Gaul. Later on, by the harbourside, Columcille (Columba) can be heard talking in Latin with the merchants before walking to Dunadd to greet the High King of the Gaels in his native tongue.

Chapter 21

Midland Valley

When we had a king, and a chancellor, and parliament-men o' our ain, we could aye peeble them wi' stanes when they werena gude bairns – But naebody's nails can reach the length o' Lunnon.

Mrs Howden's strictures in Walter Scott's *Heart of Midlothian*, as carved on Dalradian Easdale Slate from Argyll set in the Canongate wall of the Scottish Parliament building.

These stones speak a level language
Murmered word by word,

A speech pocked and porous with loss,
And the slow hungers of weathering.

Rachel Boast *Caritas* (St Andrews Cathedral) in *Pilgrim Flower* (2013)

The Midland Valley's mostly Upper Palaeozoic bedrock (Fig. 21.1A–D) defines a diverse landscape between the Grampian Highlands and the Southern Uplands. Warmer and drier than these, often below 300m elevation, its coating of glacial deposits from Highland-sourced ice provided fertile soils when enriched by dung

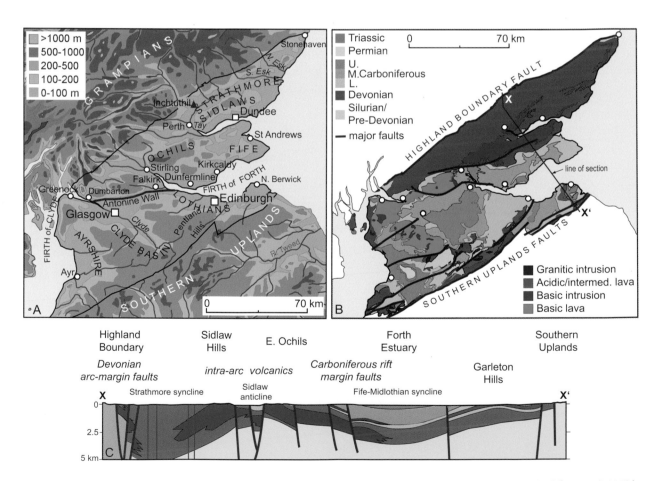

Figure 21.1 Maps and geological section of the Midland Valley. Data generalized from British Geological Survey (2007b), Ordnance Survey (1991, 2011) and *The Times Atlas of the World* (1987) sources.

and seaweed and de-acidified by locally sourced lime. The flat, reclaimed estuary-margin 'carse' landscapes are the prime arable land. Since early medieval times, together with the eastern Borders, the region has been Scotland's prime agricultural centre, early on attracting Anglian farmer-settlers into the Lothians. Pastoral areas are famous for milk cattle (think Ayrshires), beef (Angus) berry fruit (raspberries) and orchards. Sheep rough-graze tracts of fell country (Ochils, Sidlaws, Pentlands) whose summits are higher than most in the Pennines (Fig. 21.2). Remarkable local eminences include the huge protected oil-shale waste tips (bings; Chapter 12) of West Lothian (Greendyke and Five Sisters) and 'living' raised peat domes like Fannyside Muir in North Lanarkshire.

Another obvious feature is the region's isthmus-like character. Two prominent firths, Clyde and Forth, cut well into it, a bare 45km separating the inboard Atlantic from the tidal North Sea, its two largest cities on either side. Glasgow faces westwards to Ireland and the north Atlantic seaways, its glacially scoured estuary deep enough (after removal of igneous outcropping bedrock in the nineteenth century) for major portage (Greenock) and ship-building (Govan). Edinburgh's port of Leith was a notable trading partner with other eastern British ports and the medieval Hanseatic League, to which William Wallace sent friendly requests for trade after his rise to power.

The varied landscapes are home to two-thirds of Scotland's population. It grew in two surges. First, after mid- to late-eighteenth-century agricultural and industrial improvements (scientific agronomy, canals, water-power, early steam), its many nascent urban centres were peopled by an initial stream of refugees from ongoing Highland troubles and clearances. Then by industrialization, initially much facilitated by the mid-eighteenth-century Forth–Clyde canal link, notably in enterprises like Falkirk's Carron Ironworks. Subsequent growth was due to onset of the industrial steam age proper, initially enabled by Greenock's most inventive son, James Watt. Heavy industry used the region's rich reserves of Carboniferous iron ore, coal and oil shale. Thus were created the densely populated industrial lowlands centred on the coalfield basins of the Lothians, Fife, Lanarkshire and Ayrshire, with Glasgow on the Clyde rapidly becoming the undoubted working capital of the country, regarded (locally at least) as the 'Second City of Empire'.

The region's more rugged landscapes are defined by the distribution of volcanic rocks now found outcropping in large open folds cut by many faults. Such is the north-east- to south-west-trending Strathmore syncline and Sidlaw anticline whose Devonian andesitic volcanic rocks make up the Sidlaw and Ochil Hills (Fig. 21.2; highest peak around 720m). Carboniferous volcanism first featured fissure eruptions that created the lava plains and plugged volcanic necks of the Clyde Plateau Group. These envelop the Clyde basin, perhaps most definite in the stepped 'trap' country of the Renfrewshire Hills. Ongoing subsidence saw equatorial swamps peppered by volcanic intrusives (Castle Rock, Edinburgh, coastal Fife) and lava eruption (West

Figure 21.2 View to the north-east of the Ochil Fault line taken from the Wallace Monument, Stirling (56.138773, –3.917889). Photo: ©Vaughan Melzer. The fault-line separates the lava flow scarplands of the Ochil Hills from adjacent carseland. The Midland Valley specializes in such vivid and sudden juxtapositions of contrasting relief, slope and land-use.

Figure 21.3 Dumbarton Rock (55.936479, −4.563012). Photo: ©Vaughan Melzer. Vertiginous polygonal cooling joints feature on the right buttress of this basaltic volcanic intrusion of Permo-Carboniferous age. Situated on the north bank of the Clyde in the outer estuary, the castle (mostly rebuilt in the C17th) is supposedly the fortress with the longest recorded history in Britain, mentioned in a 4th-century letter written by St Patrick. It was the lynch-pin of the maritime defences of the British (Brythonic) kingdom of Strathclyde.

Lothian's Bathgate Hills, Edinburgh's Arthur's Seat and Calton Hill, East Lothian's Garleton Hills). From the Iron Age onwards such resistant volcanic outcrops have provided innumerable steep-sided eminences for fortified settlements and castles: Stirling, Edinburgh, Dumbarton (Fig. 21.3), North Berwick Law/Tantallon and Traprain Law to name but a few. Others such as Dunnottar Castle, Stonehaven are built upon Old Red Sandstone cliffs. From certain such vantage points were viewed pivotal battles in the Wars of Independence (e.g. Stirling Bridge, Falkirk, Bannockburn).

The region's sharp boundaries are long-lived faults, the Highland Border Fault being the most emphatic. Swift-flowing Highland rivers run across it onto the chequered pasturage of Strathmore's ('large valley') broad, rolling landscape founded on Lower Old Red Sandstone. After the battle of Mons Graupius (Chapter 20), Agricola built strategic forts and watchtowers along the dolerite dykes of the Gask Ridge just west of Perth to guard the main Highland valley exits. The legionary complex of Inchtuthil by the Tay south-west of Blairgowrie was at the centre of these 'glen-blocking' fortlets, but it was abandoned soon after completion as Agricola was recalled to Rome. Later the Antonine Wall, a 'new build' by Hadrian's successor, briefly closed the Forth–Clyde isthmus. It was in turn abandoned and the remaining Picts of the Highland Borders left to themselves. They eventually prospered and became part of a distinctive pre-Gaelic culture for several hundred years (Fig. 21.4; Chapter 20).

By way of contrast to the northern borderlands, not only are the Southern Uplands of lesser relief, but

Figure 21.4 Aberlemno No. 1, a 1.9m high Pictish carved stone, the 'Serpent Stone' (56.692099, −2.781294). Photo: Wikimedia Commons ©Catfish Jim and the soapdish. It is flat-laminated Old Red Sandstone, probably locally quarried from the Scone Sandstone Formation: one of a group of carved stones at Aberlemno, Angus, in the kirkyard and by the road between Forfar and Brechin. The symbols here include a snake, a double-disc cut by a z-shaped rod, a ?mirror and comb. Bronze Age cup marks on the reverse side suggest it is recycled. The smoothly curved lower side looks specially masoned; it would be interesting to know its buried extent and shape.

subsidiary and parallel bounding faults to the main Southern Uplands Fault cause the topography to change more gradually, in stepped fashion. Also, discontinuous but resistant Siluro-Devonian andesitic volcanic outcrops just to the north lessen the impact of the faulting. Their emphatic round-topped and craggy-sided uplands, most notably the Pentland Hills (highest peak 580m), Auld Reekie's breezy lungs, extend north east-south-west for 25km or so. In southern Lanarkshire and eastern Ayrshire from Tinto (700m elevation) to New Cumnock lies another billow of higher ground caused by resistant Silurian and Devonian sedimentary and igneous rocks.

Eighteenth- to nineteenth-century Glasgow grew up around the Clyde on a series of glacial drumlins. These low hills controlled both the course of the Kelvin River and the street layout, their summits chosen for many Victorian public buildings. These and the distinctive tenement blocks and kirks are built of lovely brown-red weathering freestone: Permian sandstone from Mauchline, Ayrshire. The eighteenth-century light-industry and manufacturing was replaced in the mid-nineteenth century by the emphatic steel-based Clydeside heavy industry of ship and locomotive building (Fig. 21.5). The city was heavily bombed during the Second World War, with 1960s planners splitting it in two along an elevated

Figure 21.5 *River Clyde Shipyard Study* (2011) by Adam Kennedy (b. 1987). Mixed media. 65x60cm. ©Adam Kennedy. The artist grew up next to the Clyde; he describes the atmosphere in Glasgow as a feeling you'd recognize if you live there: 'The clouds of Glasgow are so heavy and wet...'. His paintings are nostalgic and dream-like with a ghostly atmosphere in greys and blacks: '...colour would be too much.'

motorway. Since the 1970s it has become a post-industrial city with all the advantages and disadvantages that this implies. Yet, like Liverpool, it is a tough city full of wit and energy, a working-place whose industrialists, scientists and artists nowadays look outward towards Europe more than to England, Ireland and Empire as it once did. Novelist Nicola Harding suggests that the vibrant arts scene is due to 'the collective egalitarian feel of Glasgow, the multitude of practices and groupings, the respect for hard work, the now.' Fine teachers like David Harding nourished influential 'Art and the Environment' courses at the Art College, focusing on particular environmental issues in Scotland and the linkage of peoples and places. The decline of heavy industry on the Clyde has been a recurring environmental sculpture theme.

Unique amongst major British cities, Edinburgh's architecture and urban development down the ages has played spectacular riffs on its geological heritage and the proximity of the Pentland Hills and Firth of Forth; it is a cityscape of striking views. Most obvious is its medieval castle perched on vertiginous crags formed by black-jointed dolerite that plugged a Carboniferous volcanic neck intruded into more easily weathered sedimentary bedrock. Glacial ice spreading from western Highland catchments streamed round the volcanic edifice, eroding and sharpening its upflow margins and spreading an easterly tapering 'tail' of morainic detritus down its leeside. On this elongated and raised cone was built the medieval High Street down to Holyrood Palace: the Royal Mile, 'Whose ridgy back heaves to the sky' (Scott). On both sides there gradually grew up the little streets (closes) with their tall tenements sheltering a high-density population hemmed in by surrounding bogs.

Architects and planners have since been kind to Edinburgh's physicality, beginning with the numerous Neo-Classical public buildings of Calton Hill and the restrained domestic and public buildings of the grid-patterned Georgian New Town built upon the drained wetlands. Here are street upon street of buildings in fine, durable Lower Carboniferous sandstone from the Craigleith quarries, creamy-yellow ashlar blocks whose subtle weathering stains and internal decoration stand comparison with Bath. Today the city is the second largest financial and administrative centre after London, and its outstanding modern buildings are respectful of the medieval core. Notably these include two contemporary round-towers: the colourful and curvaceous National Museum of Scotland (2011, Gareth Hoskins) built of variegated pink, grey and pale-yellow Permian desert sandstone from specially re-opened quarries at Hopeman on the Moray coast, and the dumpy International Conference Centre (1995, Terry Farrell). The jagged skyline of Scotland's Parliament Building (Enric Miralles, 1999) nods to the heritage of igneous rock and mountain in Scottish landscapes, especially in its exterior decoration. In the wider region, plentiful building stone has gifted a rich architectural heritage: Norman Dunfermline Abbey and the late-medieval Rosslyn Chapel in Midlothian with its elaborate sculptures are noteworthy.

During the nineteenth century, national sentiment demanded public statues of famous Scots. John Steel's Scott Monument in Edinburgh (1840–46) was carved out of a 30 ton block of Carrara marble. Pioneering Scottish female sculptor Amelia Hill's striking statue of David Livingstone is in Princes Street Gardens, Edinburgh. Romantic paintings of the sublime Scottish landscapes became *de rigueur,* a tendency reinforced by the contrasting literary outputs of authors like Scott, Carlyle and Stevenson. An early example is *The Falls of the Clyde* by Jacob Moore, a view also drawn by Joseph Turner. Alexander Nasmyth is taken to be the founder of the Scottish Landscape school; a friend of Robert Burns (whom he painted), he was fascinated by engineering and is best known for his *View of the City of Edinburgh* (1822). Other notable nineteenth-century landscape artists include Horatio McCulloch (Glen Coe, Trossachs), Joseph Farquharson (snowscapes and sheep), William McTaggart (clouds and stormy seas) and Edwin Landseer (Highland scenes with no people, stags).

Chapter 22

Formerly Volcanic Islands of the Inner Hebrides

Còmhlan bheanntan, stòitteachd bheanntan, còrr-lios
bheanntan fàsmhor…
('A company of mountains, an upthrust of mountains, a
great garth of growing mountains …')
Somhairle MacGill-Eain (Sorley MacLean) *Ceann Loch Aoineart*
(*Kinloch Ainort*) (1932–1940)

Many a Mortal of these days
Dares to pass our sacred ways
Dares to touch audaciously
This Cathedral of the Sea …
John Keats, from verse in a letter to his brother Tom after visiting Staffa
on 31 July, 1818

Of the three score or so inhabited islands that make up the entire Hebridean archipelago, a dozen qualify as 'Volcanic Islands': partly or wholly comprising early Cenozoic (c.60Ma) igneous rocks. Four of the chief are shown in Figure 22.1A (we include Arran for convenience, but geographically it is in the outer Firth of Clyde). The magmatism was of two types, creating distinctive landscapes. Extensive lava tablelands (northern Skye, west Mull, Morvern peninsula) subsequently carved out by ocean, ice and river are dissected leaving darkly layered remnants with bold scarps and often huge coastal landslips. The geological gift of the lava fields is not just their own physical ruggedness and regularity; during eruptions the flows also smothered and preserved remnants of Mesozoic strata. These had previously coated the entire western Precambrian oldlands as they slowly foundered during Mesozoic crustal subsidence. The thorough-going early Cenozoic uplift and erosion after the cessation of volcanism (Chapter 7) has removed almost all traces of these strata where not protected by lava cover. The remnants contain phosphate and lime which, along with nutrients from weathered basalt, add much-needed fertility to otherwise acid soils. The highest peaks of the islands comprise central sub-volcanic intrusions (central complexes) formed by foundering of lava into huge caldera-like magma chambers, so forming distinctive ring dykes. Seen from the distance the dark gabbroic examples that make up the Cuillins' frost-shattered and jointed, jagged ridges are like the fins on a Chinese dragon's back (Fig. 22.2). Such rocks and intrusive granitic plutons are the corried sources of Ice Age valley glaciers, also on Ben Resipole in the Ardnamurchan peninsula, Ben More of east Mull, and Goat Fell of north Arran.

In prehistory, the islands were first home to Mesolithic colonizers who used and traded sharp, glassy volcanic pitchstone for use in cutting and fishing artefacts. Neolithic and Bronze Age expansion witnessed monumental and funereal installations in maritime settings at the foot of mountains. During Iron Age times the islands were dominated by imported Gaelic culture (Pictish symbol stones are very rare here) with boat-borne missionaries from Iona bringing the new religion. Then arose the fiercely independent Gael–Norse maritime culture, with its fleets of highly-manoeuvrable galleys (birlinns), under the suzerainty of the Lords of the Isles. After the late fifteenth century came feudalistic fealty to the Scottish kings. The common people eventually numbered tens of thousands on each of Mull, Skye and Arran. They survived in their feudal chains with no rights of perennial tenure until the gales of change from the 'Age of Improvement' swept the majority of them away. Their fate was in the balance until the passing of the 'Crofters' Holdings Act' by Gladstone's penultimate administration. The Crofters' Union in recent years has been prominent in its fight to let communities buy their own lands back from the laird's estates, for example on Eigg (also in Assynt).

It is with this sometimes dark heritage in mind that the visitor looks down to the once-populous coastal lowlands from the Cuillin mountains of central Skye. Once visible from here were numerous crofting hamlets with islanders numbering over 20,000 – the island at the very centre of the movement for crofting reform.

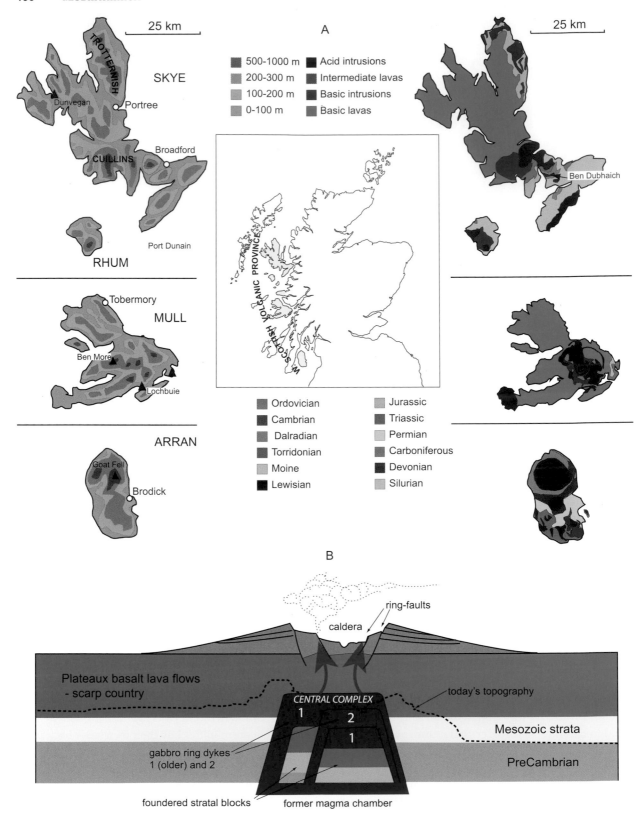

Figure 22.1 Maps and a geological section reconstructing idealized Hebridean volcanic/plutonic relations; the lava fields may once have been around 1000m thick. Data generalized from British Geological Survey (2007b), Ordnance Survey (1991, 2011) and *The Times Atlas of the World* (1987) sources. Section after Trewin (2002).

Figure 22.2 The unforgettable jagged skyline of the Cuillin Hills, Skye. Photo: ©Vaughan Melzer, camera at 57.148600, −6.106966. View is north-west from Elgol foreshore on Loch Scavaig, Strathaird peninsula. The highest peak to the left is Sgurr Alasdair (1009m).

Today tourism, crofting agriculture, forestry, fishing and distilling provide livings for around nine thousand permanent inhabitants including retirees, artists and writers. The tourists include many walkers and climbers whose assaults on the redoubtable Cuillin ridges are to be celebrated in an outdoor bronze sculpture of Norman Collie and John McKenzie, two pioneer climbers, by sculptor Stephen Tinney. Julie Brook is a Skye land artist who usually chooses desolate physical locations for her works (Jura was an early one): 'the artist draws the landscape, with the landscape, in the landscape.' She uses natural materials that can become eroded or burnt: like ourselves and all cultural matter, they are ephemeral and decay. Artist Diana Mackie paints landscape scenes of turbulent mountain and coastal weather in her Sligachan series. From Caroline Bailey's work (see Fig. 22.5) we are aware that atmosphere, light and colour are of prime importance on any island where land meets sea.

Music sings out on Skye in the history of pibroch, brought to its later genius by the MacCrimmons, hereditary pipers to the MacLeod clan chiefs at Dunvegan for 800 years. Much earlier, around 300 BC, the inhabitants of nearby Dun Beag ('small fort'), now a 2m high stump of a broch with views over Loch Bracadale, would have listened to a different music. A wooden lyre bridge found in hearth deposits at High Pasture Cave is the oldest remains of a stringed instrument in western Europe. The cave system is in the southernmost outcrop of the Cambro-Ordovician Durness Limestone (Chapters 18,

19), on the flanks of Beinn Dubhaich, east of the granitic 'Red' Cuillins: the limestones famously metamorphosed by the heat from the granite intrusion, a classic example of contact metamorphism.

Mull is the mountainous portal to the Firth of Lorne and Loch Linnhe. Like northern Skye, its western peninsulas comprise extensive 'trap' country (Fig. 22.3), as does Morvern to the north across the Sound of Mull. The famous fossil Tertiary tree in West Mull stood bolt upright even as it was enveloped by a new flow of glowing lava. Three overlapping volcanic centres make up the main mass of East Mull's glaciated core of Beinn More and Beinn Talaidh. Both here and on the Ardnamurchan peninsula to the north are found arcuate and concentric ridges made by outcrops of both basic and acidic composition whose intrusive contacts consistently dip inwards towards a central focus. First discovered here, they are aptly known as cone sheets; think of an inverted stack of traffic cones. Mull's more southerly location is evident in the remains of a large (32 metres long, 11 wide) Neolithic Clyde-type cairn (cairn with forecourt) overlooking Port Donain. At the eastern end of gulf-like Loch Buie ('yellow loch') in southern Mull a small granite-stoned circle stands in perfect harmony below the twin peaks of Ben Buie and Creach Beinn on its square kilometre of raised beach platform and alluvium.

Like Shetland, Arran has been called 'a Scotland in miniature'. Its northern mountain core is of Tertiary granite intruded into Dalradian metamorphic rocks;

Figure 22.3 View of the lava tableland ('trap' country) in western Mull. Photo: Shutterstock 228536599©Spumador. The layered traces of successive basalt flows are clearly visible above the steep lower slopes.

its Old Red Sandstone/Carboniferous periphery and southern hilly central belt have volcanics, minor basic and acidic intrusions, and Permo-Triassic strata. The highest peak is granitic Goat Fell (Gaothe Bheinn, 'Mountain of Winds'), which features rugged, jointed castellated ridges. The savagely glaciated and rugged outer part of this, the largest of the Hebridean Tertiary granites, is cut through by spectacular ice-carved U-shaped valleys (Glens Rosa and Sannox) with their valley moraines and hanging-valley tributary waterfalls. The lower parts are well stocked with many Clyde-type cairns and stone circles (six on Machrie Moor alone). Modern islanders are heavily tourist-dependent through proximity to urban Clydeside and Ayrshire by ferry boat. A significant part of this influx is of landscape tourists; the place is a haven for lovers of physical landscape and geology. One of James Hutton's three unconformities is here, the first-discovered: Old Red Sandstone resting on Dalradian.

Turning to smaller islands, the noble slabs, scarps and slopes of Askival (814m) and Hallival in Rùm's mountainous central complex are layered gabbros that formerly accumulated as hot, sticky, crystal-sediment in magma chambers. The island was Norse till the Treaty of Perth in 1266. Even by Highland standards the Clearances here were ruthless. The entire population of 400 in nine hamlets who had farmed and fished from its raised beaches for countless generations were removed at a year's notice. Today the island is a nature reserve with an embryonic crofting community once more. All 5000 inhabitants on Eigg's 17 square miles of basalt and Jurassic strata were also forcibly removed. The island had a previous bloody history. In 617 a Pictish Queen massacred the monks of St Donnan's monastery. In 1577 visiting Macleods massacred by fire and smoke the entire population who had taken refuge in a coastal cave; James Boswell saw human remains in the cave 200 years later. He and Samuel Johnson were only two of many notables who have visited the 82 acre island of Staffa (a name from Old Norse for stave or pillar; Fig. 22.4) whose basalt cooling joints and columns inspired both John Keats (*This Cathedral of the Sea*) and Felix Mendelssohn (*Hebrides Overture*). Since all aspects of island life, culture, history and tourism depend so fundamentally on the ferry system centred at termini like Mallaig, we leave the reader with an evocative image by artist Caroline Bailey (Fig. 22.5).

Figure 22.4 Fingal's Cave, Isle of Staffa (56.431358, –6.341364). Photo: Shutterstock 152634794 ©Steve Allen. These *c.*3m high vertical polygonal joints (the 'organ pipes') formed in the interior of a stationary basalt lava flow as it slowly cooled and contracted after eruption had ceased.

Fig. 22.5 *Sailing out of Mallaig 12* (2013–2014) by Caroline Bailey (b. 1953). Mixed media. 76x86cm. ©Caroline Bailey. The artist shows the CalMac ferry bound for Skye. She says: 'I always thought they look like great seabirds plying their way across the waters…they are a lifeline to the islands and provide an integrated transport system…never leaving port until an expected bus has arrived.'

Chapter 23

Southern Uplands and Galloway

Kirkcudbright County is very beautiful, very
wild country with craggy hills somewhat in the
Westmoreland fashion ... Our landlady of yesterday said
very few Southrens passed these ways. The children
jabber away as in a foreign Language – The barefooted
Girls look very much in keeping – I mean with the
Scenery about them...

John Keats, writing to brother Tom from Auchencairn in the first week of
July 1818

I would hope that collectively these arches are a
celebration and monument to the Scottish people and
the travels they have made, and that they will act as a
connection between those who have left and those who
have stayed here...

Andy Goldsworthy speaking of his landscape sculptures, *Striding Arches*,
located in the hills around Cairnhead Glen, Dumfriesshire

This round-peaked, glacially dissected plateau founded
mostly in Lower Palaeozoic rocks (Fig. 23.1A–D) is
mountainous in places and much travelled through; it
must be crossed to reach or leave Scotland by land. For
northward invaders and colonizers – Romans, Angles,
Normans and English – it was the necessary land route
into wealthy Midland Scotland. The alternative before
turnpikes and railways was by boat along the eastern sea-
board, the route a sickly John Keats took back down to
London from his Scottish hike in the autumn of 1818.
The steep-cliffed coast here is due to resistant Silurian
outcrops, a challenge for construction of this branch of
the rail link from Edinburgh to London. From Dunbar to
Cockburnspath and again from Burnmouth to Berwick
between its numerous cuttings the traveller gets glimpses
of spectacular coastal views: folded and faulted Lower
Palaeozoic greywackes and the vivid red-brown hues of
the Old Red Sandstone – perhaps even a hopeful glance
at Pease Bay, towards the site of Hutton's unconformity
at Siccar Point (Fig. 6.3).

Dunbar itself, like Berwick (Chapter 24), had a
strategic position; with its castle on a volcanic outcrop

it was a much fought-over part of Anglian Northumbria
for several centuries. Neighbouring East Barns featured a
notable Mesolithic site. Here, prior to a quarry extension,
geophysical surveys and rescue excavations discovered
post-holes for a 6m-diameter roundhouse whose hearth
deposit was dated to 10ka, the earliest preserved house
in Scotland.

The highest ground over 600m is in the Lammermuir,
Ettrick, Lowther and Galloway Hills, all deeply glaciated, the rounded summits with blanket peats, val-
ley-fringing crags and boulder fields. As in Grampian,
the Galloway massifs of Criffell, Cairnsmore-of-Fleet,
Merrick and Cairsphairn are rooted in resistant granitic
and intermediate intrusions (Fig. 23.2). Their corries
hosted the nuclei of Quaternary ice that flowed out into
the Irish Sea and away southwards, carrying within them
the tell-tale igneous stones that became glacial erratics.
Away from the granites, the lower landscapes are pastoral country with scattered NE–SW trending Siluro-
Ordovician outcrops – gorse-covered, rock-slabbed
ridges that end in skerried shores or coastal cliffs, many
of which carry Bronze Age cup-and-ring markings (Fig.
23.3). The foothills of Cairnsmore feature the striking
hillside Neolithic court tomb of Cairn Holy, its massive
dark exhumed greywacke slabs fanning open to the sky
like brutally broken bad teeth. The cliffed fastness of
Criffel's southern faulted margin, with its vivid-pink
felsitic and porphyritic dykes, hosts the Iron Age broch
of the Mote of Mark. The Novantae here were subdued
by Agricola in AD 82. From the Rhinns of Galloway in
the far west, he briefly (perhaps optimistically) contemplated a single-legion invasion of Ireland.

The early conversion of the Southern Uplands
Brythonic tribes to Christianity was aided by the construction of huge masoned preaching crosses in local
stone that stood near main routes or early churches. The
largest and unique survivor is the early-seventh-century
Ruthwell Cross (Chapter 4; Fig. 23.4) located close to
the coast a few miles east of the Nith estuary. This 5.5m
high carved Permian sandstone block has panels that

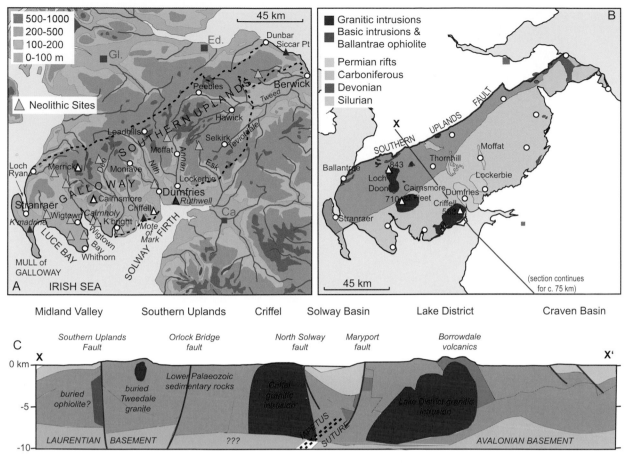

Figure 23.1 Maps and a geological section for the Southern Uplands in its regional context. Data generalized from the British Geological Survey (2007b), *The Times Atlas of the World* and other sources.

Figure 23.2 Bruce's Stone and Glen Trool, Galloway (55.089493, −4.496750). Photo: Shutterstock 150822881 ©Kevin Eaves. In 1306/7 Robert the Bruce took refuge here as a fugitive in the rugged landscapes of western Galloway. His memorial stone is below the rocky granitic crags that lead vertiginously up to the peak of Merrick, at 843m.

Figure 23.3 Cast (*c.* 40x25cm) of exquisite cup-and-ring petroglyphs exhibited outside Kirkcudbright's Stewartry Museum. This example is extraordinarily geometrical and may depict a central henge monument with surrounding close-packed hut circles.

feature both figurative work (e.g. shepherd with lamb, birds eating grapes on the vine) and decorative themes. Most famous of all are the runic lettered inscriptions discussed in Chapter 16. Cultural and architectural scholar Nicholas Pevsner regarded the Ruthwell Cross, together with that at Bewcastle (Chapter 24), as 'the greatest achievement of their date in the whole of Europe…' (but he knew nothing of Ireland).

The closeness of Ulster led to the emergence of both Ulster Scots and Galloway Irish as cultural groupings in modern times. Such links originated after Elizabeth I's 'plantations' and later during the Jacobite troubles and the Irish Famine. A stage-coach service subsequently wended its slow way from Stranraer across once-notorious roads to Carlisle and southwards, amply documented in the letters of Viscount Castlereagh of Castle Stewart in Down to his wife Emily, staying with her sister at Blickling Hall, Norfolk. By 1859–61 the rail link from Carlisle to Stranraer via Dumfries opened and today the trucks of Ulster carriers roar along the

only partially upgraded A75. They bypass places like Tongland, up the River Dee from Kirkcudbright with its striking Telford bridge and a beautiful concrete Art Deco hydro-electric station.

The most efficaceous routes to and from Glasgow and the western Midland Valley are provided by geological structure: low ground along the faulted Permo-Triassic rift basins of Dumfries, Lockerbie and Thornhill. These linked the earliest Roman marching routes, the former a rich source of fine red-brown building and monumental sandstone, notably at the Locharbriggs quarries. The M74 now screeches through the Lockerbie basin into the Lower Palaeozoic core via the Beattock pass, with its nod to glaciated Moffat valleys and headwaters of the Clyde. It was here in the Lowther Hills of south Lanark and northern Dumfriesshire that lead, zinc and gold were worked around Leadhills and Wanlockhead (the highest Scottish villages, around 450m); gold in the sixteenth century and lead/zinc from the early seventeenth century to the 1930s. The lead was blasted and shovelled

from vein ores, and the gold panned from valley-floor alluvial deposits. The latter provided coinage and crowns (2 kilograms in total weight) for James V and his Queen, Mary of Guise; the metal may still be panned (by extreme optimists) under licence. In their spare time the lead miners set up an early working-men's library in the 1740s, supposedly the oldest subscription library in the Island.

From late-medieval times the narrow north–south glens and valleys of the central Southern Uplands hosted important 'clan' groups (Humes, Dixons, Kerrs, Johnstons, Scotts, Elliots, Irvings, Maxwells and others). They were cattle-breeding pastoralists with a marked tendency to brigandage and feuding, both with kinsfolk, neighbours and over the border (see also Chapter 24). Pacification by James VI led eventually to a largely legal cattle trade along the drove roads of Nithsdale southwards. One popular route led the skinny but hardy black Galloways from Carlisle far south-east to fatten up on the water meadows of East Anglia prior to sale at Smithfield.

A few decades after Union, numerous market towns were planned and (re)built. Factory industries developed, including boot-and-shoe, weaving and bobbin-making, powered by water and steam. Dumfries and Stranraer were and are the largest towns. Dumfries, the old town constructed from Locharbriggs sandstone, is the cultural and administrative centre of central Galloway. It lies snugly in the hollow of its Permo-Triassic rift basin, shaltered below the smooth leeside slopes of Criffell. Its bloody history, occupied and/or plundered by the English six times between 1300 and 1570, began when the Bruce is said to have murdered his rival the Red Comyn in front of the Greyfriars altar. The town was where philosopher and cleric Duns Scotus took his orders. A supporter of the Bruce, a student in Paris and resident of Cologne, his sarcophagus in that place's cathedral bears the splendidly European inscription: *Scotia me genuit. Anglia me suscepit. Gallia me docuit. Colonia me tenet.* ('Scotland brought me forth. England sustained me. France taught me. Cologne holds me.'). Better known is the poetic genius of the Scottish Enlightenment, Robert Burns, who spent the last third of his life in the town as an exciseman; the thirsty fieldworker can still take a 'haf-and-a-haf' in his honour at the bar of his favourite howff.

The port town of Stranraer on Loch Ryan is the administrative centre for Western Galloway, the ferries to Ulster now served from Cairn Ryan. Like Dumfries, its location and anchorages are determined by its location in a fertile lowland Permo-Triassic rift basin. The axe-head shaped Rhinns peninsula of resistant Ordovician greywacke to the west is the flanking upland to the double embayments of Loch Ryan and Luce Bay. At nearby Kirkmadrine the oldest surviving Christian standing stones in Britain are fifth to early sixth century. One has the theological inscription *Initium et Finis* ('Beginning and End'); a pronouncement that Playfair and Hutton would doubtless have profoundly disagreed with. Across Luce Bay on the Whithorn peninsula the poorly documented Saint Ninian

Figure 23.4 The Ruthwell Cross inside its kirk in the village, 15km SE of Dumfries (54.993055, −3.408912).

lived and worked in the early Christian era, founding the first church in Scotland, the *Candida Casa* (White House) around AD 400.

By the late eighteenth century newly planned settlements and extended medieval towns prospered in the region, from Wigtown in the west to Peebles and Galashiels in the east. The region's wilder places (Lowthers, Lammermuirs, Solway) grew famous through Scott's 'Waverley novels', a literary output that, with Burns's poetry, can fairly be said to have created the image of Romantic Scotland, perhaps to its detriment. Once linked into the national rail network, the region became popular amongst the more discerning and wealthy as a touristic and recreational destination. Its towns still have a special relationship with their surrounding agricultural hinterlands and proudly hold on to their special identities, many through the annual fiestas of 'Common Riding'. Here in the nineteenth-century mill-towns with terrace-rows developed, eventually specializing in weaving fine woollen and silk yarns: Hawick and Galashiels survive; Peebles just; Langholm and others not.

Medieval Kirkcudbright on the Dee estuary became an Edwardian and later artist's colony associated with the 'Glasgow Boys' school of colourists and others, notable residents being E.A. Hornel, Jessie M. King, E.A. Taylor and C. Oppenheimer. Attracted by the quality of coastal light and varied topography, they were part of a more general phenomenon, the trend towards Impressionism and *plein air* composition. Artists still live and work in the town and its hinterlands, producing landscapes showing play of light and colour from coastal greywacke, craggy granite hillslopes, and estuarine shores and marshes; our favourites are works by Helen Campbell and Linda Mallett. The latter writes: 'Throughout my working life I have been drawn to border and threshold situations, such as the littoral between land and sea, and the liminal between states of consciousness.'

Landscape sculptor Andy Goldsworthy's *Striding Arches* (Fig. 23.5) are located on the summits of Ordovician-age greywacke hills around Cairnhead Glen near Moniaive, a glaciated corrie and valley in the central Southern Uplands. Begun in 2002, they are four Romanesque-style barrel arches; 4m high dressed red-brown Permian sandstone from Locharbriggs, each visible from one other. They have inspired further collaborative efforts in landscape art: Alex Finlay, Alexander Maris and Susan Maris celebrate in poetry, photography and sound-sculpture the riverine confluences of Dalwhat Water that runs through Cairnhead Glen.

Figure 23.5 One of Andy Goldsworthy's four *Striding Arches*. Photo: Wikimedia Commons©Andrew Bowden. The stonemason's art, the stone itself and the bridging theme are all palpable. There is another arch on the distant hill framed in the arch space.

Chapter 24

Scottish–English Borderlands

I curse their head and all the hairs of their head, I curse their face, their brain, their mouth, their nose, their tongue, their teeth. May the thunder and lightning which rained down upon Sodom and Gomorrah, rain down upon them ...

A small part of the world's longest curse (1068 words), written by the early-sixteenth-century Archbishop of Glasgow, Gavin Dunbar, against the lawless Border reivers.

Turner's tour ... Through Derbyshire, Yorkshire and Durham ... is mostly the old stones of the antiquarians ... Gradually in the borderlands of Northumberland and the Tweed valley this subject-matter expands to include the broader sweep of their setting and a sense of his own and their exposure to the elements under broad skies, battered by waves on wide sea strands and bold headlands or suffused by the sun in quiet inland valleys.

Biographer and archivist David Hill, on 22-year old J.M.W. Turner's 1797 excursion in northern England (1996)

Here (Fig. 24.1A–C) are some of the choicest agricultural lands of Scotland, and some of England's wildest and most deserted fell-country. They feature Neolithic inscriptions on outcrops, monastic ruins, a rich heritage of folk legends and ballads, and a savage history of invasions, feuding and outlawry. They also contain the most fixed and final physical statement of the practical limits to Roman imperial power: the Roman Wall, sited by wily Hadrian's engineers in Carboniferous scarp

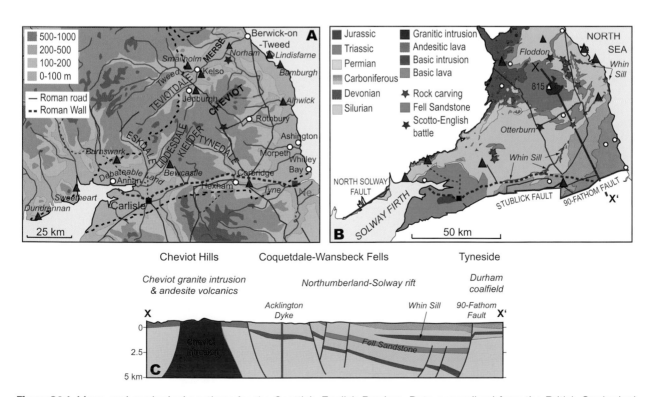

Figure 24.1 Maps and geological sections for the Scottish–English Borders. Data generalized from the British Geological Survey (2007a, b), *The Times Atlas of the World* and other sources.

country, well south of the modern border. It is built upon resistant limestones and sandstone crags intruded by the jointed black dolerite of the Whin Sill, forming one of the Island's most iconic outcrops, uniquely combining geology with architecture and world history. As Rome's most Hellenophile emperor, we imagine Hadrian's ghost now double-comforted: first by the sight of his still-standing wall after nearly two millenia, with its forts and fortlets lovingly excavated; second by the knowledge that deep in the crust under it runs the trace of the Greek-named Iapetus: the lost ocean 'discovered' by late-twentieth-century geologists (Chapters 6, 7).

To the west, fringed by the Solway Firth, are fertile and often arable lowlands, a mix of Carboniferous and Permo-Triassic marls and sandstones. Here, around the deepest indentation of the Island's coastline north of the Severn, portage by coastal ketch and steamer into often tiny anchorages was for long the most rapid and dependable mode of transport. To the east is Carboniferous fell country; the now-forested uplands of Kielder drained by tributaries of the North Tyne. The region's geological boundary to the south has no significant topographic expression; it is taken along the Maryport–Stublick–90 Fathom fault lines from the northern Lakes, south of Tynedale to Corbridge and north of Tyneside to the North Sea coast.

The pastoral lands along the modern border feature much volcanic-sourced rock: spreads of basalt in 'trap' country (lavas of Birrenswark, Glencartholm, Kershope, Cottonhope, Cockermouth, Kelso) and scores of volcanic plugs and necks forming pleasingly conical and rounded gorse-covered hills (Eildons, Maiden Paps). These rise strikingly above the general wide-rolling topography eroded out of Upper Old Red and Lower Carboniferous sedimentary rocks. In the late nineteenth century, Scottish geologist Archibald Geikie (he who named the Dalradian), recalling his travels in volcanic Auvergne, could write of it as 'Scotland's Puy Country'. Many of the formerly volcanic edifices – Birrenswark and the Eildons are notable examples – have Iron Age and Roman settlements, and many more feature younger fortresses and fifteenth- to sixteenth-century fortified tower houses ('peel' towers; Fig. 24.2).

Figure 24.2 Smailholm peel tower (55.604328, −2.576164). Photo: Shutterstock 349192334 ©Georgethefourth. The late-15th- to early-16th-century structure is built on a handy crag of early Carboniferous basaltic Kelso Lavas. Irregular blocks of the dark rock make up the majority of the building. These set off local Old Red Sandstone ashlar blocks used for the quoins, mullions and a pretty Romanesque-style arched doorway.

The rather monotonously rounded and often treeless Cheviot massif has glacial corries cut into its poorly exposed Lower Devonian volcanic surrounds; its core is of granite. Geologically it is an outlier of the Southern Uplands, munched over by multitudinous sheep and cursed by fell-walkers for its heathery, peaty hillocks. The Upper Old Red Sandstone envelops almost its entire western margins; in many out-of-the-way places the modern fieldworker can, like Hutton before them, 'with eagle eyes', find it resting unconformably upon Lower Palaeozoic rocks. In the northern Border fells it outcrops in a string of narrow gorges almost to Ecclefechan in the southwest. A favourite is just north of Langholm, 'Bloody Cleuch Linn'; the red-cliffed gorge.

End-Carboniferous compression bulldozed the eastern Borders strata into notable fold structures, in an arcuate belt that swings around the Cheviot buttress. The anticlinal folds (Berwick, Holborn, Lemmington, Bewcastle) set resistant sandstone outcrops of rough-grazed and treeless scarps amongst more general grazed pastureland of adjacent Cementstone vales. The outcrops gift notable sculptural heritages to the region. There are over 1000 Neolithic 'cup-and-ring' petroglyph marks on natural outcrops (Fig. 14.5). These are carved into the ultra-resistant silica-cemented sandstone scarps of the Fell Sandstone that define the upland fringes of fertile valleys. North Cumbria has recorded carvings on Barrons Pike and Hazelgill Crag that, with Black Stantling, The Whams and Crying Crags, overlook the pastoral grazing lands of the Bewcastle inlier. From one of the first two vividly named outcrops was later taken the immense raw slab lovingly carved by early-eighth-century Anglo-Saxon masons into the Bewcastle High Cross (Fig. 24.3). Petroglyph-bearing crags in Northumberland have evocative names: Roughting Linn, Hare Law Crags, Doddington Moor, Wheetwood Moor, Berwick Hill, Snook Bank, Millstone Burn and Horseshoe Rock.

Geology has made north–south lines of communication avoid the Cheviot's 800m-plus summits. The easternmost route via the Great North Road was easiest along the Northumberland and Berwickshire coastal stretch, where the soft Cementstone bedrock was scraped clean and lowered by Cheviot-sourced ice. Two subsidiary through-routes run either side of the mountain. The eastern strikes north-west over the Wansbeck at Morpeth, then via Cornhill-on-Tweed and Coldstream across the chequer of fields that comprise the rich arable and pastoral lowlands of the Merse of Berwick, classic drumlin country, to the Lauder valley. Along this route,

Figure 24.3 Bewcastle High Cross in its churchyard (55.063417, –2.682345). Photo: Wikimedia Commons©Doug Sim. The 4.4m stone has its cross-parts missing, presumably destroyed in the 17th century. The right-hand face in this view has intertwined swirls of vine with birds eating grapes; a motif also pursued by the Ruthwell sculptor (Fig. 23.4). The surrounding Permian sandstone grave slabs and church burial records reveal the surnames of the chief English Border families: Armstrongs, Elliots, Forresters, Nixons, Routledges, Telfords, Waughs.

but from opposite directions, came the Scottish and English armies during the Scotto-English Wars of the sixteenth century, notably to Flodden in 1513. Flodden Ridge itself marks the northernmost extent of the Cheviot lava pile. The over-confident Scots disastrously abandoned it to fight from the lower Branxton Ridge, a fluvio-glacial feature. The result is history: 'The flooers o' the forest are a' wede away'.

The Romans pioneered the western route, later known as Dere Street, north-west from the legionary centre of Corbridge in Tynedale up to Lauderdale through Rochester, Jedburgh and Melrose. It was probably the route taken by Agricola and his invading legions in AD 80. The Tyne's riverine hinterland stretches far westwards, almost into Cumbria, and includes Redesdale. It was along this valley, the Cheviot shimmering in the distant east, that Hotspur raced to intercept Earl Douglas's Scottish army at Otterburn on a hot summer's afternoon in August 1388.

Security in the Borders depended on how effectively kings, queens, earls, dukes and border lairds pulled their levers of power. Former governance of the region's western marches was chiefly from Carlisle. Originally a strong Roman fortress town at the western end of the Roman Wall, by the late eighteenth century it had become a mill town based on calico weaving. Earlier the pink Permo-Triassic sandstone of its battered castle and cathedral had fended off many reiver raids, such was the region's lawless state in the C16th. The eastern marches were governed from Berwick-upon-Tweed, whose precariously exposed position on the main route to and from Lothian led to frequent sieges and sackings. Finally, Henry VIII built its still-complete encircling walls from local Fell and Scremerston Sandstone. Their immensely thick zig-zag bastions laid every approach open to enfilade cannon fire, though the design was never thereafter tested by siege. Other castles scattered over the region include Alnwick, Bamburgh, Dunstanburgh, Norham and Caerlaverock.

The lowlands and dales were also peacefully endowed – witness the spread of medieval monastic ruins whose local building stones meld into a landscape rich in its own folklore and legend. Dundrennan Abbey is an eclectic mix of grey Criffell granite, pink-grey Carboniferous grit and brown Locharbriggs sandstone. It was the place of Mary Stuart's last night in Scotland before passage by ship across the Solway Firth to Maryport. Sweetheart Abbey, the 'New Abbey', was founded by Dundrennan monks after a founding charter by Lady Devorgilla – the last Cistercian abbey to be built in Scotland. It is pure Locharbriggs Sandstone, probably fetched hence down the Nith by coastal barges, the brown-red ashlar unforgettably rich in low morning sunshine (Fig. 24.4). Here amongst the woods and

Figure 24.4 Sweetheart Abbey (54.980275, –3.619176). Photo: Shutterstock 130439072 ©Tutti Frutti. A long view west down the six-bay nave from the presbytery where the High Altar once stood.

streams of New Abbey, Lady Devorgilla and the heart of her husband Lord John de Balliol were eventually interred together at the High Altar just before the Wars of Independence. Further east are the abbey ruins (all special in their own ways) of Hexham, Morpeth, Melrose (below the Eildon Hills), Jedburgh (astonishingly complete), Dryburgh, Kelso, and unforgettable Lindisfarne. All suffered pillage, sacking and burning numerous times in the many troubled years of Border conflict. Finally, as everywhere, all were brutally, but legally, destroyed Henrician-style.

In 1797 the 22-year-old J.M.W. Turner undertook his most ambitious sketching tour yet; to the north of England. He travelled along the windswept beaches of the Northumberland coast as far as Berwick and up the Tweed valley sketching *en plein air* and subsequently painting much-loved landscapes: sometimes repeatedly and after many years had passed. They include Norham castle above its rocky Tweed gorge (Fig. 24.5), ruined sandstone abbeys (Jedburgh, Melrose, Dryburgh) and remote coastal castles on Whin Sill dolerite crags (Dunstanburgh and Bamburgh; Fig. 8.1).

South of the wild Berwickshire coast, from Ashington to Seaton Delaval, coal-bearing Upper Carboniferous strata crop out. This part of the Northumberland coalfield supplied London and the south-east with some of its 'sea coal' for over 600 years, the majority coming from lower reaches of the Tyne, also painted by Turner (Chapter 25). Smaller and older deposits were mined at Scremerston, south of Berwick (the oldest workable coals in Europe), at Canonbie and along the trace of the Stublick Fault just south of the Roman Wall.

Figure 24.5 *Norham Castle, Sunrise* (*c*.1787) by Joseph Mallord William Turner (1775–1851). Graphite, watercolour and body colour on paper. 50.1x70.5cm. Acc. No. TW0017 ©Trustees of the Cecil Higgins Art Gallery, Bedford. Turner must have sketched from a mid-channel bar/island with one foot in Scotland and the other in England to get this view downstream. The castle, once a key border fortress, now a ghostly ruin, sits on an early Carboniferous crag above a wide ford. The artist wryly catches the contrasting dress, activities and variable prosperity of the left and right banks – the fat English cattle, for example.

Chapter 25

Lakeland, its Surrounds and the Isle of Man

[there were] shows in these mountains of great store of copper ores and rich leaders. If these were ordered and the several natures of the ores known; how to smelt them with their several additaments, these mines would in time become as famous a mineral town as any in Germany...

A mid-sixteenth-century report to Elizabeth I on the potential mineral richness of the Lakes.

With Kirkby Roundheads on the roof
Purple as polyanthus, proof
Against the flocking, mid-March weather,
When the wind's wings and the gull's feather
Fly screaming off the sea together.

Norman Nicholson, *Coniston Flag* in the sequence of poems *The Seven Rocks* (1954)

The Lakeland core and its Isle of Man offshoot (Fig. 25.1) are geological inliers of much-loved wild and mountainous country formed on Lower Palaeozoic rocks. The former is well protected as one of our oldest and largest National Parks. Its surrounds are tourist-neglected lowlands of interest and beauty in their own right: on three sides a maritime periphery grounded in Carboniferous and Permo-Triassic bedrock that nurtures urban populations (Barrow, Whitehaven, Workington), historic regional centres (Maryport, Cockermouth, Penrith, Kendal, Brough) and numerous other sturdy market towns. Since the onset of the Romantic Movement, the core has become one of England's most noted centres of landscape and geological tourism; William Wordsworth and geologist Adam Sedgwick wrote their popular Complete Guide to the English Lakes in 1842. Wordsworth grew up in Hawkshead, and it was he who first suggested that Lakeland should become a 'sort of national property...' Influential Beatrix Potter at Hill Top Farm championed its becoming a national park, and left her estate to that end.

Yet at one time, the Lakes rang with the thumps, rumbles, scrapes and explosions associated with mining – this by courtesy of German mining engineers brought from Saxony to Coniston, Catbells, Keswick and elsewhere. The smelted copper ore (together with Cornish tin) made the long-barrelled bronze cannon that helped defeat the Spanish Armada. Other treasures once mined here include the unique Keswick graphite (which started the pencil industry worldwide) and the ores of lead, zinc and tungsten. In the lowland periphery, Workington/Whitehaven coal and the rich haematite iron-ore of Egremont and the Furness lowlands served to concentrate the Victorian/Edwardian population that emigrated to these parts from all points of the compass. Barrow's steel industry has gone, but the naval shipyard still thrives, offshore wind turbine construction helping to diversify the region's military-industrial base. The country market towns nowadays seek their share of an ever-increasing touristic cake through the arts, crafts and heritage. The nation's entire nuclear waste is stored at the Sellafield/Windscale repository: a wide-ranging geological inquiry into its future safety is under way at the time of writing.

How this division between Lakeland core and periphery came about concerns (once again) the development and demise of the Iapetus Ocean, but also the much later Cenozoic uplift and erosion that climaxed with Quaternary glaciation (Chapter 7). In the later Ordovician, both Lakeland and Man were in muddy oceanic depths, the former the site of island-arc volcanism as subduction began. Its foundations were intruded by voluminous upwelling granitic magma that solidified to become the $1500km^2$ Lake District batholith. Its cupolas poke up as the eroded masses of Ennerdale, Carrock Fell, Eskdale, Broad Oak and Threlkeld; Shap to the east is younger, a 'Newer Granite'. These cupolas and their volcanics sourced the Lakes' mineral deposits (bar the iron ore). Once exposed to the elements, the rocks

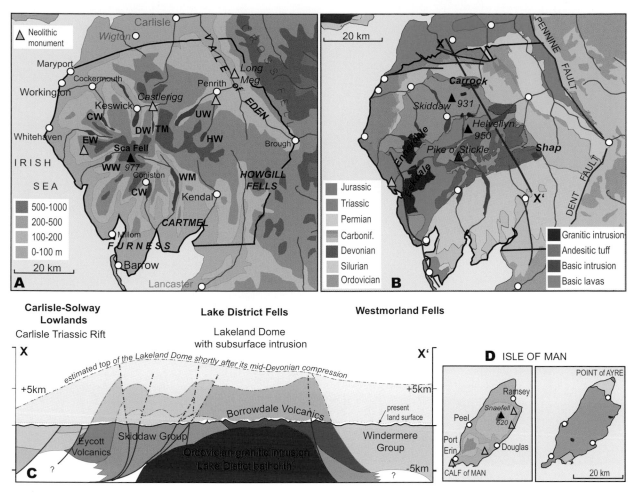

Figure 25.1 Maps and geological sections for Lakeland, its surrounds and the Isle of Man. Data generalized from British Geological Survey (2007a, b), Ordnance Survey (1991, 2011) and *The Times Atlas of the World* (1987) sources.

had a natural resistance to erosion, eventually 'sharpened' by the thorough scouring of repeated Quaternary glaciations. Despite the prominence of the Lakeland peaks and ridges with their arêtes and corries, it is also the U-shaped valleys, hanging valleys, morainic ridges and lakes that give pleasure to natives and visitors alike (Fig. 25.2). Wordsworth imagined viewing them from the air between Great Gable and Scafell: the eight major valleys 'diverging…like spokes from the nave of a wheel.' This divergence is also a consequence of Cenozoic uplift and erosion when the core's cover of Permo-Triassic and younger Mesozoic rocks was entirely removed so that a pre-glacial radial river system was slowly 'let-down' onto the diverse Lower Palaeozoic basement.

Lakeland has three distinctive belts of mountain and fell scenery. The main peaks north of the latitude of Keswick (Causey Pike, Catbells, Skiddaw-Blencathra,

Grasmoor) and along the pastoral valleys of Bassenthwaite, Derwentwater, Crummockwater and Buttermere are sculpted in the slaty Skiddaw Group. The ridges are smooth when eroded parallel with the generally north-east-south-west stratal strike, but rockier on the tops where glacier ice has scoured across them: thrilling fell-walking along the sweeping ridges. Skiddaw is one of the five Lake District giants above 900m, its broad ridge-top plateau littered with late-glacial frost-shattered debris. Across the Irish Sea the Isle of Man rears up as a wriggle of hills to the peak of Snaefell (620m), its grey silhouette clearly visible from both the Lakes themselves and Galloway to the north. Long an important Norse centre, its geology and scenery are akin to the northern Lakes. The Cambro-Ordovician Manx Slate group forms the Island's core and its prodigious coastal clifflines. Like the Lakes, its glaciated mountain, hills

Figure 25.2 Wastwater (54.427181, −3.319393) looking north-east along one of Wordsworth's 'wheel spokes'. Photo: Shutterstock 314475872 ©GrahamMoore999. This classic U-shaped valley hosted Pleistocene valley glaciers sourced from Great Gable (788m) and its surrounds. Here at the south-west end of the lake are outcrops of the Upper Ordovician Ennerdale granite. Middle Ordovician andesite lavas and the spectacular Wastwater screes are to the right with Yewbarrow's ash-flow volcanics in the crags to the left. Great Gable is in the centre distance with the lower slopes of Scafell.

and valleys are bordered by Lower Carboniferous and Permo-Triassic rocks to north and south, making up more fertile lowland outliers.

Just south-east of Keswick, on slightly elevated pastureland above Derwentwater and the Greta valley, is the late-Neolithic stone circle of Castlerigg (Fig. 25.3). To the modern visitor this place seems like landscape theatre. Grounded in the Skiddaw Group, it borders the central volcanic belt: the site is at the northern pastoral limits of the arc of mountain core. In detail, the complexity of the central volcanics creates a myriad of local microterrains. Thick andesite lava piles of the northernmost volcanics give rise to regular and obvious scarplands, as seen along the flanks of High Rigg (Fig. 25.3) north of Thirlmere, Border End and Scoat Tarn. The serious peaks of the central belt are Scafell, Scafell Pike, Great End and Bow Fell, with Great Gable, Langdale Pikes, Helvellyn and

Pillar in support. All were formed by crustal foundering into magma-grounded calderas. This caused explosive outpourings of hot ash-flows and the settling out of stratosphere-high ash columns. The resistant ash-flows now form thrillingly sharp and glacially scoured ridges in the landscape. The ash fall-out indurated and partly metamorphosed to form Pike O'Stickle's fine greenstones, source-material for the widely exported Neolithic ceremonial handaxes (Chapter 2).

The southern Lakes fells are rooted in Windermere Group sedimentary rocks comprising many fine turbidites. As they descend southwards their gentler slopes and wooded valleys are reassuring after descent from the rumbustious central peaks. Across glaciated Lunedale and north of Garsdale the dissected flat-topped and treeless plateaux of the Howgill Fells in the hanging wall of the reverse Dent fault seems more substantial,

Figure 25.3 Castlerigg looking to the south (54.602824, −3.098402). Photo: Shutterstock 222969199 ©Dave Head. The 40 or so stones (the tallest is 2.2m) are caught in a burst of sunlight with the central volcanic mountain core dark across the southern skyline. The stepped 'trap' country of High Rigg, formed of Birker Fell andesite lava flows, is clear in the middle distance.

providing unhindered fell walking and the delights of unspoilt market towns and villages like Kirkby Stephen, Ravenstonedale and Brough.

Lakeland periphery's Carboniferous and Permo-Triassic strata rest unconformably upon the Lower Palaeozoics of the core. The former are limestone-rich in Furness, Cartmel and Arnside with karstic scenery, cave systems and numerous historic lime kilns. The northern coastal coalfield of Workington–Whitehaven continues under the Irish Sea. Its Coal Measures are overlain unconformably by Permo-Triassic successions in the Carlisle basin and the Vale of Eden (surely the Island's most beautiful English place-name). These were Permo-Triassic rift basins, developed along the Maryport and Pennine Faults respectively. The formerly active faulted eastern margin features breccias known locally as 'brock-ram', derived as alluvial fans from the Permo-Triassic

ancestor to the modern Cross Fell scarp. Desert dune sands accumulated across the basin floor and gained thin skins of iron oxide and then silica cements during their shallow burial. They are now the pink-red Penrith/Lazonby Sandstones – widely used throughout Cumbria and beyond as fine sculpting and building stone (Fig. 4.8). The intermontane desert sand-sea later became a Triassic playa lake, accumulating the salty muds now mined for gypsum and anhydrite.

The central Eden Valley was a major focus for late-Neolithic monuments with numerous henges and stone circles; many Langdale stone axes and Bronze Age beakers, urns and cup-and-ring carved stones are found hereabouts. Most notable of the Neolithic/early Bronze Age circles is 'Long Meg and Her Daughters' north-east of Penrith (Fig. 25.4). The site has over-views across the fertile vale to the distant mountain

Figure 25.4 Long Meg (54.727946, −2.667336). Photo: Shutterstock 186456855 ©Dave Head. She is a 3.6m high block of pinkish-red Permian sandstone, shown here with some of her daughters in the background. They have varied pedigrees, including acid volcanics and white quartzes.

with two more marking the south-west entrance. Long Meg herself has carved rings and a spiral, and stands outside her oval of over 60 daughters, positioning the sunset at winter solstice. The stone carvings and others in the region are considered similar to the outstanding Irish Boyne Valley grave art. A modern re-connection with this Neolithic monumental richness are the already-cited (Chapter 4) Eden Benchmarks. The most upstream, and perhaps most spectacular of this notable chain of sculptural works, is at the head of Mallerstang Dale (Fig. 25.5) on the Pennine borders.

Roman forts and cavalry stations in the region are mostly related to the coastal continuation of Hadrian's Wall along the southern shores of the Solway Firth. There is Norse influence throughout the region: in place names the endings -by (settlement) and -thwaite (clearing); short and emphatic watery words like gill (stream), fell, force (waterfall); as carving on crosses; hogsback grave markers and grave-goods. It is thought that the Norse influx began when residents of Viking Dublin fled here as refugees (also to the Wirral, Chapter 28) around AD 900. The carved stonework generally shows a mingling of Viking, Pagan and Christian styles, as on the tenth-century Gosforth Cross.

core. It is adjacent to an earlier ditched enclosure on a Penrith Sandstone terrace above the River Eden. Two of the largest daughter stones are to the east and west,

Figure 25.5 *Water Cut* (1996) by Mary Bourne (b. 1963) at 54.382001,−2.331745. Photo: ©2016 Val Corbett Photography. All Rights Reserved. The sculpture, 2.2m of Salterwath Limestone from near Shap, stands alongside the ancient green road known as 'Lady Anne Clifford's Way'. Mallerstang Dale contains the headwaters of the Eden, to whose meandering course the sculpture's aperture pays homage, and from where the sculpture stands clear on the skyline of Mallerstang Edge above it. The fells here have the scarp-and-flat form typical of the Yoredale and Stainmore Groups in the North Pennines (Chapter 26). The succession begins down in the valley as the Great Limestone with scattered disused lime-kilns. The prominent scarp on the right skyline is formed by the Pickersett Edge Grit.

Chapter 26

North Pennines

I see the native of my kind
As a locality I love,
Those limestone moors that stretch from Brough
To Hexham and the Roman Wall,
There is my symbol of us all.

W.H. Auden, *New Year Letter* (1941)

In Garsdale, dawn;
At Hawes, tea from the can.
Rain stops, sacks
Steam in the sun, they sit up.
Copper-wire moustache,
Sea-reflecting eyes
And Baltic plainsong speech
Declare: By such rocks
Men killed Bloodaxe.

Basil Bunting, *Briggflatts, Part 1* (1966)

The name of the region (Fig. 26.1) and its southerly sibling (Chapter 27) might suggest a Roman origin, perhaps a nod to their Apennines. In fact the prefix 'Pen' is native British (Brythonic; Welsh) for a mountain head, as in Yorkshire's Pen-y-Ghent. From the excerpts above it is clear that poets W.H. Auden and Basil Bunting loved the region's landscapes deeply. Auden kept an OS map of it on his study wall in New York City during his days there. The itinerant Bunting returned from Persia in the early 1950s to live and joyously write in and about his homeland: once the larger part of Anglo-Saxon Northumbria, a kingdom that included Lothian and East Yorkshire.

From Cross Fell, at 850m the highest point, the region forms an eastward-declining plateau cut by east-draining glaciated dales. Structurally it is set within an array of fault lines, and in general the sedimentary strata also slope gently eastwards. The major massifs of Alston and Askrigg and intervening Stainmore are underlain by buried granitic plutons of the 'Newer Granite' suite and bounded by east–west trending structures, part of the wider Carboniferous rift province. To the west is the prominent zig-zag of the Pennine–Dent faults, ruptures formed later during Hercynian compression. To the east wiggles the Permo-Triassic unconformity that marks the western limits to the eastern Mesozoic lowlands (Chapter 34).

Below Quaternary glacial moraines and tills, almost everywhere has Carboniferous sedimentary bedrock. This is notably intruded by the Whin Sill along Cross Fell; the resistant dolerite also pops up, due to folding and faulting, to form the High Force waterfall and Cauldron Snout's cataracts in upper Teesdale. In the west, inliers of Lower Palaeozoic rocks occur along the foot of Cross Fell and to the south along the North Craven Fault from Ingleton eastwards to Malham (in Chapel-le-Dale, Crummack Dale and Ribblesdale). Thick limestones of the Lower Carboniferous Great Scar Group with their karstic pavements and cave systems are widespread (Figs. 2.2A, B) and rest upon the Lower Palaeozoic with spectacular unconformity (Fig. 6.4). Above are the interstratified limestone, sandstone and mudrock members of the Yoredale Series (Fig. 5.3). These are succeeded with slight unconformity by thick sandstones of the Stainmore Group, which cap notable North Yorkshire peaks like the isolated monadnocks of the well-known and much-loved 'Three Peaks': Ingleborough, Pen-y-Ghent and Whernside.

From the moorland drainage divide in the higher and wetter west begin the long courses of rivers contained in a series of pastoral stone-walled dales, all clearly valley-glaciated. With Old English and Norse roots, their mostly monosyllabic, curt and resonant names fall easily off the tongue; Tyne, Wear, Tees, Greta, Swale, Ure, Nidd and Wharfe. Here are the ancient lead-mining centres of Alston, Weardale, Swaledale (Fig. 26.2) and Grassington/Greenhow. The walls and roofs of sturdy valley farmhouses, barns, market towns and villages are invariably of local Yoredale Series stone (Fig. 2.2), the resistant limestones and sandstones quarried from between weaker mudrocks. Stepped hillslope scarps are cut by tributary streams that flow quickly through

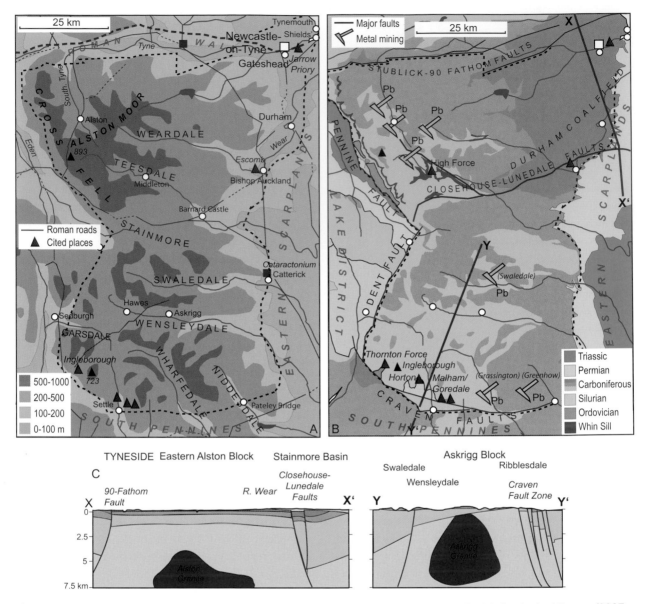

Figure 26.1 Maps and geological sections for the North Pennines. Data generalized from British Geological Survey (2007a, b), Ordnance Survey (1991, 2011) and *The Times Atlas of the World* (1987) sources.

wooded gashes to broad, glaciated U-shaped valley floors below (Fig. 5.7). Their startling waterfalls and rock pools (Hardraw Force near Hawes is a notable one) are hemmed in by steep bounding crags. The generally eastern stratal dip eventually brings the Coal Measures to outcrop, upon which the Northumberland–Durham coalfield developed.

The west-to-east pattern of North Pennine structure and drainage determined that north–south travel and communication depended upon sturdy bridges to cross the wide and fast-flowing rivers and their sea-fringe of long, wide estuaries. Modern road and rail visitors

to the north-east arrive via bridges and viaducts that offer splendid urban views. There are strong similarities between Newcastle-on-Tyne and London in this respect. Both have Roman origins, each nucleated along a large river with clusters of iconic buildings joined by distinctive and much-loved bridges. On the 200m wide Tyne, at a site close to the Norman castle, was the first Roman bridge, 'Pons Aelii' (Emperor Hadrian's family name); it led to adjacent left-bank coastal regions. The city's rapid industrial growth engendered iconic bridges: Robert Stephenson's emphatic High Level Bridge of 1849 carries both road and rail traffic on two separate

Figure 26.2 Remains of Old Gang Smelt Mill along the mine-waste choked channel of Mill Gill/Old Gang Beck, a left bank tributary to the River Swale, North Yorkshire (54.400121, −2.040429). Photo: Shutterstock 119605924 ©Andrew Fletcher. The smelter was fed by lead ore from several mines in the vicinity, and its flue extends 160m higher to the SE margin of Pinseat Hill. The east-north-east trending galena-bearing veins cut the Lower and Middle Carboniferous rocks of the area; the Great Limestone forms a prominent bench above the right bank.

levels, its four cast-iron arches sitting on Coal Measures sandstone piers. Later bridges are the Tyne (Fig. 26.3) and Swing Bridges and, more playfully, the tilting Gateshead Millennium Bridge.

Further emphatic reminders of past ecclesiastical and military identity are seen in Durham City's Norman cathedral and castle perched upon the incised meander loop of the River Wear. Bridges built of local sandstone by its early bishops join this medieval core to the wider township; both Framwellgate and Elvet had shops built along them. The Norman cathedral is the region's defining architectural masterpiece. Its exterior shell (Fig. 26.4) features yellow-brown weathering Coal Measures sandstone quarried from the walls of the adjacent Wear gorge. Inside, on notably massive carved sandstone pillars (Fig. 13.6) are early Gothic arches. In the 1272 Chapel of Nine Altars the slender pillars of Upper Weardale's own Frosterley Marble, a dark

coralliferous limestone from the Lower Carboniferous, frame the deluge of coloured light entering through stained glass: pure Gothic. At outcrop in upper Weardale, naturalist Phil Gates describes viewing this beautiful black limestone *in situ*: 'Splashed with water from the burn the wet rock matrix darkens to reveal the coral's white filigree.' The sausage-shaped fossils are a species of the extinct rugose coral genus *Dibunophyllum*.

Far older than Durham cathedral and with arguably even more ecclesiastical clout is the seventh-century Anglo-Saxon priory of Jarrow-on-Tyneside (Monkwearmouth, its sister foundation, is by Sunderland's Wearside). Along with the exquisite Escomb church near Bishop Auckland, its stone recycled from nearby Binchester Roman fort, they are the oldest surviving stone-built churches in the land, beautifully weathered and still solid after 1400 years' exposure to the elements. Protected by the military might of Northumbrian kings

Figure 26.3 The iconic Tyne Bridge of 1928 with the great civic buildings of Newcastle in the background (54.968000, −1.606062). Photo: Shutterstock 74142781 ©Shaun Dodds. The girdered arch and throughway are of Middlesbrough steel supported by two multi-storeyed end-piers of Cornish granite. The latter stone seems out of joint with both the tough sandstone framework of Stephenson's adjacent High Level Bridge pillars and with the old cityscape itself.

Figure 26.4 Durham Cathedral (1093–1133), the most physical statement of Norman ecclesiastical character in Britain; construction began only shortly after William I's 'Harrowing of the North'. Photo: Shutterstock 114217864 ©Kevin Tate.

like Edwin, Oswald and Oswiu since the early seventh century, the abbeys served as regional centres for early Christian missionary and scholarly activity. The earliest coherent history of post-Roman Britain, Bede's *Ecclesiastical History of the English People*, was written here. One missionary was Bede's protégé, Cuthbert: both men now entombed at Durham. This 'golden age' of North Pennine ecclesiastical culture was brutally terminated by Norse incursions and eventual invasion.

In later medieval times began the systematic and lucrative trades that lifted the region's moorland and estuaries from mostly pastoral agriculture. This was coal-mining and the winning, refining, smelting and fashioning of metals from minerals found abundantly in the Carboniferous bedrock. The thick High Main coal seam and others supplied London out of Tynemouth for 700 years; keel boats (lighters) brought the coal from mines and adits along the valley to the coastal transports docked in droves at the Shields in the outer estuary (Fig. 15.8). To the south, Durham City, Chester-Le-Street, Crook, Spennymoor and Bishop Auckland became nuclei for scores of surrounding pit villages. The Royal County Hotel in Durham City was the place for visiting socialist and union dignitaries to view the parades of the Durham Miners' Gala, an annual celebration of the might of coal and its miners. In the mid nineteenth century, shipyards like Jarrow on Tyneside used iron and steel plate from new towns like Consett on the western edge of the coalfield: the coking coal, iron ore and limestone mined locally from the Coal Measures and older Carboniferous.

The lead veins of the richest orefields – Alston Moor, Weardale and Swaledale – may have been worked by the Romans. From the mid-eighteenth century they were vigorously exploited by the Quaker-financed London Lead Company. Sturdy upland villages and the market town of Alston (England's highest) grew up, the orefield dominating the Island's lead production until the early twentieth century.

Late-twentieth-century de-industrialization saw the extinction of the region's coal mining, steelmaking and shipbuilding, though there is nowadays a resurgence of vehicle manufacture. The older legacy is celebrated in Anthony Gormley's 20m high steel sculpture Angel of the North (1998; Fig. 26.5). Perhaps the Island's most startling and most-viewed outdoor sculpture, it looks huge and comforting set within a wide valley; its captive audience the travellers driving daily by Gateshead on the A1. Gormley says he consciously had the Tyneside vernacular architecture of shipyards and bridges in mind when creating the sculpture on a site where coal was once mined. By way of contrast, Katharine Holmes's artwork depicts her timeless native limestone country of the Yorkshire dales. Using mixed media and in situ materials like river gravel, mud and grass, her often large-scale paintings (notably of Goredale Scar; Fig. 5.4.) depict geological features like bedding and jointing and the changing colour of the weathered rock outcrops at different times; they are full of texture and reveal only traces of humans, in tracks or fences.

Figure 26.5 *Angel of the North* (1998) by Anthony Gormley (b. 1950) (54.914122, −1.589494). Steel. 20m tall, wing span 54m. Photo: Shutterstock 307645835 ©DavidGraham76.

Chapter 27

South Pennines

we found the houses thicker, and the villages greater in every bottom; and not only so, but the sides of the hills, which are very steep every way, were spread with houses, and that very thick ... we could see that almost at every house there was a tenter, and almost on every tenter a piece of cloth, or kersey or shalloon ...

Daniel Defoe describes his arrival into the Halifax area of West Yorkshire in the 1720s.

On my first visit to the industrial north I rode on the top of a tram all the way from Leeds to Batley and all the way I rode through urban streets ... but as darkness closed and a few smoky stars soothed and extended my thoughts, the lamps going up in innumerable little houses restored the contours of hill and dale in shimmering lines of light.

Jacquetta Hawkes in *A Land* (1951)

The South Pennines (Fig. 27.1) are founded upon one of the Island's most emphatic geological structures: an upfold (anticline) whose arcuate Carboniferous outcrops defined a powerhouse of the Industrial Revolution. Today some 6 million people live within an hour's drive from its central core of Millstone Grit, which stretches from Pendle Hill and Ilkley Moor in the north-west and north-east, to Stoke-on-Trent and almost to Derby in the south-west and south-east. This 60km long, 15–25km wide moorland plateau, generally around 3–400m elevation, has plentiful craggy 'edges' and deeply incised glaciated dales; Airedale and Calderdale in the north are the largest, rising far to the west. The gritstone edges are up to 50m above intervening wide 'flats' where thinner and softer intercalated mudrocks have been eroded out. Always noteworthy and noble is the widespread Kinderscout Grit; other local developments in Yorkshire and Lancashire are the Pendle, Almscliff–Warley Wise, and Bramhope Grits; in Derbyshire, the Ashover, Chatsworth and Roaches Grits. The youngest of all is the widespread Rough Rock, particularly striking at its type locality around Halifax, the town built on its gently

east-dipping slope. All of these strata have provided first-class building stone and roofing stone-slates (the famed Elland Flags are from the lowermost Coal Measures) from the twelfth century (as at Kirkstall Abbey) to the present day.

The treeless nature of the southern Pennine moorland tops was not always thus. Pioneering pollen analysis of cores from Rossendale, since confirmed elsewhere, show that until the Iron Age the moorland flats and tops were largely wooded. A combination of pastoral expansion, timber exploitation and blanket peat growth in cooler, wetter climates did for them; adjacent steep-sided valleys became woodland refugia, in our time preserved as much-loved nature reserves, as at Hardcastle Crags near Hebden Bridge. Yet the rarity of Neolithic monuments, and particularly of outcrop inscriptions, throughout the gritstone edge country is striking.

The gritstone moorland core is bounded to east and west by the softer landscapes of the Yorkshire–Nottinghamshire and Lancashire Coal Measures. These are lower (usually <200m elevation) and less jagged because of fewer gritstone members. That the Pennine fold is asymmetric is evident in the contrast between its gently dipping eastern and steep-dipping western limbs. The former takes the Coal Measures under the Mesozoic cover of England's eastern scarplands (Chapter 34). The thrust-faulted western limb faces like a gigantic comma into the Cheshire and West Lancashire Permo-Triassic basins. These fertile lowlands became intimately linked to the industrial fortunes of their Pennine hinterland, not least for Wakes week rail excursions to coastal resorts like Blackpool, Formby, Scarborough and Filey. To the north the folds of the Ribblesdale–Craven Belt bring rumpled Lower Carboniferous limestones and mudrocks emphatically to surface, forming rich pastoral landscapes enclosed by gritstone moors. Southwards the central grit-country bifurcates east (Hathersage to Ashbourne) and west (Buxton to Leek) to define Derbyshire's own Dark Peak moorlands (Fig. 27.2). Within them are dales carved from soft Edale Shales

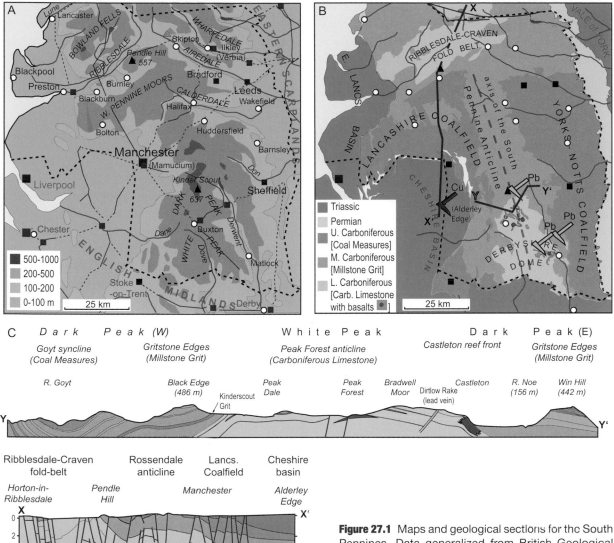

Figure 27.1 Maps and geological sections for the South Pennines. Data generalized from British Geological Survey (2007a), Ordnance Survey (1991, 2011) and *The Times Atlas of the World* (1987) sources.

Figure 27.2 Walking on the wilder side of South Pennine gritstone landscape: a boulder field on the western slopes of Kinderscout (53.386152, −1.878933). Photo: Shutterstock 285999734 ©Richard Bowden. View looks down to Kinder reservoir and beyond to the pastoral landscapes around New Mills, Derbyshire.

whose dammed valleys like the upper Derwent contain large reservoirs. Between is the Derbyshire White Peak; a domed anticlinal inlier of Carboniferous Limestone that forms well-drained, sweet-grassed and much-mineralized pastoral karstlands, as in the Castleton area. They are cut by incised wooded gorges with precipitous tors and cave levels with linear mined-out lead and fluorite veins ('rakes') visible across the landscape.

The region's influence on the Island's political and economic destiny started in Roman times with subjugation of the stroppy Brigantes. A millennium later it was part of stiff northern resistance to Norman rule, savagely suppressed in William I's 'Harrowing of the North'. In later medieval times the 'Pilgrimage of Grace' by recusant Catholics was ruthlessly put down by Henry Tudor. Farms and hamlets built of local sandstone subsequently grew up within the upland landscape whose climate, relief and communications were exactly suited to the production of wool. Along the many springlines, water is abundant everywhere in Grit country, where a prosperous cottage economy was established, as witnessed by Defoe in our chapter motto. The thrifty, independent hill- and dale-folk were part-farmers, part-spinner/ weavers whose local masters and squires built spacious sandstone halls and houses, themselves building nonconformist chapels. Adjacent small communities were joined by steep gritstone-cobbled packhorse routes connecting to larger towns with cloth warehouses and markets; Piece Hall in Halifax is the prime surviving example. Later came trans-Pennine horse-powered transport: silent barge-craft gliding along canals like the Leeds–Liverpool and Rochdale, transporting the raw materials of coal, wool and cotton into warehouses in innumerable canal basins.

Valley-bottom towns grew up and became early centres of industrialization of the woollen cloth trades (Fig. 27.3). A mix of hard-worked rivers, canals and later railways were strewn with mines, factories, mills and row upon row of terraced housing – the first truly mixed industrial landscapes. Yet above all this the bounding crags, flats and moorland tops gave physical freedom to the inhabitants who walked them on Sundays; they were (and are) the recreational 'lungs' of the landscape. Conurbations grew up: Manchester, Leeds, Bradford and Sheffield. Manchester, world-renowned 'cottonopolis', burgeoned over the East Lancashire plain, fed by canals, railways and roads: notably that wonder of the Victorian age, the Ship Canal that linked it and Salford to worldwide hinterlands. Leeds/Bradford became the Island's worsted, linen and woollen capitals, their mechanized looms eventually destroying ancient rural hand-loom and domestic spinning economies across the entire country

Figure 27.3 Autumn in Hebden Bridge, Calderdale, West Yorkshire. Photo: Shutterstock 163940420 ©peterjeffreys. Sandstone-built terrace dwellings, some 'back-to-back', cling to the steep SE-facing slopes of the valley side. Down below the River Calder, road, canal and railway pass within hailing distance through the gritstone landscape and town centre. Though the river is usually benign, such valley-bottom towns in upland Pennine landscapes have proven terribly vulnerable to flooding during the past decade or so.

Figure 27.4 *An Iron Forge* (1772) by Joseph Wright of Derby (1734–1797). Oil paint on canvas. 121.3x132.0cm. Acc. No. T06670 ©Tate, London 2015. Wright was *the* painter of the early Industrial Revolution. Here the iron-forger's family are bathed in industrial light, secure under the proud eyes and strong arms of the *pater familias*. The light comes from the white-hot iron bar under the gigantic water-powered trip-hammer poised to begin its forging. The scene is obviously compressed by the artist but his intent is clear: the bright lights of industry are only comforting.

from Mendip to Norfolk. Sheffield's steel industry came from master-cutler origins, the original water-powered iron forges here and in Derbyshire (Fig. 27.4) using iron smelted from locally sourced clay-ironstone ores (Chapter 11). Many other urban centres accreted from pre-industrial manufacturing nuclei or were entirely new-builds. By the mid-nineteenth century an all-enveloping coal-based mechanized culture sustaining millions of people had rooted itself in the South Pennines. Now the mining, steel and much of the textile industries are extinct.

Yet the people, moors and dales remain. Industrial heritage centres and museums record and remind us of past glory and misery. Since passing of the Clean Air Acts and the long-drawn-out decline of smoke-stack industry there has been an increase in people's willingness to commute by rail or road in order to live in the region's upland fringes. Previously the South Pennines was perceived to have limited scenic value due to a legacy of industrial pollution, made clear by smoke-blackened buildings and outcrops, thousands of acres of post-industrial desolation, and eroded moorland peat hags. Commuters now prize mullion-windowed terraced houses (urban cottages), surveys consistently emphasizing the value placed on local uplands giving a deep sense of landscape-belonging, tranquillity, and escape from the stresses of modern urban life. There is now awareness of environmental factors, ecology and habitat. Moor-top sites (flat, remote, high) are favoured for wind-farm projects, yet surviving moor peats and mosses are valuable parts of a fragile ecosystem and are also potent agents for ecological balance and carbon storage. In recent years the large-scale project 'Moors For The Future' has seen cultures of sphagnum moss spread over Kinderscout and adjacent moorland in the Peak District National Park and at many other locations in an attempt to reverse the harm done by 150 years of industrial pollution (smoke and acid rain), oxidation by drainage and peat fires.

The physicality of moorland tops, craggy edges and the stepped tablelands of intervening dalesides continue to inspire countless writers, artists, sculptors and outdoor enthusiasts of every sort, notably climbers and walkers. The gritstone setts of old packhorse routes now rattle with the boots of recreational walkers. Chalk patches in climbers' handholds zig-zag up gritstone crag and edge. Victorian and Edwardian sandstone buildings in former milltowns and villages are scrubbed up, weathering naturally at last. Yet the rocks quarried in the region still provide essential raw materials: building stone, cement, aggregates and minerals.

In countless museums, galleries, sculpture parks and wild places the carved stonework and cast bronzes of celebrated South Pennine sculptors like Henry Moore and Barbara Hepworth (Chapter 14) and their numerous successors amaze the visitor and comfort the locals with their sense of timelessness, space and integrity – likewise the literature set in industrial/rural Pennine landscapes by often verbally gritty authors such as J.B. Priestley, Alan Sillitoe, John Braine, Stan Barstow, Keith Waterhouse, Glyn Hughes, Ted Hughes, Tony Harrison and Alan Bennett. And no artist has painted wet urban flagstones by moonlight with more sense of 'home' than Atkinson Grimshaw did in Park Row, Leeds (Fig. 27.5).

Figure 27.5 Park Row, Leeds (1882) by John Atkinson Grimshaw (1836–1893). Oil on Canvas. 63.5x76.2cm. Acc. No. LMG100030 ©Bridgeman Images/ Leeds Museums and Art Galleries/City Art Gallery. Moonlight dominates the weak gas lamps, illuminating wet Elland flagstones and gritstone cobbles; a horse-drawn carriage and tram are in the distance. The building to the right is fairly new, the delicate pink brick and yellow sandstone detailing as yet free of soot.

Chapter 28

English Midlands

The glowering carnival: nightly solar-flare
From the Black Country; minatory beacons
Of ironstone, sulphur. Then, greying, east-northeast,
Lawrence's wasted pit-villages rising early,
Spinning-wheel gear-iron girding above each
Iron garth; old stanchions wet with field-dew.

Geoffrey Hill, CXXXVII, *The Triumph of Love* (1998)

But the coals were murmuring of their mine,
And moans down there
Of boys that slept wry sleep, and men
Writhing for air.
And I saw white bones in the cinder-shard,
Bones without number.

Many the muscled bodies charred,
And few remember...

Wilfred Owen, *Miners* (1918)

A region much travelled through, often not included in the north/south debate; signs on the M1 leaving London indicate 'The North', yet the HS2 railway will (eventually, perhaps) join Birmingham to the metropolis. A glance at the maps (Fig. 28.1) reveals this northernmost part of lowland England anchored in mainly soft, erodible Triassic marls and drained by the headwaters of great rivers: the Trent, Severn and Avon. Its smooth bosomy hills are mostly below 150m elevation: much was under rotation as open field in medieval times, the characteristic

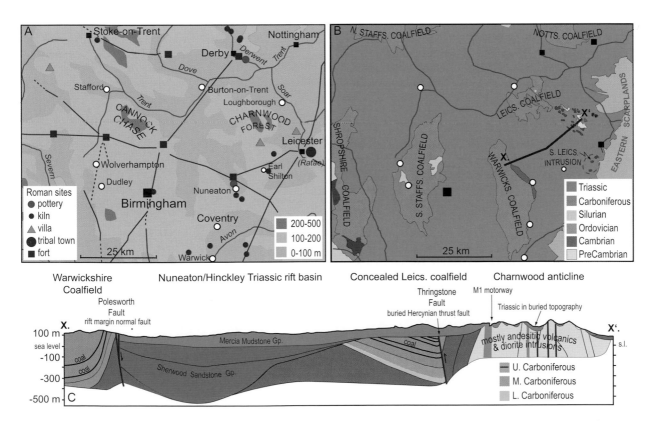

Figure 28.1 Maps and geological section for the English Midlands. Data generalized from British Geological Survey (2007a), Ordnance Survey (1991, 2011) and *The Times Atlas of the World* (1987) sources.

ridge-and-furrow often clearly seen in today's pasture-land. Defoe described the Midlands as being 'empty and dreary' while according to Walpole, 'an open country is but a canvas upon which a landscape might be designed.' Both opinions seem perverse. The land rises higher when more resistant and permeable Sherwood Sandstone crops out. Such is the preserved acid heathland landscape of Cannock Chase, Staffordshire. In Nottingham this same rock formation is less well cemented, hence the excavation of caves that riddle its ancient core. Other elevated ground makes up the densely populated Upper Carboniferous inliers of what were once five coalfields: Leicestershire, Warwickshire, North and South Staffordshire and Shropshire. Charnwood Forest to the east is late-Precambrian terrane, the most rugged ground of all. This core landscape, rich in volcanic rocks and upland wildlife, was disgracefully bisected in the 1970s by the M1 motorway (when it could have gone eastwards along the plains) and whose constant traffic noise defiles it.

Standing today on one of Charnwood's crags, it takes no great imagination to reconnect with its fiery past between 600 and 700Ma. Such rocks were part of a volcanic arc around the Avalonian terrane on the fringes of Gondwana. They slowly subsided beneath the waters of Iapetus during the Lower Palaeozoic, picking up sediment on the way. First were shallow marine Cambrian sandstones, now landscape-forming quartzites, notably along the narrow Nuneaton ridge. These are followed by thick mudrocks with the trilobite fossils of Gondwanan affinities (Chapter 7) first discovered by Charles Lapworth.

Voluminous mid-Ordovician igneous intrusions, roughly contemporaneous with those of Snowdonia and Lakeland, peep out from younger strata or are penetrated by deep boreholes and traced from geophysical soundings. Several outcrops form hillier country where huge quarries like Mountsorrel (Fig. 4.6) and Croft (Fig. 28.2) exploit tonalites and diorites for hardy angular aggregate. To the west is the slightly more elevated South Staffordshire coalfield. Here, around Dudley, Silurian strata are again represented, with the common trilobite genus *Calymene* known as the 'Dudley Locust'. There are also Silurian limestones whose coral patch reefs with abundant crinoid fossils match those of Shropshire's Wenlock Edge (Chapter 29).

Long after its late-Caledonian deformation, Avalonia was slowly covered by deltaic sediment in Coal Measures times, with thick peat accumulations that gave rise to such monstrous coal seams as North Staffordshire's 'Thick Seam', a source for the country's first 'million-ton' pit. It was the disaster here in 1917 at Minnie Pit that sparked Wilfred Owen's dark poem quoted above and discussed in Chapter 16. After end-Carboniferous

Figure 28.2 Essence of the English Midlands: looking west from Croft Hill summit rising 130m or so above the River Soar floodplain (52.568871, −1.243778). It gives broad views westwards to the industrial landscapes of the old Warwickshire coalfield on the horizon (Nuneaton, Coventry). Views east are to the pastoral landscapes of the rural East Midlands. Here at its summit are resistant outcrops of late-Ordovician diorite, which the immense quarry extracts for road aggregate. The working face shows red-brown Triassic Mercia Mudstone resting unconformably on the undulose, dark-weathering diorite surface. The excavation has vividly revealed a buried pre-Triassic landscape cut into the diorite. Beyond the quarry face a new landscape emerges along a linear ridge as quarry-waste is being landscaped.

tectonic compression produced folds and thrust faults (see Fig. 28.1C), sedimentation resumed in several north–south trending Permo-Triassic rift basins: Stafford, Wolverhampton, Worcester, Solihull, Burton-on-Trent and Hinckley/Nuneaton. Their faulted margins sourced alluvial fans and rivers whose sediment forms the afore-mentioned Sherwood Sandstone, a major aquifer and source of fine building stone. Later the rifts became interior drainage basins filled by the fine-grained sedi-ments of playa lakes. These Mercia Mudstones (Fig. 28.2) include Derbyshire and Nottinghamshire gypsum and fine alabaster. Their partial dissolution by groundwaters gives the non-carbonate permanent hardness that makes Burton-on-Trent a real-ale haven.

Important Roman roads traversed the region: the Fosse Way joined distant Isca (Exeter) to Lindum (Lincoln) and another, Chester to London via Ratae (Leicester), a large tribal market town (oppidum), later walled and with the usual Roman amenities. Greater Mercia developed after the earliest Saxons settled and ploughed the claylands and gravelly terraces of the Forest of Worcester. Penda's Angles conquered them, moving into the east Midlands along the Tame and Trent. Their capital at Tamworth was founded in 573 on a Sherwood Sandstone ridge between Bromsgrove and Lichfield, perhaps just one of several bases for a generally itinerant court. The Mercian kingdom expanded under Penda so that by the eighth century, King Offa, a correspondent of Charlemagne, ruled all England south of the Humber between Wales and East Anglia. His kingdom subse-quently fell to the Danes in 875.

The Midlands' agriculturally-rich medieval landscapes spawned subsequent cottage industries based on animal hides and wool, some of which survived into modern times alongside mining in the Leicestershire coalfield. The west Midlands also entered the eighteenth century as a largely agricultural society with cottage industries. It left it as England's foremost manufacturing centre using metals, and for pottery. It owed this transformation to both readily available local raw materials (coal, iron ore, various clays) and the undoubted native ingenuity and inventiveness of its inhabitants. Since Birmingham was twenty miles from the nearest navigable water, far-sighted manufacturers and the likes of Josiah Wedgwood of Burslem realized the full potential of canals for transport. Canalization was time-consuming on account of prob-lems with landowners, leases and logistics; eventually an extensive network was completed by 1777, founded on the Trent–Mersey link, later with the Grand Union network southwards. Wedgwood was a major investor and energetically pushed through the Trent–Mersey link. This joined his own works at Etruria with the Duke of Bridgewater's pre-existing waterway, the pioneering Bridgewater canal of 1761 that connected coal mines at Worsley with the Mersey. Over 150km of canals are still navigable in the region (Fig. 28.3).

Wedgwood's skill as a potter was noticed at an early age. He was apprenticed to Thomas Whieldon (as was Josiah Spode), the leading workshop potter of his age. His artistic talent, wide reading around classical themes, and acute business acumen were combined with a notably practical geological and chemical bent

Figure 28.3 A narrow boat navigates the Trent and Mersey canal past a typical Potteries bottle-kiln and its attendant wharves, workshops and warehouses: Price and Kensington Potteries, National Teapot Works, Longport, Burslem (53.055743, −2.209864). Photo: Shutterstock 367822 ©Tim Large. The Longport works are an early example of a fire-proof industrial building, all brick and iron. Out of the thousands of kilns formerly in the Potteries, 47 proudly remain today, listed buildings all. Canals supplied all the raw materials to most kilns, also transporting the delicate finished goods to ports and (eventually) railway distribution centres.

concerning coloured glazes and clay-mixes from different local geological strata. Following John Astbury before him, he added pulverized bone and flint to Upper Carboniferous Etruria Marl clays (also sourced for bricks and earthenware) to help whiten them after firing. He also invented a pyrometer to accurately measure *in situ* kiln temperatures. He travelled to Cornwall in 1775 to oppose attempts to monopolize the exploitation of St Austell's china clay (Chapter 31), eventually purchasing Carloggas pit, and selling direct to the Staffordshire industry. Within his factories humans, not machines, painted the delicate and subtly coloured visual images that characterize his decorated products, notably the famous Neo-Classical Jasperware. The Pottery towns were thus major employers not just of manual workers, but also of moulders (using Nottinghamshire gypsum), firers, artists and skilled decorators.

Well before the Industrial Revolution the West Midlands had a flourishing indigenous iron and brass industry centred around Dudley. The metals were cast in moulding sands obtained from the more friable levels of local Sherwood Sandstone. Its Coal Measures iron ores were smelted with charcoal and fluxed by local Wenlock Limestone. The first Newcomen steam engine pumped water from the Conygree coal mine on Lord Dudley's estates in 1712. By the seventeenth century Birmingham's artisan workers were already established metalworkers, making cutlery, nails, swords, guns, horse harness, buttons and much else. Gradually steam-powered machinery took over as the canal link with Staffordshire and Worcestershire was completed in 1772.

In 1774 James Watt accepted financial sponsorship from the risk-taking Birmingham manufacturer Matthew Boulton. Watt travelled south from Edinburgh with his geological friend James Hutton, bringing along the prototype of his revolutionary Kinneil steam engine. He soon became immersed in West Midlands intellectual life, becoming a notable member of the Lunar Society, a nonconformist scientific/literary coterie of friends who met monthly at full moon. Fellow Lunaticks (including Erasmus Darwin, Charles's grandfather) could more than match the scope and influence of the Edinburgh Enlightenment natural philosophers and savants, of whom Hutton, as we have seen (Chapters 3, 6), was one. Chemist members like Priestley (Fig. 28.4) and Keir were also practical men, the former famous for his discovery of oxygen. Keir developed Birmingham's glass manufacturing industry until, by the mid-nineteenth century, it supplied the entire one million or so panes

Figure 28.4 One of nine carved and masoned slabs of red sandstone, The Moonstones, that celebrate the chief members of the Lunar Society at the Asda Queslett Superstore car park, Great Barr, Birmingham (52.551707, –1.909235). Photo: Wikimedia Commons©Andy Mabbett. Joseph Priestley's stone here shows chemical apparatus with the full moon above. All designs were by Steve Field, carved by Malcolm Sier and Michael Scheurmann (1998).

for the Crystal Palace. He also experimented on the rapid cooling of molten silicate liquids to glass, much to Hutton's volcanic delight.

Today, Birmingham, Coventry, Nottingham and Derby and other towns across the region are still inventive manufacturing centres, each with a proud tradition of 'making': Rolls-Royce engines, Jaguars, Land Rovers, Royal Worcester, JCBs and Marmite. Josiah Wedgwood's name survives (just) in pottery from Stoke-on-Trent, still importing fine Cornish clay: Stoke natives will always turn over an old cup or plate to check its manufacturer's name, nodding sagely as they do so. Leicestershire remains in the memory of one of us (HJL) as pure pleasure, seen through a thrown-down four-barred gate with wildflower meadows and a distant bell-tower (Fig. 28.5) – this especially so since the inception of the National Forest project in the 1980s, which has already led to afforestation of much industrial wasteland, and with plans to include much more in the formerly industrial Midlands.

Figure 28.5 *Distant Lutterworth* (1994) by John Lines (b. 1938). Oil on board. 47.5 x 55.5cm. Acc. No. LC028 Rugby Museum and Art Gallery ©John Lines. A painting of memory. Leicestershire countryside: its small fields, abundant hedgerows and ancient trees.

Chapter 29

Welsh–English Borderlands

On Wenlock Edge the woods in trouble;
His forest fleece the Wrekin heaves;
The gale, it plies the saplings double,
And thick on Severn snow the leaves.

'Twould blow like this through holt and hanger
When Uricon the city stood:
'Tis the old wind in the old anger,
But then it threshed another wood.

A.E. Housman, *Poem XXXI* in *A Shropshire Lad* (1896)

Mrs Cotton (softly): 'It sounds a sad piece.'
Professor (quietly): 'Yes, it is. A kind of long farewell.
An elderly man remembers his world before the war
of 1914, some of it years and years before perhaps –
being a boy at Worcester – or Germany in the 'Nineties
– long days on the Malvern Hills – smiling Edwardian
afternoons …'

From J.B. Priestley *The Linden Tree*, Act Two, Scene One (1947) [The
chief characters hear Dinah Linden playing the second subject of the First
Movement of Elgar's Cello Concerto in an adjacent room.]

Geological boundaries along the Wales/England border
are much fragmented by folds, faults and unconformities,
so that the region's rocks (and thus its scenery) are wonderfully various (Fig. 29.1). The landscape often changes
over just a dozen kilometres or so: rough-grazed upland
tops over 400m elevation have Precambrian and Lower
Palaeozoic (often volcanic) foundations (Clwydian
Range, Malverns, Wrekin, Caradoc, Long Mynd); Silurian limestones and mudrocks form the rich-grazed
grassland slopes of Wenlock Edge; valley-floor orchards
line numerous tributaries to the Dee, Wye and Severn;
brown-soiled Hereford, Ludlow and Clee are of Siluro-Devonian Old Red Sandstone; Coalbrookdale, the
Forest of Dean and Wye's incised valleys also have Old
Red but with Carboniferous strata rich in coal and iron
ore. The gradual decline of slopes to the English lowlands
begin on Triassic Sherwood Sandstone in the Wirral
and scarp-hilled northern Cheshire. The Cheshire and

Worcestershire plains are founded on Mercia Mudstone,
usually covered by thick spreads of Quaternary glacial
detritus from Scottish–Lake District ice lobes that swung
into the region from the Irish Sea.

On very many natural rock outcrops and several
artificial mounds the Iron Age peoples, as elsewhere in
the west, erected their hill forts (Fig. 29.2). After the
Roman penetration of Wales the defeated British hill
tribes were ringed by garrisoned settlements, campaign
(vexillation) fortresses and marching roads, from Deva
(Chester) and Viroconium (near Wroxeter) in the north
to Glevum (Gloucester) and Caerleon (near Newport)
in the south. The later Anglo-Saxon surge of conquest
against the Welsh had its westerly boundary emphatically
defined by the lengthy ditched earthwork that is Offa's
Dyke. It follows the margin of the Cheshire basin and
then defends the headwaters of the Severn and Teme in
the Silurian plateau between Mongomery and Kington.

The major populous centres of the Wirral, Deeside
and Merseyside grew up as maritime outlets linked by
canal and then rail for South Pennine Lancashire and
Midlands industry (Chapter 28). The Wirral was a major
Norse centre, witnessed by place names, archaeological finds (hogback tombs, Viking crosses) and written
sources. Ingimund's Saga tells of the Norse expelled from
Dublin in 902 travelling *en masse* to the Wirral (also to
coastal Lakeland; Chapter 25). That same year, at Cross
Hill, Thingwall, the Island's first recorded democratic
parliamentary debate took place. It was possibly also
on the Wirral (though there is a counter-argument for
Northumberland, near Bamburgh) that King Athelstan
finally defeated a Norse army gathered from far and wide
at the battle of Brunanburh (937), thereby unifying the
English state. According to the Anglo-Saxon Chronicle:
'Five lay still on that battlefield – young kings by swords
put to sleep – and seven also of Anlaf's earls, countless
of the army, of sailors and Scotsmen.'

Major developments on the eastern Wirral during
the Industrial Revolution included shipbuilding at Birkenhead and, in the 1880s, relocation of

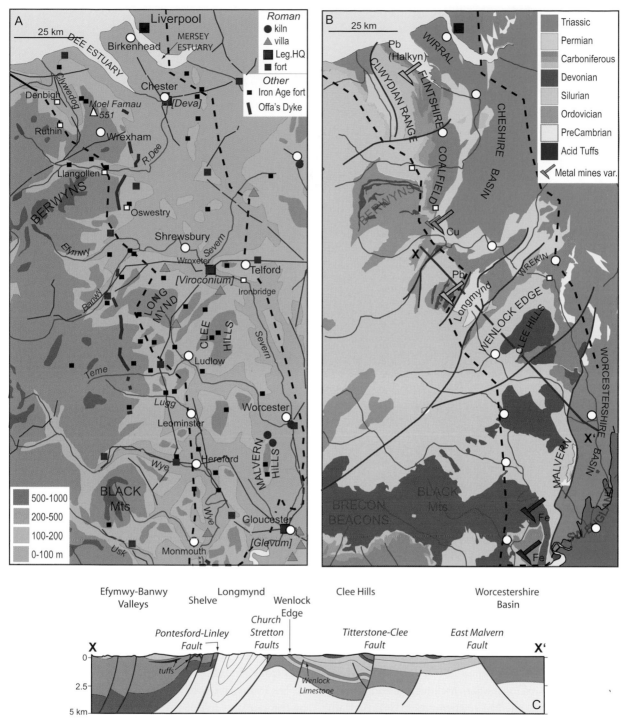

Figure 29.1 Maps and geological section for the Welsh–English Borders. Data generalized from British Geological Survey (2007a), Ordnance Survey (1991, 2011) and *The Times Atlas of the World* (1987) sources.

industrial-scale soap manufacture from Warrington and Tipton in the West Midlands to Port Sunlight. The Lever Brothers' model village took ideas inherent in the Arts and Crafts movement, William Morris especially: social engineering at its best. Across the Mersey, Liverpool grew originally as a trading port with Norse Dublin. Facing west, with much Welsh and Irish immigration, it boomed from the late 1600s. This was largely on account of the burgeoning North American and West Indies trade that it shared with Bristol – whence came

Figure 29.2 Old Oswestry Iron Age hillfort with its elaborate multiple ditches and ramparts (52.872046, −3.048577). Photo: ©Webb Aviation. The gulley-like cut in the centre left leads into the innermost upper living space.

what we see now as a great infamy, the 'golden triangle': Lancashire and Midlands goods traded for West African slaves, the survivors traded on for West Indian sugar. Today the past wealth and elegance indicated by the city's surviving early Georgian brick-built squares and fine public buildings (many in Sherwood Sandstone) signify an era of human exploitation, suffering and loss: something the striking modern architecture and restored dock basins along the Mersey front seem only partially to compensate for.

Chester's position at the head of the Dee estuary was initially strategic; the Romans chose Sherwood Sandstone to build their fortress-port of Deva there; the hard-worked XXth Legion attempted to control both Welsh and Pennine hill tribes. Paulinus took the legion eastwards from the shores of Menai to achieve the last-ditch defeat of Boudicca's army in AD 60/61. By

late-medieval times Chester was a conduit for Cheshire's rich agricultural and mineral hinterland, a heritage visible in the town and county inventory of fine late-medieval timber-framed buildings. The trade included rock salt from the Mercian Mudstones of Nantwich (mined and brined, it is still our chief indigenous source) and lead ores from Flintshire.

The industrial northern borderlands are epitomized by the rise and fall of the Flintshire coalfield and the Wrexham iron industry, both sourced in Coal Measures strata. A remarkable outlier of pottery manufactory grew up as a cottage industry and flourished at Buckley until 1946; intense faulting had juxtaposed there a variety of Upper Carboniferous red, buff and white clays. To the west, pastoral Carboniferous Limestone country culminates in the long east-tilted scarp of the Clywdian Range whose rocks sit unconformable on a Silurian pediment; the range features cairns, tumuli and Iron Age forts. East of its peak, Moel Famau, at Bryn yr Ellyllon (Fairies' Hill), a wondrous Bronze Age gold ceremonial cape (the 'Mold cape') was discovered in a cist grave (Chapter 11). Today, Anthony Gormley's Koros-like bronzes at Blundellsands on the outer Mersey (Fig. 14.1) eye the Clywds out of the corner of their eyes just twenty or so miles away – a reminder of both Liverpool's significant Welsh heritage and the universal and timeless power of metallic imagery.

Between Shrewsbury and Hereford are other ancient wool-based market towns: Church Stretton, Ludlow, Leominster. Shrewsbury itself nestles on Carboniferous and Permo-Triassic bedrock in the lee of the Precambrian Longmynd plateau and overlooked to the east by volcanic Wrekin. It hosts an unlikely piece of architectural history in the five-storey late-eighteenth-century Ditherington Flax Mill, the world's first entirely iron-framed building. Designed to reduce fire hazard, it is the structural ancestor of all modern skyscrapers. Coalbrookdale's early longwall mines provided the coal that Abraham Darby first coked to reduce local iron-ore and to heat the various tile kilns and potteries that grew up here. Thomas Telford used the local iron for his bridge across the Severn gorge (Fig. 29.3). Its heritage survives today in more than name: the Ironbridge

Figure 29.3 Telford's bridge spanning the Severn gorge at Ironbridge, Shropshire (52.627379, −2.485467). Photo: ©Authors. Massive piers of local Upper Carboniferous sandstone support the riveted archwork and iron-railed inflection of the deck. The inset shows the central decoration with the date 1779. For the last 40 years a thick, arched bed of submerged concrete has prevented the piers from sliding riverwards.

Snowdonia was wooded up to a treeline at 600m in Neolithic times. From the top of Snowdon today one can see several distant ranges bordering the Irish Sea, a discontinuous mountainous periphery: all have concentrations of Neolithic sites in the adjacent lowlands. Two fine examples are Harlech's Dyffryn Ardudwy burial chamber, a very early (*c*.3500 BC) double portal dolmen and Capel Garmon, originally a passage grave covered by a cairn, with spectacular mountain views. Pottery fragments from Neolithic to Bronze Age suggest prolonged use.

At the very north-eastern margin of Snowdonia, within sight of the cream-white Carboniferous Limestone of Great Orme, is the important Neolithic ceremonial axe factory of Penmaen Mawr ('Big Stone Headland'). The stone for the widely traded polished axes comes from the western slopes of Cwm Graiglwyd, site of a late-Ordovician subvolcanic intrusion. The dark, hard, speckled and compact microcrystalline diorite is nowadays quarried for use as more prosaic (but essential) aggregate.

In the nineteenth century the dramatic landscapes of the region began to attract landscape tourists *en masse*. Following in Richard Wilson's (Fig. 30.5) and Turner's footsteps came later landscape artists like Sidney Richard Percy; his limpid 'Llyn-y-Ddinas', a lake north of Beddgelert in the Nant Gwynant Pass, shows exquisite details of stratification and cleavage in the Snowdon Volcanics. In the twentieth century Gwilym Prichard from Criccieth in Lleyn, who taught in Anglesey, said 'I paint the land and farms, the sea and the sky because they shaped my life.' His approach is intimate and emotional rather than analytical.

Figure 30.5 *Snowdon from Llyn Nantlle* (1765/67) by Richard Wilson (1713/14–1782). Oil on canvas. 103.6x126.3cm. Acc. No. WAG 2429 ©Walker Art Gallery, Liverpool. The Walker catalogue reads: 'Because Wilson sought a sense of harmony this view is not accurate topographically. It is a scene of order and tranquillity. Nature seems benign. It is a peaceful and happy scene with people fishing in the foreground. The view, eastwards from the western end of Nantlle lake, was later visited by Turner who painted the same scene.'

Chapter 31

Southern Wales

I have been taught the script of stones, and I know the tongue of the wave.

Vernon Watkins, *Taliesin in Gower* in: *New Selected Poems* (1954)

(Used as the epitaph on his memorial at Hunt's Bay, Penrhyn Gŵyr)

The force that drives the water through the rocks
Drives my red blood; that dries the mouthing streams
Turns mine to wax.
And I am dumb to mouth unto my veins
How at the mountain spring the same mouth sucks ...

Dylan Thomas, *The force that through the green fuse drives the flower* (1934)

Here the Cambrian Mountains (Fig. 31.1) curve in an arc north-east from Carmarthen to an apex in the Tywi Forest, thence to the dissected plateau of Pumlumon Fawr (752m) behind Aberystwyth. In the west, Dyfed's stubby peninsula fingers well out into the Irish Sea, separating the St George and Bristol Channels. Its wave-resistant igneous core borders the concave sweep of the less resistant hinterlands of south-central Cardigan Bay. The Cambrian range, with its glaciated northern scarps, forms a drainage divide; to the west the Teifi descends to Cardigan, to the east Tywi to Caernarvon's estuary. The Wye and Severn tributaries, with their many modern reservoirs, flow east to the Borders. Down them all came Quaternary ice streams and outwash. During the last deglaciation, ice in the Tywi valley paused and laid across a moraine that ponded Lake Tregaron, its icy waters warming, infilling with detritus and organic fines. Subsequent overflow breached the lake but

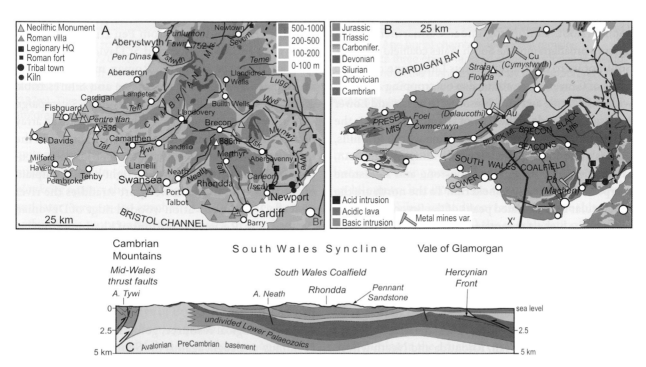

Figure 31.1 Maps and geological section for Southern Wales. Data generalized from British Geological Survey (2007a), Ordnance Survey (1991, 2011) and *The Times Atlas of the World* (1987) sources.

Chapter 32

South West England

Field threaded with flowers
Cools in lost sun.
Under furze bank, yarrow
Sinks the drowned mine.
By spoil dump and bothy
Down the moor spine
Hear long-vanished voices
Falling again.
Now they are all gone, says bully blackbird.
All gone.

Charles Causley, *Trethevy Quoit* in *A Field of Vision* (1988)

So the landscape was the common factor for all of us, a presence of perpetual power which in its transitoriness reminds us of our own ... any pathway we followed, over moors, or down the shafts of mines, or along the corridors of gales, led only to oneself.

From the personal memoir of David Lewis, architect, in the catalogue to the Tate Gallery exhibition *The Art of St Ives* (1985)

England's longest peninsula tapers irregularly south-westwards, dividing incoming Atlantic wave and tide between the Celtic Sea and English Channel (Fig. 32.1). At its tip, the Land's End headland has low concordant peaks set back from a dissected coastal platform above rugged, castellated cliffs of bare-jointed granite (Fig. 2.2D). Neolithic monuments and thick-walled farm buildings are set in a maze of small Iron Age stone-walled fields. Around St Just, long-abandoned engine-houses still stand, sentinel-like, on the skyline; the old miners worked rich veins of tin and copper ore here from great depths.

Bounded by thrust faults, South Cornwall's geology nods to the Brittany peninsula and Galicia's Cape Finisterre. All share a common geological heritage – born in a marginal seaway to the Rheic Ocean north of the nascent Hercynian mountain chain. The southern headlands comprise older resistant igneous and metamorphic masses: Lizard's dark gabbros and softer black-green serpentinite define it absolutely as an ophiolitic remnant; Dodman and Start Points' silvery micaceous schists witness an early consummation of the Hercynian deformation. The Lizard gabbros sourced widely traded Neolithic ceremonial polished stone axes. Tell-tale gritty fragments remain in local clays dug for the throwing of tough, round-bottomed pots distributed widely through the lands. Start's low cliffs reveal the coastal wave-cut platform of the Last Interglacial preserved a few metres above the modern beach (Fig. 7.16).

The Upper Palaeozoic successions of the rest of Cornwall and Devon define a regional syncline, albeit much deformed and dismembered, the imprinted cleavage forming Delabole's famed roofing slates. North Devon has the northern limb of the syncline, its coastal cliffs cut into resistant Old Red Sandstone. Modern Exmoor's steep valley-frilled and often bare-topped plateau was once a Royal Forest. In the 1390s Geoffrey Chaucer managed what remained of it, as Petherton Forest in the lowland lee to the Quantock Hills. Exmoor's open moorland was enclosed wholesale in 1818.

The marine Devonian deposits in their type localities contrast with the similar-aged continental Old Red Sandstones of Wales, the Borders and elsewhere in the Island. They include the spectacularly coralliferous Torquay and Plymouth limestones and the ammonoid-bearing mudrocks of Padstow and Chudleigh. The synclinal core has Carboniferous successions traditionally called 'Culm', whose resistant sandstones form cliffs and headlands around Bude Bay. Despite its Upper Carboniferous age the Culm yields no coal; deposition occurred from turbidity currents debouching into a deep lake basin.

The large Cornubian granite batholith determines the peninsula's shape and form. Its unyielding substrate bears five main cupolas that define the high moorland massifs of Dartmoor, Bodmin, St Austell, Carnmenellis, Land's End and the remnants that are the Scillies. In the unparalleled warmth of the Paleogene period the hillier granite outcrops stood proud and forested, experiencing

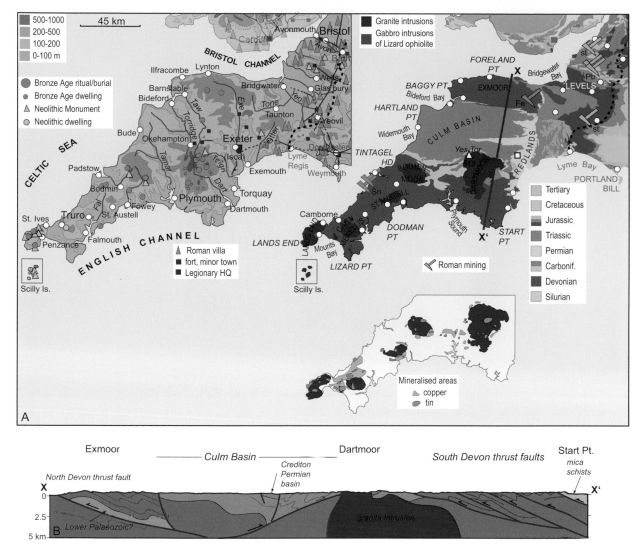

Figure 32.1 Maps and geological section of Southern England. Data generalized from British Geological Survey (2007a), Ordnance Survey

intense acidic weathering under a hot, wet climate. The moorlands, entirely unglaciated, have V-shaped and spurred valleys, gently sloping interfluves (no arêtes) and headwaters (no corries/cwms). Ill-sorted periglacial deposits, formed during the later cold intervals of the Quaternary period, are known locally as 'head'; these moved slowly downslope during brief summer thawing of tundra topsoil. The many rounded, convex hilltops are usually crowned by iconic and isolated granite tors (Fig. 10.1), remnant masses of *in situ* rock. The periglacial location of the granitic uplands preserved long-weathered mantles of detritus, with tin minerals richly concentrated in the many valley terraces.

The Devon 'Redlands' border the region's eastern transition to the scarplands of southern England, their

red-brown Permo-Triassic bedrock giving that rich colour to soil and cliff, which is seen to abundance in the spectacular Dawlish coastal cuttings of the Great Western Railway, and along the coastal outcrops from Budleigh Salterton to Sidmouth (Fig. 32.2). Sandstone and cob-built farms and villages (cob is an adobe-like mix of clay and straw) are at the rich agricultural heart of a region of small fields, high fuschia hedges, winding sunken lanes and clotted cream teas.

To the north, between the Quantocks and Mendip along the sharply indented south-eastern margin of Bridgewater Bay, are the flood-prone Somerset Levels. Here, mid-Holocene coastal plain wetlands accumulated alluvium and peat on a substrate of soft, low-relief Triassic (Mercia Mudstone) and

Figure 32.2 The ' Redlands' coast of Devon. (50.659064, –3.279020). Photo: Shutterstock 78130198 ©L. Trott. Cliffs of Otter Sandstone; Triassic river-deposited sediments at Ladram Bay, Otterton.

Liassic mudrock at the outfall of the rivers Tone, Parrett and Brue. Over them to the east rise the 'whaleback' anticlinal folds of the Mendip massif (Blackdown, North Hill, Pen Hill and Beacon Hill). These asymmetric structures, their Carboniferous strata slightly overturned to the north along north-throwing thrust faults (see Chapter 6) were generated during Hercynian compression. Their limestone bedrock forms a cave-ridden karstic landscape with notable dry valleys like Cheddar Gorge. It is cut by dozens of fabulously rich lead/zinc veins mined since Roman times (Fig. 9.2). As in Glamorgan (Chapter 31) the Mendip Carboniferous inliers are ancient pre-Triassic hills, now ringed by vales and low scarps of unconformably overlying Triassic breccio-conglomerates and Liassic mudrocks and limestones. Ham Hill Limestone west of Yeovil has been quarried as exquisite building stone since Roman times (Fig. 9.3). Additionally, outliers of Middle Jurassic limestones are famously unconformable (with oysters and molluscan borings) on the Carboniferous Limestone east of Leigh-upon-Mendip. The Doulting Stone from near Shepton Mallet is a famous and gorgeous building stone of bioclastic limestone with bands of shelly detritus rich in crinoidal fragments ('sea lilies'). It was used in the masterpiece that is the frontage to Wells Cathedral, splendid in the Triassic lowlands below Mendip's lower slopes. On a much smaller scale but just as precious, it was used for Cheddar's elegant arched surround to the town's fifteenth-century Market Cross, recently expertly reconstructed with the same stone after a traffic accident.

In addition to Mendip, other folded Carboniferous inliers occur across Avon, including the now mined-out north Somerset coalfield (Fig. 3.1) once so important to Bristol. This west-facing port-city was south-west England's industrial and commercial centre, surpassing distant and coal-less Exeter and naval Plymouth (the first British slaving port) in Georgian times, though eventually eclipsed by Liverpool. There remains a strong maritime heritage in the region's many other sheltered ports and anchorages, mostly along the southern estuaries. Mesolithic adventurers and Neolithic farmer-traders arrived via these estuaries to change its landscapes for ever. Cultural (and presumably also kinship) links with

Armorica (also rich in monumental works) came to a final flourish during the Bronze Age. In the benign pre-Iron Age climate there was much deforestation; upland Dartmoor in particular became the centre of an upland agricultural nexus (Fig. 7.20). During the Iron Age the Cornish language developed from Brythonic roots, related to Breton and Galician. It comes as no surprise to learn of visits by Hellenistic explorers like Pytheas and the suspicion (never proved) that the islands known to ancient geographers as the 'Cassiterides' (off Brittany) in fact sourced their tin from Cornwall (Chapter 11).

The Romans established Isca (Exeter) as their major south-western legionary centre, settling widely in the fertile and climatically benign Durotriges tribal lands (modern East Devon, Dorset and Somerset). In the latter they began draining the Levels and mining Mendip ores (Fig. 9.2). By way of contrast only a single villa has been discovered in Cornwall (near Camborne, on the coastal platform), with marching forts only as far west as Okehampton. This, and milestones excavated at Tintagel, imply the Dumnonii of Cornwall were peaceful and friendly, as befitting international traders. Celtic Christianity flourished in 'The Age of the Saints', the chief monastic centre being St Petroc's, Bodmin. The 'Saint's Way' is an overland pilgrimage route between the Camel estuary and Fowey that avoided maritime passage around treacherous Land's End.

Into the lost Bronze Age landscapes of the western uplands gradually came the determined makers of a later-medieval and modern palimpsest: alluvial tinners, small upland smelt masters (using pollarded valley willow for fuel), mineral prospectors and mine captains. With them came packhorse routes, leats, shafts, pits, narrow-guage railways, multitudinous spoil tips and 'hushed' floodplains. Granite from the moors was used for everything: gate posts, field walls, buildings, stamping anvils, smelt furnaces, metal moulds. Today's landscapes feature surviving iconic engine houses, sturdily built of local granite, their architecture grand and timeless. They bear silent witness to the immense depths of the mine shafts sunk in the mid-nineteenth century. Coastal locations around St Just (Fig. 32.3) and St Agnes hint at the extent of subsea workings. Mining communities passed from boom to crash as individual mines shut; almost the whole mineral field had failed by the early twentieth century. Mining the deep ore became too expensive to compete with foreign competition; the mines were deep, hot, wet, insanitary (hookworm flourished in Dolcoath) and

Figure 32.3 Granite-built engine houses on the coastal platform at Botallack near St. Just, Land's End (50.134508, –5.686791). Photo: Shutterstock 282708971©Helen Hotson. Wheal Owles is in the foreground, the Crowns to the far left, the Count House in the far right and the metal structure in the distance is Allen's Shaft.

labour-intensive; furthermore their mine engines were dependent on expensive coastal importation of Welsh coal.

The southern Dartmoor and St Austell granites are notable for their huge volumes of completely kaolinized (Fig. 5.2) granite, extending hundreds of metres below the present land surface. The china-clay industry began with William Cookworthy, a pioneering Falmouth porcelain ceramicist. In 1775 Mr Wedgwood rushed down from Staffordshire (Chapter 28) to view the clay and buy his own pit. Today the quartz- and mica-rich spoil tips west of St Austell glare and glint stark-white against surrounding hillsides with their ancient field patterns and early Christian parishes (Fig. 32.4). Yet clay-digging was once an integral part of an ancient dual economy. Around the 1850s many of the 7000 or so employees were also long-term leaseholders of smallholdings: a few acres of farmland. They were paid by day-wage and could only be 'encouraged', not forced or threatened, to work regularly and efficiently in the pits.

It was into this depressed and hungry Cornish landscape of cliff and moor and its multilayer palimpsests of former working environments that St Ives grew from its fishing heritage into a major resort, after the railway arrived in 1877. The St. Ives School of Artists became internationally significant with the arrival of Bernard Leach in 1920. This artist-potter was interested in the revival of craft work but also saw pottery as a form of abstract art with the same

Figure 32.4 View from the east of the 15th-century St Michael's Chapel on the 20m high tor known as Roche Rock (50.407303, −4.833962C). Photo: Shutterstock 183146237 ©Helen Hotson. The igneous rock is part of a smaller intrusion (a dyke?) to the north of the main St Austell granite. It comprises quartz and disseminated crystals or clumps of prismatic black tourmaline (common in SW England, rare elsewhere); the assemblage known as 'schorl'. The china-clay pits of St Austell lie just to the south (left).

formal elements as sculpture and painting. In 1939, Barbara Hepworth, Ben Nicholson and Nuam Gabo moved here; their creative presence ensured that St Ives produced some of the most serious and exciting work of the mid-twentieth century. It addressed the technical and ethical issues inherent in modernism: not parochial yet often rooted in place. Thus Barbara Hepworth's sculptures were intended for outdoor display (Fig. 32.5) and whilst abstract they are also rooted in landscape. She says of Sculpture with Colour (1943) that it expresses 'the tension felt between myself and the sea, the wind or the hills.'

Figure 32.5 Part-view of '*Two Forms (Divided Circle)*' (1969) by Barbara Hepworth (1903–1975). Bronze. 237x234x54cm. Acc. No. BH 477 Barbara Hepworth Museum, St Ives ©Tate. Photo: Shutterstock 1065358 ©9548315445/ Shutterstock.com. The artist said that her sculptures were always about figures in landscape, embedded in it, and that there are strong parallels between them and megaliths.

Chapter 33

Southern England

Towards winter flowers, forms of ecstatic water,
Chalk lies dry with all its throats open.
Winter flowers last maybe one frost
Chalk drifts its heap through billions of slow
sea-years...

Alice Oswald, *Sonnet,* in *Woods etc.* (2005)

The star-filled seas are smooth to-night
From France to England strown;
Black towers above the Portland light
The felon-quarried stone.
On yonder island, not to rise,
Never to stir forth free,
Far from his folk a dead lad lies
That once was friends with me.

A.E. Housman, LIX, The Isle of Portland in *A Shropshire Lad* (1896)

This unglaciated region of Mesozoic lowlands, scarplands and low ridges (Fig. 33.1) has few summits over 300m and a dryish mild climate: little wonder that it is fertile and populous. Its long south-facing coastline is a straggle of cliffs and landslipped bays (Lyme, Kimmeridge) eroded from sands and clays. Longshore drift (eastwards in the Channel, westwards from the southern North Sea) feeds sediment along it to estuary mouths (Solent, Medway, Thames), coastal marshes (Romney, Pevensey), elongate shingle spits (Chesil) and sandy, beach-ridged plains (Dungeness). Intervening high-cliffed headlands from Portland Bill via Beachy Head to the Forelands are held up by resistant chalks, limestones and sandstones, prone to spectacular slippage when underlain by clay (as at Folkestone and Hastings).

The Mesozoic/early Cenozoic geological processes that produced the region's landscape (Chapter 7) culminated in active fold-thrust tectonics, just like southern Iran today. This produced the step-like monoclinal folds and thrusts of Dorset's Jurassic Coast; the scarp and dip slopes of the Cotswolds and Wessex chalk downlands; the soft dome of the Weald between the North and South Downs; and the broad sag of the London basin syncline between the gently tilted Chilterns and North Downs.

The Thames runs through it, from its Cotswold headwaters to the southern North Sea, its tributaries cutting through ancient breaches in the chalk ridges on the way, like the Goring, Guildford and Dorking Gaps. The long estuarine inlet is finally corralled below the sheltering North Downs, its wide and terraced valley witness to progressive incision through the Cenozoic sediments of the London Basin. The region's history reflects this spatial accessibility, rooted long before the founding of Roman Londinium.

Neolithic and early Bronze Age monumental landscapes (Avebury, Silbury, Stonehenge, Woodhenge) signify wealth, power and common purpose. The Iron Age newcomers cropped well-drained lowland river terraces with their sophisticated ploughshares made from Wealden iron. Tribal rulers in *oppida* like Dorchester, Wheathampstead, Colchester, Chichester and Silchester were in trading contact with the Roman world well before the Claudian invasion of AD 43. Of the thirty or so Iron Age forts in Dorset, multivallate Maiden Castle is as impressive in its chalky landscape as are stone-built Tiryns and Mykene of Mycenaean Greece; though there was no Homer (or any other writer) to glorify it.

To the west, the Cotswolds Middle Jurassic limestone scarp faces emphatically over Severn's wooded Liassic lowlands. It features Neolithic long barrows and Iron Age forts that exploited defensive positions. Around Bath and Stroud are deep valley bottoms of swift-flowing tributaries to the Avon, exploited during the Industrial Revolution with mill chimneys, canals and terraced cottages in the surrounding landscape. Bath's seven hills and steep-sided wooded valleys host its famous warm springs; in AD 60 the thermal-loving Romans built Aquae Sulis as an exquisite bathing complex. In the seventeenth century the curative properties of the waters were widely advertized and Bath became the greatest spa in the land, a fashionable meeting place for the well-off. John Woods and his son designed and built

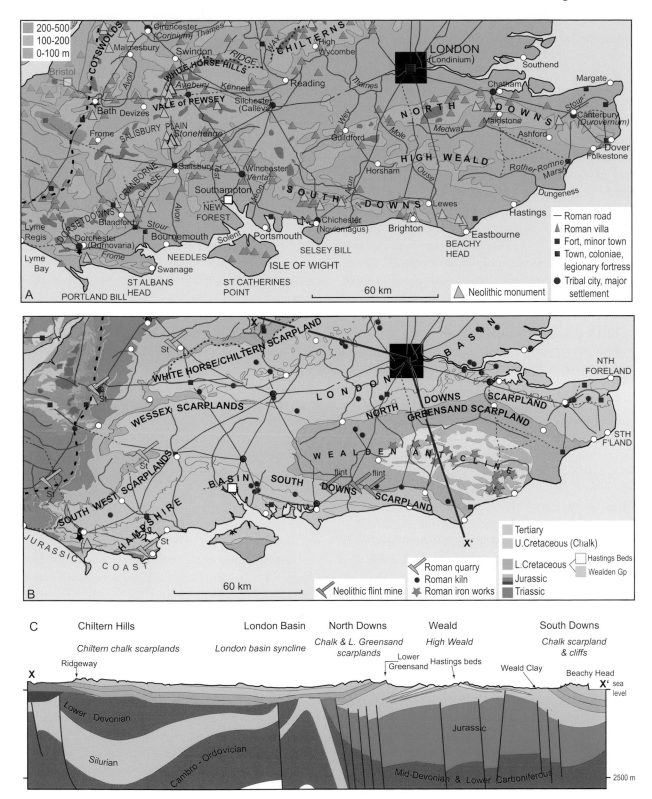

Figure 33.1 Maps and geological section of Southern England. Data generalized from British Geological Survey (2007a), Ordnance Survey (1991, 2011) and *The Times Atlas of the World* (1987) sources.

much of the town in cream-yellow weathering oolitic limestone from the Combe Down and Bathampton mines. The Royal Crescent, Circus, Assembly and Pump Rooms show off a pleasing mix of weathered stone and Georgian proportions. Above the town a broad plateau slopes gently eastwards with the stratal dip of the Jurassic limestones. To the north, wide valleys along the dip-slope are tributary to the upper Thames. Here the soil gets deeper and seventeenth- to eighteenth-century cloth mills at Dursley, Chalford and Painswick testify to the former supremacy of sheep and wool. The Cotswolds today are intensively arable within the 6000km of drystone walls built during eighteenth- and nineteenth-century enclosures. The long-weathered honey-coloured limestone unifies many Cotswold villages in an intimate connection with local geology, church, manor house, farm, barn, field walls and cottage all sharing the same stone.

Yet despite all this Jurassic splendour, the iconic landscape of the region as a whole is chalk downland, the site of monumental complexes and human-cut engravings of horses and giants on a thrilling scale (Fig. 33.2). Thin-soiled and free-draining, the scarps were carved naturally by periglacial erosion into concave coombes and ephemeral bournes, as in the several Winterbournes of Dorset and Wiltshire. They rise surely from the spring-lined lower slopes of clayland vales and basins. These transitional places were the favoured sites for Roman villas, mostly estate farms, which lie thickest on the ground in Hampshire, Dorset, Wiltshire and Sussex.

The downs are now mostly landscapes of large fields, mixed grassland and arable in the west and arable with scattered ancient woodlands (particularly over the Chilterns) in the east. The Normans introduced Forest Law, with still-wooded areas like Cranbourne Chase becoming royal hunting grounds. There is much parkland on the lower slopes bordering the London basin, with race-courses like Epsom on the downs. The valleys are generally more intimate, with sunken winding holloways, lanes and wooded coombes. Many valley villages grew up around the medieval trade in sheep, built along springlines and in the bournes where the permeable Chalk and Greensands overlie impermeable Gault Clay.

Acid heathlands characterize the spreads of Cenozoic sands and gravels at the gently tilted margins to both the Hampshire basin and the London basin in Hertfordshire and Essex. Formerly mixed-forest, they were cleared to heath, and in the last century much built-upon around the fringes of the metropolis. The heath/

woodland terrains of Epping Forest and the New Forest that we walk or ride on today remind us of the mixed landscapes we have lost. Yet they were partly human constructs: the creation of William I's New Forest in 1079 involved the abandonment of thirty-six parishes. Nearby Bournemouth was founded as a late-Georgian resort town on deserted coastal heathland, the Duke of Rutland noting, 'on this barren and uncultivated heath there was not a human to direct us.'

The High Weald is mostly Hastings Sandstone (Lower Cretaceous), the surrounding vales cut into older Weald Clay. Following Iron Age and Roman workings, the well-wooded Weald became the most important industrial (iron-producing) area in Britain in the sixteenth and seventeenth centuries (Chapter 11). A few kilometres inside the chalk rim of the North Downs is the prominent but discontinuous chert-capped Greenstone scarp, the Surrey Hills, that extends in a convex arc by Haslemere, Dorking, Sevenoaks and Maidstone (Fig. 33.3). Its summit at Leith Hill (294m) is SE England's highest point.

Kent's gentle landscape includes a scattering of brick oast houses for hop-drying, ancient oak-framed Wealden houses, and the gaiety of peg-tiled roofs. Over the millennia it has had the most varied juxtapositions of architectural styles, stones and cultures imposed on it. First was the Roman victory arch at Richborough, done originally in Italian marble. Next came Norman Canterbury, built with flashy imported Caen limestone. In the early twentieth century rose steel winding-towers and spoil tips at four of England's deepest (c.900m) coal mines: Tilmanstone, Betteshanger, Snowdown and Chislet. The chalk landscape at Folkestone now features the Channel Tunnel, incoming travellers emerging from its dark concrete-lined depths into sunlit birdsong.

Roman Londinium was a north-bank settlement at the first bridgeable Thames crossing point across from right-bank tidal marshland (modern Southwark). The site was about 12m above flood level on the loamy-soiled brickearth surface to the well-drained sands and gravels of the late-Pleistocene Taplow Terrace, lowest and youngest of the dozen or so river terraces formed by the downcutting Thames during the late-Quaternary. Initially, from AD 43–60/1, the wood-built settlement stretched about 500m eastwards from the left bank of the Walbrook tributary and the same distance westwards along the main road out towards the banks of the River Fleet. All that remains of the original settlement is a thin layer of black ash left after the Boudiccan revolt of

Figure 33.2 *The Vale of the White Horse* (*c*.1938) by Eric William Ravilious (1903–1942). Graphite and Watercolour on paper. 45.1x32.4cm. Acc. No. 5164 ©Tate, London, 2015. The Tate caption includes: '...the low viewpoint makes the image disconcerting. This emphasizes the mass of the hill and provides an unusual view of the White Horse cut into the chalk at Uffington in Berkshire...'

Figure 33.3 View in early autumn from Surrey's North Downs chalklands to the wooded Lower Greensand scarp that is prominent west of the Dorking Gap and eastwards over neighbouring Kent. Photo: Shutterstock 333957863 ©RubinowaDama. The wooded slopes in the centre include Leith Hill to the right and run east–west, the strata dipping gently towards the viewer.

AD 60/1. The historian Tacitus describes Londinium on the eve of destruction as being 'very full of businessmen and commerce'. Stylus writing tablets found in Walbrook landfill dated AD 62 (requesting 20 loads of provisions from St Albans) show the settlement was rapidly restarted after the rebellion (Tomlin 2016) to eventually become a Europe-wide centre of trade. By AD 200, in common with other towns threatened by Saxon depredations, it was comprehensively enclosed by a 5km-long, 6m-high wall constructed in Kentish Ragstone.

Christopher Wren was one of the commissioners who formulated new building regulations (e.g. no thatch allowed) after the Great Fire of 1666. He pioneered use of the tough and durable Portland Stone for the ecclesiastical and public buildings of London. The Upper Jurassic oolitic freestone was dug out by quarrymen and, in later years, by convict-labour in the royal quarries of the Isle of Portland (Fig. 4.10) to be shipped to stoneyards on the Thames frontage below the City. Within fifty years Neo-Classical Georgian brick terraces and squares began to spread out northwards and westwards from Covent Garden. The wealthy faced their houses with Portland Stone; the upper middle classes rendered their brick walls white to mimic it. Tasteful and socialistic iconoclasts of the Pre-Raphaelite movement like William Morris

later moved out altogether into their exquisite brick new-builds (Fig. 33.4) in the countryside.

Immigrants have always begun their new lives in London's central parishes, the successful then moving outwards. As each generation became more prosperous they in turn did the same. Today it is the opposite: London's inner housing is bought up as investment property, forcing indigenous folk further out to the borough fringes.

Figure 33.4 The Red House (1850) by Philip Webb (1831–1915) and William Morris (1834–1896) (51.455450, +0.130311). Photo: Wikimedia Commons ©Tony Hisgett. The Morris residence in Bexley Heath, Kent, has a Neo-Gothic exterior inspired by 13th-century medieval: fine, deep red-brown brickwork, baked roof-tiles and hipped dormers. The house faces north and was apparently notoriously cold during the Morris' residency. The interior is pure Arts and Crafts with murals by the Pre-Raphaelite Edward Burne-Jones.

Chapter 34

Eastern and Central English Scarplands

I used to know the landmarks on this route
The industries of Britain left and right.
Once I'd know exactly where we were
From the shapes of spoil heaps and from winding gear
Spinning their spokes and winching down a shift
Miles deep into this sealed and filled-in shaft...

Tony Harrison, *Cremation eclogue* in *Under the Clock* (2005)

Then across the marsh it comes,
the sound as of an endless
train in a distant cutting,
the god working his way back,
butting and shunting,
reclaiming his territory...

Kevin Crossley-Holland, *Dusk, Burnham Overy Staithe* (1972)

The geological map of eastern and central England (Fig. 34.1) has sinuous tracts of colour: resistant strata forming scarps, intervening vales sculpted from softer stuff. The North York Moors are the most emphatic landforms; vertiginous mid-Jurassic sandstones cap Roseberry Topping above Teeside (Fig. 34.2), with Round Hill on Urra Moor the region's highest peak at 454m. In contrast, Norfolk's 105m Beacon Hill on the morainic Cromer-Holt Ridge is the lowest 'highest point' of any British county. Numerous and populous settlements developed throughout the region along ancient communication routes through rich agricultural districts and at sites for extraction of mineral resources. The most populous, wealthy and ecclesiastical districts of the Domesday and Henrician reckonings are here: in Lincolnshire, Norfolk, Suffolk and Essex, all once in the old Danelaw. Lucrative trade through its east-facing ports with the Hanseatic League, chiefly wool and then worsted cloth, explains the region's dense scattering of medieval buildings. Earliest land communication was along scarp ridges: the Icknield Way to the Wash and the Roman route north to Lincoln and York (Chapter 8). Waterborne traffic was important along the Tees, Ouse, Trent, Great Ouse and Yare, their wide, shallow estuaries also providing easy access for invading longships. Later, Viking settlers traded at riverside sites in York, Norwich, Colchester and Cambridge. The Anglo-Saxon and Scandinavian cultures dominated till 1066, with numerous stone-built churches, the larger abbeys, monasteries and cathedrals of timber. These were abandoned (Elmham) or rebuilt (Canterbury) by stone-skilled Norman masons.

Working from old-to-young, generally west-to-east, from Wearside down to Nottingham, Permo-Triassic rocks rest unconformably on the Carboniferous of the Pennine fringes. They begin their scarp-length in Sunderland's suburbs, the Magnesian Limestone reef front at Tunstall Hills. Here you can collect *Phillipsia*, one of the last extant trilobite genera in Britain. Roman engineers first spotted the suitability of Tadcaster's Lower Magnesian Limestone for building stone and used it widely in York (Chapters 10, 13). Disused quarries in Durham and North Yorkshire are now nature reserves, hosting lime-loving plants that are rare elsewhere. Southwards the low Magnesian Limestone scarp separates concealed and exposed Coal Measures: the gentler east-facing dip slopes dotted with reclaimed pits and afforested spoil tips. The scarp is cut by prominent joints and minor faults whose enlargement by solution produced Creswell Crags and other Upper Palaeolithic cave sites in the region (Fig. 34.3). To the south, the sandy dolomitic limestones are quarried for building stone as 'Mansfield White' and (formerly) 'Mansfield Red'. Another low scarp appears to the east along the Triassic Sherwood Sandstone outcrop: a good aquifer here, but not so good for building stone as in the Midlands (Chapter 28).

Further east, a prominent vale in red-brown Triassic Mercia Marls and grey Lias claystone stretches from Teesside down the Vale of York to Nottingham. The Ouse and Trent rivers follow its N–S trend before entering the Humber estuary. Into this came Roman, Anglian and

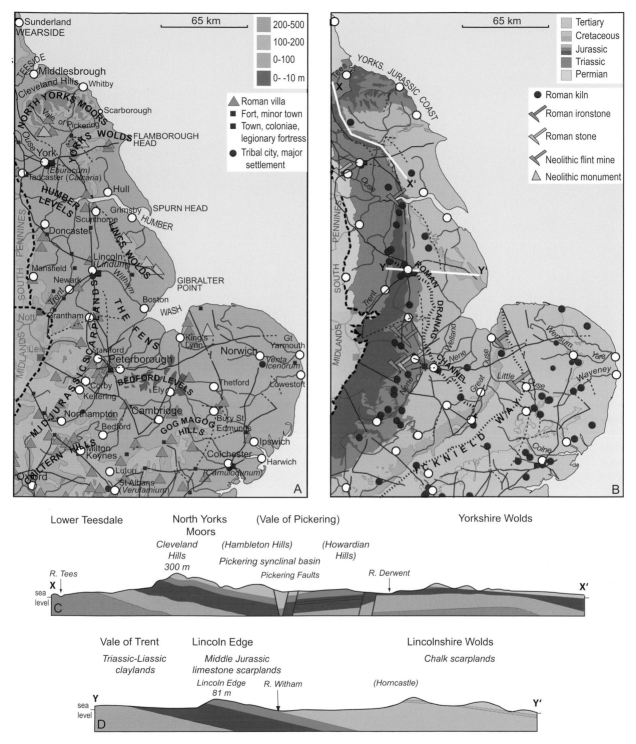

Figure 34.1 Maps and geological sections for the Eastern Scarplands. Data generalized from British Geological Survey (2007a), Ordnance Survey (1991, 2011) and *The Times Atlas of the World* (1987) sources.

Viking raiders, each in turn settling the fenland margins and rich terraced claylands, the former constructing the Fosse Dyke drainage system. A huge Danish Army overwintered by the lower Trent at Torksey in 872;

pottery finds indicate use of the potter's wheel, a craft lost since Roman times.

South-tilted Jurassic outcrops curve around the North York Moors, thick sandstones and ironstones

Figure 34.2 View east of Roseberry Topping, North Yorkshire (54.505485, −1.107493). Photo: Shutterstock 258590336 ©Michael Stubbs. This outlier of Middle Jurassic river-deposited sandstone stands upon a concentric plinth of Liassic Cleveland Ironstone and Whitby Mudrock. The ironstone was worked in a ring of bell-pits. At Hasty Bank, c.10km to the south on the main north-facing escarpment to the North York Moors, finer-grained Middle Jurassic strata of floodplain origin yield exquisite plant fossils.

Figure 34.3 The cross-stratified oolitic and shelly dolomitic limestone forming Creswell Crags (53.263491, −1.193529), Britain's premier Mesolithic cave site. The joint fractures enlarged to cave spaces by solution are clearly seen. The gated and locked entrances to Dog Hole and Pin Hole caves are to the left.

rising up from the Triassic marls of Teeside and the Vale of York. The high-cliffed coastal stretch from Redcar to Filey is a 60km-long Yorkshire 'Jurassic Coast', quite equal in both scenery and heritage to the one in Dorset. Jet occurs in the Lias as sporadic thin seams around Whitby, with alum previously sourced between Saltburn and Ravenscar. Several Norman monastic houses were founded in coastal and valley sites within or around the periphery of the heathery moorlands: Whitby itself, Rievaulx (where the monks smelted iron), Byland, Kirkham and Mount Grace. Their huge flocks of sheep grazed pasturelands beneath the rugged edges and tops. Well-cemented quartzose and iron-stained Middle Jurassic sandstones and Upper Jurassic (Corallian) limestones and calcareous sandstones provided robust building and paving materials. Local Aislaby sandstone makes up Whitby's pantiled and cliff-huddled cottages, the modern harbour's foundation blocks (Fig. 10.2) and the steep connecting cliff path to the abbey where Abbess Hild hosted the notable Synod of Whitby (AD 657). East of York the Jurassic scarps diminish southwards as the strata thin, and by Market Weighton near the Humber they are absent, the Chalk resting unconformably on the Lias.

Ironstone from the Cleveland Hills and north-west Lincolnshire (Frodingham Ironstone) led to rapid mid-nineteenth-century industrial expansion at Middlesbrough (once a single farm) and Scunthorpe. Railways linked Stockton, Darlington, the Durham coalfield (Chapter 26) and the shipyards of Wearside to Middlesbrough's sixty or more blast furnaces. Here Dorman-Long specialized in the manufacture of iron bridges: Middlesbrough's unique Transporter Bridge, Newcastle's Tyne Bridge, London's Lambeth Bridge and Sydney Harbour Bridge, amongst many others worldwide, proudly bearing the plaque 'Made in Middlesbrough'.

Across the Humber, from Lincoln Edge to Stamford, a vigorous mid-Jurassic limestone scarp reappears. Along it ran the Roman marching road to Lincoln and York from the junction fortress of Durobrivae at the crossing of the Nene. The modern A1 is a palimpsest on the Roman road and its successors, medieval Ermine Street and the Great North Road. The stone-built towns and villages signposted from it record historic limestone quarrying centres: Ancaster, Clipsham, Barnack (Fig. 34.4) and Ketton. The ultimate one-stone town is Stamford; like other former woollen towns on the route it had numerous coaching inns.

Figure 34.4 Barnack Church, Cambridgeshire (52.632585, –0.406617). Anglo-Saxon stonework in Barnack Rag limestone makes up the cream-white tower below the c.AD1200 spire and turreted belfry. A beautiful interlaced carving done out of a single piece of the limestone and intricate windows feature on the south face below the gilded clock. The irregular limestone blocks that form the stonework are rather crudely laid, but it is clear from the protruding and perfectly geometrical ashlar pilasters that they were originally meant to be covered with plaster (presumably also rendered white).

Southwards the limestone is replaced by Northampton Sands Ironstone, an ore smelted by the Romans around Kettering. From the 1850s, opencast extraction supported ironworks at Corby, Kettering and Wellingborough. Many of the workers recruited for Corby's integrated steel-making in the 1920s-30s came from Midland Scotland and Ireland – the mass unemployed of the last Great Depression. At Hornton (type locality of 'Green Hornton Stone', Chapter 14) and in the Banbury area generally, iron-rich sandstones are important traditional building materials, the dark green iron mineral, chamosite, weathering to pleasingly variegated mixes of ochreous hues.

To the east Upper Jurassic mudrocks, the Oxford and Kimmeridge Clays and the Lower Cretaceous Speeton Clay, form the bedrock to wide vales overlain by Quaternary glacial lake deposits and Holocene alluvium deposited by rivers like the Ancholme, Witham, Welland, Nene and Great Ouse. Most important are thick Anglian tills, decidedly chalky in the south and east. These are remnants in the landscape, local terraces bounded by erosional slopes and younger gravel terraces. The drained wetlands of the Humber Levels and the wider Lincolnshire Fens were once natural areas of ecological richness, dotted with 'islands' of habitation and settlement (Axeholme, Ely), many dating back to the Bronze Age.

In East Yorkshire and Lincolnshire the billowing low chalk hills of the Wolds (*wold* is Old English for a wooded upland, as in modern German, 'wald') run west from Flamborough Head. They define glacial Lake Pickering's southern shoreline, part of Star Carr's wider Mesolithic landscape (Chapter 8), the eastern limit to the Vale of York and the west-facing escarpment of the Lincolnshire Wolds. The original woodlands gave way to medieval sheep pasture, nowadays mostly large, thin-soiled arable fields, their chalk substrate showing patchily through after autumn ploughing. Along their base and on the lower slopes, settlements nucleated at spring lines. The hilltops have Neolithic, Bronze and Iron Age monuments, settlements and cemeteries. During the later Iron Age the area was inhabited by Belgic immigrants. Unique in Britain, these Parisii buried members of their elite (men and women) with dismantled chariot pieces, horse trappings and possessions, the metalwork in La Tène style.

Across the Wash, Chalk scarpland topography in west Norfolk and Suffolk is subdued by glacial erosion (Chapter 7), reappearing in the Gog Magog Hills and northern Chilterns. The remainder of East Anglia has subdued but definite relief, properly described as 'gently undulating' rather than Noel Coward's 'very flat'. The fabulous discoveries of extinct mammals from the Cromer Forest Bed (Chapter 7) led Jacquetta Hawkes to write in *A Land* (1951),

'I wish they might have their resurrection day, step from the cliffs and process through the town like circus beasts.' Her wish was realized 63 years later when Cromer town centre witnessed a procession that included a mock-up of the gigantic steppe elephant found there in the 1980s. The coinage of the Iron Age Iceni included notable animal images, the horses a testament to the pastoral and arable wealth of its varied and fertile soils. Roman Colchester was built on a terrace at the Colne's navigable limits; as at London (Chapter 33) and St Albans, a dark subsurface cinder layer marks the total destruction wrought by the Icenian revolt.

The richness of late-medieval East Anglia was founded on wool: rural spinning and hand-loom weaving in countless villages, towns and in the core city of wealthy and populous Norwich. The small group of weavers, 'thirty Douchmen of the Low Countreys of Flaunders …', invited by Elizabeth I in 1566 were followed by several generations of Protestant immigrants and refugees. By the mid-seventeenth century their descendants constituted a third of Norwich's population. These 'Strangers' introduced new cloths, contributing to the popularity of 'Norwich Stuffs' from Restoration to late-Georgian times. The industry crashed as worsted manufacturing moved to the mechanized looms of the West Riding; East Anglia had no swift-flowing water or local coal to compete.

The legacy of links with the Low Countries influenced regional culture and landscape directly – drained wetlands (Fig. 34.5), Dutch gables and the 'Norwich School' of landscape painting (Chapter 15). Proximity to industrial north Germany finally led to East Anglia's flattish interfluves being the premier sites for scores of Second World War American bomber airfields. Today, the coastline lies at the mercy of raw nature rather than enemies; equinoxial storm-surge is an increasing threat (see Kevin Crossley-Holland's verse in our chapter motto).

To end our account on a bibulous and optimistic note, our warming climate is emulating medieval conditions (Fig. 34.6), for flavoursome dry white wines are now harvested in scores of vineyards from the Chilterns' Essex chalklands to the depths of chalk-clayed Norfolk.

Figure 34.5 Holme Fen, Cambridgeshire (52.492541, −0.232709) contains approximately 5 hectares of rare acid grassland/ heath and a hectare of raised bog. Amongst silver birch woodland, Joy Lawlor is viewing the companion marker post to the original that was sunk to ground level in the mid-19th century. Compaction has been around 4.5m since the area was drained; as a consequence the area has the lowest elevation in Britain, some 2.75m below sea level.

Figure 34.6 *September: The Labours of the Months*. (*c*.1470) 23cm diameter. Victoria and Albert Museum, London ©Victoria and Albert Museum. A late-medieval stained glass roundel of a youth harvesting grapes. Probably made in Norwich and possibly from the old parsonage at St Michael-at-Coslany in that city; the splendid Perpendicular church itself survives.

Glossary

Italics indicate a word also defined in a separate entry

A

Acadian *Mountain-building* and deformation in northern Britain during the mid-*Devonian*

Accretionary prism Off-scraped sediment wedge accreted by *plate subduction*

Acidic (felsic) Light-coloured *igneous* rocks like *granite* and *rhyolite* with >65% *silica* and common *quartz*

Adit Tunnel driven into a hillside to intersect ore-bearing *veins* or *coal*

Alabaster Finely crystalline *gypsum*

Alluvial Pertaining to rivers

Alpine *Mountain-building* by destruction of the *Tethys* ocean

Ammonite Extinct cephalopod related to modern Nautilus

Amphibole Hydrous alumino-silicate mineral with linked double-chain structure

Andesite Mid-grey/brownish finely crystalline *intermediate lava*

Anglo-Saxons Fifth- to tenth-century Germanic immigrants in variable hegemony from Northumbria to Wessex

Anhydrite Calcium sulphate, an *evaporite* salt

Anthracite Bright-burning ashless *coal* with high carbon content (>92%) and little water

Anticline Stratal up-bend; fold with crest and two outward-tilted limbs

Aragonite Rarer mineral form (polymorph) of calcium carbonate; see *calcite*

Arcadian References an idyllic Greek past with people (shepherds especially) living peacefully and in harmony with nature

Archean Eon of time 4000-2500Ma

Arête (French) Narrow-crested ridge produced by adjacent valley glaciers

Art Nouveau 1890s-early-twentieth-century applied art and architectural style; chiefly with long sinuous lines and patterns

Arts and Crafts Movement 1880s-1920s anti-industrial movement inspired by nature, folk art, traditional crafts; particularly in design

Ash Fine *volcanic* deposit sedimented from eruptive plumes or as hot, ground-hugging *ash flow*

Ash flow (aka nueé ardente) Hot, ground-hugging, turbulent flow of *ash*

Ashlar A smoothly masoned *freestone* block

Asthenosphere Upper *mantle* in convection-driven flow upon which the *lithosphere* moves

Atmospheric perspective Determines depth in painting by tone-change so distant objects appear to dissolve in light and atmosphere

Augite Iron-rich dark *pyroxene* mineral; common basic *igneous* rock constituent

Avalonia Late-Precambrian to Lower Palaeozoic Anglo-Welsh crustal terrane

Azurite Blue mineral of hydrous copper carbonate

Back-arc Located inboard a *subducting plate* or *volcanic arc*

B

Background Part of a painting that appears furthest away

Barite (aka barytes) Dense flesh-coloured mineral of barium sulphate

Baroque Late-sixteenth to eighteenth century. Emotional response to the Reformation (also promoted by the Catholic Church) for movement, tension, drama; in architecture for power, opulence, control

Barrel vault Semi-cylindrical vault unbroken by cross vaults.

Basalt Dark finely-crystalline *basic (mafic) lava*

Basaltic Of approximately *basalt* composition

Basic (mafic) igneous rock Dark-coloured *gabbro, dolerite, basalt* with 45-52 % silica comprising *olivine, pyroxene, feldspar*

Basin Area of land or sea floor *subsidence* that slowly accumulates deposited *sediment*

Beaker Folk *Bronze Age* immigrants distantly descended from the Yamnaya peoples of the Pontic steppes

Bed Layer of *sedimentary rock* bounded by *bedding planes*

Bedding plane Surfaces in sedimentary rock marking depositional pauses or contrasting *sediment* influxes

Bedding The presence of *sedimentary* beds

Bedrock Indurated outcrop or subcrop below soil, alluvium or glacial deposits

Belemnite Extinct cephalopods related to modern squid

Bioclastic *Limestone* comprising *calcitic* fossil shell fragments

Biotite Dark iron-rich *mica*

Bluestone Spotted-textured *metamorphosed dolerite* from the Preseli Mountains of Dyfed

Bølling-Allerød Warm end-*Pleistocene* climatic phase, *c.*15-13ka

Boulder clay An older name for glacial *till*

Brachiopod Largely extinct marine two-valved, bilaterally symmetrical creatures attached to a substrate

Breccia Coarse-grained *sedimentary rock* or *fault-rock* comprising angular fragments

Bronze Age Division of archaeological time; in Britain *c.*4-2.8ka

Brutalist architecture 1950s–1970s descendant of modernism with a sense of massiveness and vistas of exposed concrete

Brythonic Pertaining to the Iron Age language of Britain (not Ireland) spoken from the Midland Valley south to Cornwall, including Wales, its chief remaining enclave

Byzantine Pertaining to the religious and didactic expression of Eastern Orthodox art c. 330-1450, colourful, with flat and stiff figures, lack of realism, golden or ornate backgrounds and no sense of perspective

C

Calcite Common light-coloured calcium carbonate mineral, see *aragonite*

Calcrete Calcareous lower soil horizon formed in semi-arid climates

Caldera Large circular volcanic depression within *ring-faults* over a *magma chamber*

Caledonian Lower Palaeozoic *mountain-building* episode associated with closure of the *Iapetus* ocean

Cambrian Period of time 541-485Ma

Carboniferous Period of time 359-299 Ma

Carse(land) Drained and reclaimed former estuary-margin wetland

Cassiterite Tin oxide mineral

Celestite Strontium sulphate mineral

Celtic Catch-all word referring to the culture of *Iron Age* people in Europe

Cement (hydraulic, Portland) Made by calcining *limestone* mixed with clay

Cementstone A clayey fine-textured *limestone* for calcining to *hydraulic cement*

Cenozoic Era of time from the *Cambrian* to *Permian* periods: 66Ma-present

Chalcedony Amorphous and hydrous silica, commonly as flint in chalk

Chalcopyrite Copper iron sulphide mineral

Chalk Fine-grained *limestone* comprising calcareous *coccolith* plankton

Chamosite Mineral form of sheet-like iron silicate

Chert Light-coloured finely crystalline *silica*-rich rock

Chlorite Soft greenish sheet-like iron-rich silicate mineral, commonly metamorphic

Classic Greek art fifth century BC

Clastic Detrital sediment

Clay (aka mud) The finest grade of *clastic sediment*, grains 0.0039–0.0625mm diameter

Clay mineral Aluminosilicate minerals with a sheet-like lattice

Cleavage (mineral) Splitting plane(s) along atomic weakness in a *mineral*

Cleavage (rock) Splitting plane parallel to orientated flaky silicate minerals in *folded rocks*

Coal Carbonaceous accumulation of compacted vegetation

Coccolith Calcareous parts of planktonic algae

Conceptual Art Where the idea takes precedence; tradition, aesthetics, the artist's role all questioned

Cone sheet Inverted cone-shaped, strongly curvilinear *dykes* cutting older *igneous intrusive plutonic* rocks

Conglomerate Coarse-grained *sedimentary rock* comprising boulders to granules > 2mm diameter

Contact metamorphism *Metamorphism* by heat conduction in proximity to *intruded magma* or *lava* flows

Continental collision Contiguous continents juxtaposed along faults after plate *subduction*

Continental drift English for '*Kontinente Verschiebung'*, more correctly 'continental displacement'

Cooling-joints Polygonal columnar joints formed by contraction during cooling of intrusive magma, lava or ash flow

Corrie (Scots) Scoop-shaped excavation by the former head of a valley glacier; see also 'cwm' (Cymru)

Country rock Ambient rock which *magma* intrudes

Cretaceous Period of time 145-66Ma

Crinoid Fossil echinoderm (sea lily)

Cross bedding Inclined *stratification* by deposition on the leeside slopes of ripples, dunes, or channel bars

Crust Outer solid part of the *lithosphere* down to the *Moho*

Crustal shortening Thickening due to *thrust faulting* that forms mountain ranges

Crystal form 3-D atomic lattice arrangement in *minerals*

Cwm (Cymru) Scoop-shaped excavation by the former head of a valley glacier

D

Dacite Light-coloured finely crystalline intermediate *lava*

Dalradian Late-*Proterozoic* to Lower *Palaeozoic* rocks in Grampian and Argyll

Deltaic Pertaining to deltas

Devonian Period of time 419-359Ma

Diorite Darkish coarsely crystalline *intermediate plutonic igneous rock* comprising *feldspar* and *pyroxene*

Dip The maximum angle and direction of *tilt* of *strata*

Dolomite Double magnesium and calcium carbonate mineral

Drumlin Ovoid mound of detritus moulded by subglacial ice flow

Dyke Linear vertical to sub-vertical intrusive sheet of *igneous rock*

E

Echinoid Fossil echinoderms (sea urchins)

Enlightenment Eighteenth-century (1715-1789) secular and republican philosophical movement emphasizing order and reason

Eocene Epoch of time 34-56Ma

Epidote Green-yellow hydrous aluminosilicate mineral; an alteration product from other minerals

Eruptive Subaerial ejection of *lava* and/or volcanic *ash*

Esker Ridges of glacial detritus deposited in subglacial meltwater tunnels

Evaporite Mineral salt precipitated from brine by evaporation

Expressionism 1900-1930s movement that set out to deliberately distort appearances in order to express emotion

Extension Stretching of the *lithosphere* by attenuation and normal faulting; forms *rifts*

F

Facing Younging direction of *folds*; always upward unless overturned, then downward

Fault plane Plane of displacement along a *fault*

Fault 3D sliding-displacement along a fractured slip-plane in rock

Fault footwall The lower side of a faulted rock mass below the fault plane

Fault hangingwall The upper side of a faulted rock mass above the fault plane

Fault line A *fault* traced by juxtaposed rocks across a hidden *fault plane*

Fault rock Broken-up *breccia*, recrystallized (*mylonite*) or cemented rock along a *fault plane*

Faunal province A region containing distinctive life or fossils from coeval *strata*

Feldspar Light-coloured mineral form of mixed calcium, sodium and potassium aluminosilicates

Felsic Acidic igneous rocks rich in silica (>62%)

Felsitic Older term for *Lava* or *dyke/sill* rock rich in *feldspar*, often as visible *phenocrysts*

Fissility The rock property of easy splitting along sedimentary layers or *slaty cleavage*

Flagstone *Sandstone* splittable into centimetric-thick slabs

Flint White-rinded, grey-black lustrous and amorphous form of *chalcedony* as lumps in chalk

Fluorite (aka fluorspar) Calcium fluoride mineral form

Fold, folding Up- or down-folds; undulose stratal forms produced by shortening under compression

Form An object's appearance, its shape regardless of colour or material

Freestone Stone with no significant internal splitting planes; massive, structureless

G

Gabbro Dark, coarsely crystalline *basic igneous plutonic rock* comprising *olivine, pyroxene, feldspar*

Gaels Q-Celtic (Irish Gaelic) speaking inhabitants of Dál Riata (greater Argyll) since at least the fifth century, subsequently throughout Highland Scotland; extant in Scotland and Ireland

Galena Lead sulphide mineral

Geological map (drift) Geological map of all surface *outcrops* and all Quaternary sediment

Geological map (solid) Geological map that ignores Quaternary sediment and extrapolates solid *bedrock* between observed *outcrops*

Geological map Topographic map with projection of mapped distribution of geological features

Geological section A vertical section drawn down through part of the upper *crust* to show the disposition of *outcropping* and buried rocks as inferred from field mapping and extrapolation

Georgian Period 1714-1837

Glacial erratic Often far-travelled *rock* or *mineral* brought by glacier ice

Glacial outwash *Sediment* deposited by glacial meltwater at a glacier front

Gneiss (pron. 'nice') Strongly light- and dark-banded *metamorphic* rock formed by *regional metamorphism* at mid-crustal temperature and pressure

Goethite Hydrated iron (ferric) oxide (oxyhydroxide); source of brown *ochre* pigment

Gondwana Palaeozoic and Mesozoic mostly southern hemisphere supercontinent finally split up by Atlantic and Indian oceans opening

Gothic architecture Early is 1189-1307, bulky, heavy, becoming taller, lighter. Decorated is 1307-1327, lighter, taller, with larger windows and much decoration. Perpendicular is mid-fourteenth-early-sixteenth century with fan vaulting and window decoration emphasizing the verticals. Tudor is 1509-1603 with much new domestic building using Renaissance features and brick

Graded bedding Variation of sediment size upwards through a *bed*

Grainstone *Limestone* whose larger constituents are packed against each other, the interstices occupied by younger crystal cementing crystals

Grampian *Ordovician mountain-building* episode along the

Iapetus ocean margin in Scotland

Granite Light-coloured, coarsely crystalline *acidic igneous rock* comprising *quartz, feldspar* and *mica*

Granitic Of broadly granite composition

Granodiorite Light-coloured coarsely-crystalline *intermediate igneous rock*, compositionally between *granite* and *diorite*

Graptolite Extinct Lower *Palaeozoic* chitinous planktonic marine creatures

Greenstone Hot water-altered *volcanic* rocks coloured with greenish *epidote*

Greywacke Sandstone with interstitial mud

Gritstone Granule-grade (2–4mm diameter) sandstone with scattered pebbles

Growth fault *Fault* that actively displaces *strata* while deposition takes place

Gypsum *Evaporite* salt of hydrated calcium sulphate

H

Hadean Eon of time 4600-4000Ma

Haematite Iron oxide mineral, powders red, source of red ochre

Halite *Evaporite* salt of sodium chloride

Hercynian Late-*Carboniferous mountain-building* episode associated with closure of the *Rheic Ocean*

Highstand Advent of highest stable post-glacial sea level

Holocene Epoch of time from 11.7 Ka to present

Hornblende Commonest *amphibole* mineral; prismatic dark crystals, often as phenocrysts

Hushing Mineral exploration technique whereby ponded reservoirs release their water suddenly downslope

I

Iapetus ocean Former late-*Precambrian* to Lower *Palaeozoic* ocean between Scotland and northern England

Ice wedges Detritus-filled ground cracks formed by expansion–contraction produced by seasonal freeze–thaw of ice wedges around tundra margins

Igneous rock Formed by crystallization from *magma*

Ignimbrite Deposit of hot, ground-hugging turbulent *ash flows*

Impressionism 1870s-80s Parisian art movement using new techniques to rapidly notate changing light, colour and movement

Inlier A mass of older *strata* surrounded by younger; often due to erosion of an *anticlinal fold* or of an *unconformable* cover of younger strata

Interfluve The valley slopes and summit between adjacent drainage catchments.

Intermediate igneous rock Mid-grey with 55-65% silica including *granodiorite, diorite, tonalite, dacite, andesite*

Intrusion Body of *rock* formed by *magma* that injected into

country rock

Intrusive Formed by magma injecting *country rock*

Iron Age Division of archaeological time; in Britain 2.8–2ka, elsewhere from earlier

Ironstone *Sedimentary rock* comprising iron-bearing minerals like *chamosite, haematite, limonite* and *goethite*

Island arc *Volcanic arc* formed above the melting zone caused by *plate subduction* under adjacent *ocean plate* or continental margin

J

Jadeite Microcrystalline *pyroxene* mineral, pleasingly mottled greenish

Joint More-or-less planar or gently curved dilational fracture in a *rock* mass

Jurassic Period of time 201-145Ma

K

Kame Mounded terrace-like glacial sediment transported onto or adjacent to a valley glacier

Karst Permeable limestone landscape lacking runoff; often bare or thin-soiled crags and 'pavements' riddled with solution features: enlarged *joint* fractures (clints, grikes), solution pipes and subterranean cave systems

Kaolinite (aka 'China Clay') Mineral form of aluminium silicate; a *clay mineral* formed from *feldspar* breakdown

L

Lacustrine Pertaining to lakes

Lamination Sedimentary texture comprising millimetric lenticular to planar laminae of slightly contrasting grain size

Land Art Emerged in the 1960s in reaction to consumerism; rejects gallery/museum space for landscape and left to weather and disintegrate; responds to landscape using natural materials

Laurentian (aka Laurussian) Pertaining to the *Precambrian* oldlands and their *Palaeozoic* cover at the heart of the North American continent

Lava Flow of *magma* once out of a volcanic *vent* down a volcano's flanks

Leat Channel cut around contours to take water from underground mine workings or natural springs into ore-crushing and sorting plants

Level Tunnel at a certain depth driven outwards from a central shaft to intersect and work ore-bearing *veins*

Lewisian *Archean* to early *Proterozoic* gneiss, minor *granites* and *metamorphosed dolerite dykes* outcropping in Assynt and the Outer Hebrides

Limestone A *calcareous sedimentary rock* made up of *calcite* particles *lithified* by *calcite* or *dolomite*

Limonite Mixed iron-mineral form comprising brown-to-yellow amorphous hydrated ferric oxides and hydroxides

Quoin Large dressed stone placed at the corner of buildings for ornament and/or reinforcement

R

Radiometric dating The determination of the age of any radioactive mineral or substance

Raised beach Beach subsequently raised by uplift of the land

Reaves Low drystone former *Bronze Age* field boundaries in the Dartmoor uplands

Regional metamorphism Widespread *metamorphism* during crustal thickening by *mountain-building*

Renaissance Mid-medieval (beginning late-twelfth century) 'rebirth' and rediscovery of Classical humanistic philosophy, the arts and science fuelled by liberal attitudes towards records of past scholarship in libraries from Egypt to Spain, but particularly in Florence and Northern Italy in particular

Reverse fault Fault where rock is displaced up the tilted *fault plane*

Rheic ocean Former Upper *Palaeozoic* ocean along the line of the *Hercynian* (Variscan) mountains

Rhyolite Finely crystalline or sometimes glassy light-coloured *acidic lava*

Rib vaulting Vault using framework of slender diagonal stone ribs, spaces between filled with lighter stone

Rift Linear area of *subsidence* bounded on one or both sides by active *normal faults*

Rift; failed Continental rift that ceased functioning before evolving to a nascent oceanic rift

Ring dyke Intrusion of magma as dykes around *ring-faults*, exposed by erosion in inverted 'stacked-bucket' form

Ring fault Arcuate *normal faults* bounding a volcanic *caldera*

Rock Solid natural aggregate of *minerals*

Rococo Eighteenth-century French reaction against Baroque; an ornate/florid style, exuberant natural curves in architecture and other applied arts; artwork light, graceful, elegant

Romanticism Late-eighteenth to twentieth-century reaction against the rational Enlightenment; emphasizes individual imagination, emotion, sublime landscapes

S

Saline giant Areally extensive and thick *evaporite* deposit

Sandstone A sand-grade *sedimentary rock*, grains 0.065–2mm diameter

Sarsen Hard siliceous stone found in *chalk* downland; formed as a *silcrete*

Schist *Mica*-rich irregularly cleaved *metamorphic rock*

Sediment Detrital, skeletal or precipitated grains of any size, composition or source

Sedimentary rock Formed of sedimented detrital, skeletal or precipitated grains

Sedimentary structures Features formed by currents of water or wind, raindrop imprints, desiccation cracks, etc.

Serpentinite Hydrated magnesium silicate formed by *metamorphism* of *olivine* in an *ultrabasic* host

Shape An object's outline shape, regardless of form, texture and colour

Shelly Comprising mostly calcitic fossil shells

Silcrete Siliceous lower soil horizon developed in a warm, humid climate

Silica The mineral form of silicon dioxide, as crystalline quartz, flint (amorphous *chalcedony*) or chert

Sill Sheet-like slab of *igneous rock* intruded parallel to *stratification* of *country rock*

Siltstone A silt-grade *sedimentary rock*, grains 0.0039–0.0625mm diameter

Silurian Period of time 443-419Ma

Slate Strongly compressed *mudrock* with a pronounced rock (slaty) cleavage

Snowball Earth Theory attributing late-*Precambrian* glacial deposits to global glaciation

Sough *Adit* to drain a planned working-*level* in a mine

Sphalerite Mineral form of zinc sulphide

Stope Part of an ore-bearing *vein* extracted by digging or blasting

Storm surge Phenomenon of abnormal equinoxial tidal flooding around an intense cyclonic system

Strata *Beds*

Stratification The presence of *bedding*

Stratigraphy Study of the succession of *strata* through geological time

Stratum A *bed*

Stream tin *Alluvial* deposits in SW England containing detrital *cassiterite*

Strike The sense of orientation of a *tilted bed* in space measured normal to the tilt (dip)

Subduction Movement of *plate* down into the *asthenosphere* and lower *mantle*

Sublime Mid-eighteenth-century relation to a scene that creates a sense of awe, or terror and which 'stuns the soul' with a powerful sense of the unknowable

Submarine slide Subaqueous sediment flow as a coherent mass downslope

Surrealism Early-mid-twentieth-century movement influenced by Freudian psychoanalysis, dreams, the subconscious; specializes in surprising juxtapositions of subject matter

Syenite Coarsely crystalline pinkish *alkaline igneous rock*; mostly *feldspar*

Symbolism Late-nineteenth to twentieth century. Reaction to past objective naturalism by symbolic representation (often ambiguous), especially Art Nouveau

Syncline Stratal down-bend of; fold with trough and two

inward-tilted limbs

T

Tectonic Action of plate-induced stress to fracture and/or bend rock masses

Tenebrism Dramatic painting style where darkness predominates, with details picked out by highlights

Terrane (displaced) An area of crust distinct from a contiguous terrane, implying distinct provenance

Tethys ocean Former *Mesozoic* ocean embaying central *Pangea*

Thermal sagging Broad *basin*-shaped subsidence of the earth's surface consequent upon conductional cooling of the *asthenosphere* after termination of *lithospheric* stretching

Thrust fault A reverse *fault* with a low (<10 degrees) tilt

Till Ill-sorted mix of clay, stones and sand deposited by glacier ice

Tillite Rock deposited as till

Tonalite Coarsely crystalline mid-grey *intermediate igneous rock*

Torridonian Late-*Proterozoic sedimentary rock* sequence in Assynt

Tourmaline Hard, brittle, prismatic boron silicate mineral; commonly as black 'schorl'

Trachyte Finely crystalline *alkaline lava*

Transverse fault Fracture where rock is displaced along the *strike* of a *fault plane*

Trap country Volcanic landscapes with step-and-bench topography sculpted from *lava flows*

Travertine Deposit of calcium carbonate precipitated at an emerging hot spring

Triassic Period of time 253-201 Ma

Trilobite Extinct far-seeing marine arthropod creatures

Tsunami Fast-travelling surface wave generated by sea-bed deformation after earthquake activity

Tufa Deposit of calcium carbonate precipitated at an emerging cool spring

Tuff Indurated volcanic ash

Turbidite Deposit of a turbidity current

Turbidity Current Subaqueous sediment-laden bottom flow

Tympanum Vertical recessed face above a door arch, often sculpted upon

U

Ultrabasic (ultramafic) igneous rock Has <45% silica content; *peridotite*

Unconformity Relationship whereby older strata were tilted or folded and then eroded before deposition of younger strata

Uniformitarianism Term for the philosophy of 'present causes' to explain the rock record

Upside-down strata *Strata* inverted by *folding*

V

Vault Arched roof or ceiling constructed using masonry

Veins Vertical to sub-vertical ore-seams

Victorian Gothic Period 1837-1901

Vikings Eighth- to tenth-century Scandinavian pirates who first plundered and then colonized large parts of Europe (England, Scotland, Ireland, Normandy, Sicily)

Volcanic glass Rapidly chilled material at the margins of a lava flow

Volcanic neck/pipe Former tube-like conduit for eruptive volcanic materials

Volcanic Pertaining to volcanoes

Volcanic plug Intruded neck of an extinct or resting volcano

W

Wackestone Muddy limestone whose larger constituents are entirely supported by lime-mudstone

Weathering Physical and chemical alteration of rock and mineral by atmospheric exposure

Welded tuff Compacted and indurated lower parts of *ash flow tuffs/ignimbrites*

Whinstone Scots vernacular for dolerite, in which whin refers to gorse, i.e. a stony outcrop on which gorse grows prolifically

Y

Younger Dryas Last cold phase of the *Pleistocene* epoch, *c*.13-11.7ka

Bibliography and Further Reading

Preface and Foreword

Selected Bibliography

Anderton, R., Bridges, P.H., Leeder, M.R. and Selwood, B.W. (1979) *A Dynamic Stratigraphy of the British Isles*. London: George Allen and Unwin

Auden, W.H. (1966) *About the House*. London: Faber and Faber

Clarkson, E. & Upton, B. *Death of an Ocean: A Geological Borders Ballad*. (2010) Edinburgh: Dunedin

Drabble, M. (2009) *A Writer's Britain*. London: Thames and Hudson

Eliot, G. (1871) *Middlemarch*. (Reprinted 1999, Penguin Books)

Fortey, R.A. (2010) *The Hidden Landscape*. London: Bodley Head

Hawkes, J. *A Land*. (1951) London: The Cresset Press. (Reprinted 2012, Collins)

Hoskins, W.G. (1955) *The Making of the English Landscape*. London: Hodder and Stoughton. (Reprinted 2012, Little Toller Books)

MacCaig, N. (1993) Collected Poems. London: Chatto and Windus

Patrides, C.A. (ed.) (1977) *Sir Thomas Browne: The Major Works*. London: Penguin Classics

Schama, S. (1995) *Landscape and Memory*. London: Random House

Trueman, A.E. (1949) *Geology and Scenery in England and Wales*. Harmondsworth: Pelican Books

Further Reading: Advanced

Brenchley, P.J. and Rawson, P.F. (eds) (2005) *The Geology of England and Wales*. London: The Geological Society

Trewin, N.H. (ed.) (2002) *The Geology of Scotland*. London: The Geological Society

Woodcock, N. and Strachan, R. (eds) (2000) *Geological History of Britain and Ireland*. Oxford: Blackwell Science

Further Reading: General/Introductory

Gillen, C. (2013) *Geology and Landscapes of Scotland*. Edinburgh: Dunedin Academic Press

Harvey, A. (2012) *Introducing Geomorphology: A guide to Landforms and Processes*. Edinburgh: Dunedin Academic Press

Hunter, A. (2001) *The Geological History of the British Isles: SXR60 Course*. Milton Keynes: Open University

Park, R.G. (2012) *Introducing Geology: A Guide to the World of Rocks*. Edinburgh: Dunedin Academic Press

Toghill, P. (2006) *The Geology of Britain: an Introduction*. Marlborough: Airlife Press

Part 1

Selected Bibliography

Agricola, G. (1556) *De Re Metallica*. Frieburg. (Transl. H.C. and L.H. Hoover, New York, 1950)

Bahn, P. and Pettitt, P (2009) *Britain's Oldest Art: The Ice Age cave art of Creswell Crags*. Swindon: English Heritage

Bailey, E. (1952) *Geological Survey of Great Britain*. London: Thomas Murby

Bevan, A. and Murray, B. (eds) (2005) *The Collected Poems of George Mackay Brown*. London: James Murray

Bowden, M., Soutar, S., Field, D., Barber, M. (2015) *The Stonehenge Landscape: Analysing the Stonehenge World Heritage Site*. Swindon: Historic England

Brown, G. Mackay (2005) *The Solstice Stone*. In: A. Bevan and B. Murray (eds)

Byatt, A.S. (2005) *The Stone Woman*. In: *Little Black Book of Stories*. London: Chatto

Cullum, J. (1784) An Account of a Remarkable Frost on the 23d of June, 1783. In a Letter from the Rev. Sir John Cullum, Bart. F.R.S. and S. A. to Sir Joseph Banks, Bart. P.R.S. *Philosophical Transactions of the Royal Society of London*. 1784-01-01. **74**, 416–8

Darwin, C. (1859) *Origin of Species by Natural Selection*. London: J. Murray

Dryden, J. (1660) *Astraea Redux*. In: J. Kinsley (ed.) (1970)

Ebbatson, R. (2014) *Landscape and Literature 1830–1914*. London: Palgrave MacMillan

Fagan, B. (2000) *The Little Ice Age*. New York: Basic Books

Forman, M.B. (ed.) (1935) *The Letters of John Keats*. Oxford: Oxford University Press

Fuller, J.G.C.M. (2007) Smith's other debt: John Strachey, William Smith and the strata of England 1719–1801. *Geoscientist* **17**/7, 5–12

Gibbons, S. (1932) *Cold Comfort Farm*. London: Longmans

Godwin, F. and Hughes, T. (1979) *Remains of Elmet*. London: Faber and Faber

Goudie, L. (2016) *The Story of Scottish Art*. BBC 4 Series

Holmes, K. (1999) *Katharine Holmes* Exhibition Catalogue. University Gallery: Leeds

Hutchinson, T. (ed.) (1969) *Wordsworth Poetical Works*. Oxford: Oxford University Press

Hutton, J. (1795) *Theory of the Earth with Proofs and Illustrations*. Edinburgh

Hutton, J. *Theory of the Earth*. (1788) Transactions of the Royal Society of Edinburgh **1**, 209–304 (Facsimile reprint Kessinger Publishing, **www.kessinger.net**)

International Commission on Stratigraphy (2012) International Chronostratigraphic Chart

Jenkins, D.F. and Munro, M. (eds) (2012) *John Piper: The Mountains of Wales.* Cardiff: National Museum of Wales

Kinsley, J. (ed.) (1970) *The Poems and Fables of John Dryden.* Oxford: Oxford University Press

Lyell, C. (1830–33) *Principles of Geology.* London: John Murray (Facsimile Ed. 1990, University of Chicago Press)

MacDiarmid, H. (1932) *On a Raised Beach.* In: (1993) Complete Poems Volume 1. Manchester: Carcanet Press

MacNeice, L. (1949) *Collected Poems, 1925–1948.* London: Faber and Faber

Mattingly, D. (2007) *An Imperial Possession: Britain in the Roman Empire.* London: Penguin Books

Meinig, W. (1979) *The Interpretation of Ordinary Landscapes:Geographical Essays.* Oxford: Oxford University Press

Pevsner, N. (1956) *The Englishness of English Art.* New York: Praeger

Piper, J. *The Colour of Rocks.* In: D.F. Jenkins and M. Munro (eds) (2012)

Playfair, J. (1802) *Illustrations of the Huttonian Theory of the Earth.* London and Edinburgh: Cadell and Davies and Creech

Screech, M.A. (ed.) (2003) *Michel de Montaigne: Complete Essays.* London: Penguin Books

Sinclair, G. (1672) *A Short History of Coal.* In: *Hydrostaticks* Edinburgh

Spiegl, F. (1971) *A Small Book of Grave Humour.* London: Pan

Steno, N. (1669) *De Solido.* Florence. (Transl. H. Oldenburg, London, 1671)

Strachey, J. (1717) A Curious Description of the Strata Observ'd in the Coal-Mines of Mendip in Somersetshire. *Philosophical Transactions of the Royal Society of London,* **30**, 968–73

Sullivan, A. (1998) *The Drama of Landscape: Land, Property and Social Relations.* Stanford: Stanford University Press

Winchester, S. (2001) *The Map that Changed the World: William Smith and the Birth of Modern Geology.* New York: Harper Collins

Wordsworth, W. (1850) *The Excursion.* In: Hutchinson, T. (ed.) (1969)

Further Reading

Berger, J. (1972) *Ways of Seeing.* London: Penguin Books

Betsky, A. (2013) The Complete Zaha Hadid. London: Thames & Hudson

Bourraird, N. (2002) *Relational Aesthetics.* Dijon: Les Press Du Reel

Causey, A (2013) *Paul Nash: Landscape and the Life of Objects.* Farnham: Lund Humphries

Clarke, C. (1949) *Landscape into Art:* Harmondsworth: Penguin Books

Clarke, K. (1928) *The Gothic Revival; An Essay in the History of Taste.* Harmondsworth: Penguin Books

Dobinson, T. (2013) *The Gherkin Guide*: London: Baizdon

Dyer, A. (1998) *Henry Moore: Friendship and Influence*: Great Britain: Norwich Colour Print

Gooding, M. (2016) *Frank Bowling.* London: Royal Academy of Arts

Harris, A. (2015) *Weatherland: Writers and Artists Under English Skies.* London: Thames & Hudson

Kinsman, P. (1995) *Landscape, Race and National Identity: The Photography of Ingrid Pollard.* Area **27**, 300–10.

Lingwood, J. (ed.) (1995) *Rachel Whiteread's House (Artangel Trust).* London: Phaedon Press

Marsden, P. (2014) *A Search for the Spirit of Place.* London: Granta

Moore, A. and Larkin, N. (eds) (2006) *Art at the Rockface*: *The Fascination of Stone.* London: Philip Wilson

Morris, S. (1997) *The Landscapes of Katharine Holmes.* Framlingham: Sheerham Lock

Newell, C. (2014) *John Ruskin: Artist and Observer.* London: Paul Holberton Publishing

Oliver, N. (2012*) A History of Ancient Britain.*Great Britain: Phoenix

Pawley, M. *Norman Foster: A Global Architecture.* London: Thames and Hudson

Petry, M. (2012) *The Art of Not Making: The New Artist/Artisan Relationship.* London: Thames and Hudson

Pevsner, N. (1956) *The Englishness of English Art.* New York: Praeger

Rowland, K. (1964) *Pattern and Shape:* Ipswich: W.S. Cowell

Schulman, R. and Heathcote, E. (2016) *New Architecture London.* New York: Prestel Publishing

Uglow, J. (2012) *The Pinecone: The story of Sarah Losh, Romantic heroine, architect and visionary.* London: Faber and Faber

Walmsley, A. (1990) *Guyana Dreaming: Art of Aubrey Williams.* Australia: Dangaroo Press

Internet

Agnew, J. Landscape and National Identity in Europe: England versus Italy in the role of Identity. tercud.ulusofona.pt/conferencias/pecsrl/presentations/John_Agnew.pdf (accessed 21 September 2016)

British Geological Survey bgs.ac.uk/

Halperin, I. geologicnotes.wordpress.com (accessed 21 September 2016)

International Commission on Stratigraphy. International Chronostratigraphic Chart. www.stratigraphy.org (accessed 21 September 2016)

Generationartscotland.org/artists/ilana-halperin/ (accessed 21 September 2016)

Hancock, L. www.saatchiart.com/luluhancock-hoar (accessed 21 September 2016)

Mid-Sussex District Plan (2005) A Landscape Character Assessment for Mid-Sussex, 2005. Appendix 5 – Cultural perceptions of the South Downs (online) Available from URL: www.midsussex.gov.uk/media/2308/lca29app5culturalperceptions.pdf (accessed 7 September 2016).

Smith, A. (2016) The Sublime in Crisis: Landscape Painting After Turner. www.tate.org.uk/art/research-publications/the-sublime/alison-smith-the-sublime-in-crisis-landscape-painting-after-turner-r1109220 (accessed 21 September 2016)

Shaw, G. Tile Hill paintings gerryco23.wordpress.com/2011/12/06/george-shaw-a-sense-of-our-time-acute-and-troubled/ (accessed 21 September 2016)

Newspapers/Periodicals

Greene, G. (1938) 24 hours in metroland. *New Statesman*, August issue.

Higgins, C. (2012) *Sticks, stones and furrows in the sand: the ephemeral art of a revolutionary man. Land artist Richard Long gives a rare interview before a new show.* Observer Culture 16 June

Kingsnorth, P. (2015) *Rescuing the English.* Guardian Review Saturday 14 March

Tombs, R. (2014) *A Nation Apart.* Guardian Review Saturday 8 November

Sawyer, M. (2015) *The artist who has excavated a rich seam of Nothingness. Sculptor Katrina Palmer.* Observer New Review 26 October

Part 2

Select Bibliography

British Geological Survey (2007a) *Bedrock Geology: UK South.* 1:625 000 Scale. Keyworth

British Geological Survey (2007b) *Bedrock Geology: UK North.* 1:625 000 Scale. Keyworth

Daly, R.A. (1926) *Our Mobile Earth.* New York: Charles Scribner's Sons

Holmes, A. (1913) *The Age of the Earth.* London and New York: Harper and Brothers

Holmes, A. (1929) Radioactivity and Earth Movements. *Transactions of the Geological Society of Glasgow* **18**, 559–606

MacPhee, J. (1981) *Basin and Range.* New York: Farrar, Strauss and Gilroux

Wegener, A. (1915) *The Origins of Continents and Oceans.* Engl. Transl. Skerl, J.G.A (1924) London: Methuen

Further Reading: General/Introductory

Gilmour, I. and Widdowson, M. (1999) *Maps and Landscape: S260 Course.* Milton Keynes: Open University

Hallam, A. (1989) *Great Geological Controversies.* Oxford: Oxford University Press

Lewis, C. (2000) *The Dating Game: One Man's Search for the Age of the Earth.* Cambridge: Cambridge University Press

Part 3

Select Bibliography

Atkinson, T.C., Briffa, K.R. and Coope, G.R. (1987) Seasonal temperatures in Britain during the past 22,000 years, reconstructed using beetle remains. *Nature* **325**, 587–92

Brown, T. (2008) The Bronze Age climate and environment of Britain. www.britishmuseum.org/bronzeagereview/1 (accessed 21 September 2016)

Cassidy, L.M. and 6 others (2016) Neolithic and Bronze Age migration to Ireland and the establishment of the insular Atlantic genome. *Proceedings of the National Academy of Science* doi:10.1073/pnas.151844511

Dewey, J.F. (1969) Evolution of the Appalachian/Caledonian Orogen. *Nature* **222**, 124–9

Eide, E.A. (ed.) (2002) BATLAS *Mid Norway plate reconstructions atlas with global and Atlantic perspectives.* Oslo: Geological Survey of Norway

Gibbard, P.L. and Clark, C.D. (2011) Pleistocene Glaciation Limits in Great Britain. *Developments in Quaternary Science* **15**, 75–93. doi: 10.1016/B978-0-444-53447-7.00007-6

Kennedy, W.Q. (1946) The Great Glen Fault. *Quarterly Journal of the Geological Society of London* **102**, 41–76

Price, D.G. (1985) Dartmoor: the Pattern of Prehistoric Settlement Sites. *Geographical Journal* **159**, 261–80

Shackleton, N.J. (1987) Oxygen isotopes, ice volume and sea level. *Quaternary Science Reviews* **6**, 183–90.

Sutton, J. and Watson, J. (1951) The pre-Torridonian metamorphic history of the Loch Torridon and Scourie areas in the North-West Highlands, and its bearing on the chronological classification of the Lewisian. *Quarterly Journal of the Geological Society of London* **106**, 241–308

Torsvik, T.H., Carlos, D., Mosar, J., Cocks, L.R.M. and Malme, T.N. (2002) Global reconstructions and North Atlantic paleogeography 440Ma to Recent. In: Eide, E.A. (ed.) (2002), pp. 18–39

Further Reading: General/Introductory

Cline, E.H. (2014) *1177 BC: The Year Civilisation Collapsed.* Princeton: Princeton University Press

Fagan, B. (2000) *The Little Ice Age.* New York: Basic Books

Gillen, C. (2013) op. cit. Preface and Forewords

Hunter, A. (2001) op. cit. Preface and Forewords

Hunter, J. and Ralston, I. (eds) (2009) *The Archaeology of Britain: An Introduction from Earliest Times to the Twenty-First Century.* Abingdon: Routledge

Lamb, H.H. (1995) *Climate History and the Modern World.* London: Routledge

Stringer, C. (2006) *Homo britannicus.* London: Penguin Books

Toghill, P. (2006) op. cit. Preface and Forewords

Further Reading: advanced

Brenchley, P.J. and Rawson, P.F. (eds) (2005) op. cit. Preface and Forewords

Darvill, T. (2010) *Prehistoric Britain.* Abingdon: Routledge

Gupta, S., Collier, J.S., Palmer-Felgate, A. and Graeme Potter, G. (2007) Catastrophic flooding origin of shelf valley systems in the English Channel. *Nature* **448**, 342–6

Pettitt, P. and White, M. (2012) *The British Palaeolithic: Human Societies at the Edge of the Pleistocene World.* Abingdon: Routledge

Trewin, N.H. (ed.) (2002) op. cit. Preface and Forewords

Woodcock, N. and Strachan, R. (eds) (2000) op. cit. Preface and Forewords

Part 4

Select Bibliography

Arkell, W.J. (1947) *Oxford Stone.* London: Faber and Faber

Arkell, W.J. and Tomkieff, S.I. (1953) *English Rock Terms: Chiefly as used by Miners and Quarrymen.* London: Oxford University Press

Auden, W.H. (1966) op. cit. Preface and Forewords

Bevins, R.E., Ixer, R.A., Webb, P.C. and Watson, J.S. (2012) Provenancing the rhyolitic and dacitic components of the Stonehenge landscape bluestone lithology: new petrographical and geochemical evidence. *Journal of Archaeological Science* **39**, 1005–19

Bevins, R.E., Ixer, R.A. and Pearce, N.J.G. (2014) Carn Goedog as the likely major source of Stonehenge doleritic bluestones: evidence based on compatible element geochemistry and Principal Component Analysis. *Journal of Archaeological Science* **42**, 179–93

Fleming, R. (2010) *Britain after Rome: The Fall and Rise, 400-1070*. London: Penguin Books

Pryor, F. (2003) *Britain BC*. London: HarperCollinsPublishers

Jope, E.M. (1964) The Saxon building-stone industry in southern and midland England. *Medieval Archaeology* **8**, 91–118

MacCaig, N. (1990) *Collected Poems*. London: Chatto and Windus

Mattingly, D. (2007) *An Imperial Possession: Britain in the Roman Empire*. London: Penguin Books

Nicholson, N. (1975) *Wednesday Early Closing*. London: Faber and Faber

Parsons, D. (ed.) (1990) *Stone: Quarrying and Building in England*. Chichester: Phillimore

Wood, C. (ed.) (2003) Stone Roofing. *English Heritage Research Transactions* **9**, 1–156.

Further Reading: General/Introductory

Clifton-Taylor, A. (1987) *The Pattern of English Building*. London: Faber and Faber (4th Edn)

Cunliffe, B, Bartlett, R., Morrill, J., Briggs, A. and Bourke, J. (2004) *The Penguin Illustrated History of Britain and Ireland*. London: Penguin Books

English Heritage (2013) *Stonehenge and Avebury: Exploring the World Heritage Site* (1:10 000 scale). Swindon

Hackman, G. (2014) *Stone to Build London: Portland's Legacy*. Monkton Farleigh: Folly Books

Ordnance Survey (1991) *Roman Britain: Historical Map and Guide*. (1:625 000 scale) Southampton

Ordnance Survey (2011) Ancient Britain: Historical Map. (1:625 000 Scale) Southampton

Pryor, F. (2011) *The Making of the British Landscape: How We Have Transformed the Land, from Prehistory to Today*. London: Penguin Books

Raistrick, A. (1973) *Industrial Archaeology*. London: Granada Publishing

Rotherham, I.D. (2013) *The Lost Fens: England's Greatest Ecological Disaster*. Stroud: The History Press

Sawyer, P. (2013) The Wealth of Anglo-Saxon England. Oxford: Oxford University Press

Woodcock, N. (1994) *Geology and Environment in Britain and Ireland*. London: UCL Press

Wright, D. (ed.) (1981) *Edward Thomas: Selected Poems and Prose*. Harmondsworth; Penguin Books

Websites

English, Scottish and Welsh Heritage websites have information on stone building and conservation. The BGS website has county-by-county summaries of building stone types and their distributions.

Part 5
Selected Bibliography

Cline, E.H. (2014) op. cit. Part 3

Ixer, R.A. and Budd, P. (1998) The Mineralogy of Bronze Age copper ores from the British Isles: implications for the composition of early metalwork. *Oxford Journal of Archaeology* **17**, 15–41

Kunene, M. (1970) Zulu Poems. London: André Deutsch

Lewis, A. (1994) Bronze Age mines of the Great Orme. *Peak District Mines Historical Society Bulletin* **12**, 31–6

MacCaig, N. (1990) *Collected Poems*. London: Chatto and Windus

Nield, T. (2013) *Underlands: A journey through Britain's Lost Landscapes*. London: Granta

Phillips, W. and Conybeare, W.D. (1822) *Outlines of the Geology of England and Wales*. London: William Phillips.

Raistrick, A. (1973) op. cit. Part 4

Gerrard, S. (2000) *The Early British Tin Industry*. Stroud: Tempus Publishing

Further Reading

Bird, R.H. (1974) Britain's Old Metal Mines: A Pictorial Survey. Truro: D. Bradford

Coyle, G. (2010) The Riches Beneath Our Feet: How Mining Shaped Britain. Oxford: Oxford University Press

Falconer, K. (1980) Guide to England's Industrial Heritage. London: Batsford Invaluable Gazetteer

Raistrick, A. op. cit. Part 4

Part 6
Select Bibliography

Anon (English C8th) *The Ruin* and *The Dream of the Rood* In: *The Earliest English Poems*. Transl. M. Alexander (1965). Harmondsworth: Penguin Books

Anon. (Irish C12th) *Arran* In: *A Celtic Miscellany*. Transl. K.H. Jackson (1971). Harmondsworth: Penguin Books

Auden, W.H. (ed.) (1956) *The Selected Writings of Sydney Smith*. London: Faber and Faber

Benfield, E. (1940) *Purbeck Shop*. Cambridge: Cambridge University Press

Bevan, A. and Murray, B. (eds) (2006) *The Collected Poems of George Mackay Brown*. London: John Murray

Boland, E. (ed.) (2013) *The Collected Poems of Denise Levertov*. New York: New Directions

Bunting, B. (1965) *Briggflatts*. London: Fulcrum Press

Cobbett, W. (1830) *Rural Rides*. London: Penguin Books (2005 Edition)

Curry, N. (ed.) (1994) *Norman Nicholson: Collected Poems*. London: Faber and Faber

Daniel, G. (1963) *The Megalith Builders of Western Europe*. Harmondsworth: Penguin Books

Defoe, D. (1724-27) *A tour thro' the Whole Island of Great Britain*. Harmondsworth: Penguin Books (1981 Edition)

Fiennes, C. (1698) *Through England on a Side Saddle in the Time of William and Mary, 1698*. Cambridge: Cambridge University Press (2010 Edition)

Hale, K. (ed.) (2016) *Rupert Brooke: Collected Poems*. London: Watersgreen Books

Hall, P. (2014) *Cities of Tomorrow: An Intellectual History of Urban Planning and Design since 1880*. Chichester: John Wiley

Hardy, T. (1891) *Tess of the D'Urbervilles*. London: Penguin (1998 Edition)

Hardy, T. (1996) *The Collected Poems*. London: Wordsworth Editions

Hughes, Glyn (2011) A Year in the Bull Pen. Todmorden: Arc Publications

Hughes, T. (1995) *New Selected Poems*. London: Faber and Faber

Hutchinson, T. (ed.) (1969) *Wordsworth Poetical Works*. Oxford: Oxford University Press

Keynes, G. (ed.) (1966) *Blake: Complete Writings*. Oxford: Oxford University Press

Lawrence, D.H. (1915) *The Rainbow* Harmondsworth: Penguin Books (1975 Edition)

MacCaig, N. (1993) *Collected Poems*. London: Chatto and Windus

Maclean, A. (1975) *From the Wilderness*. London: Harper and Row

MacNeice, L. (1966) *Collected Poems*. London: Faber and Faber

Mendelson, E. (ed.) (1991) *W.H. Auden: Collected Poems*. New York: Vintage International

Moritz, C.P. (1785) *Journeys of a German in England: A Walking Tour of England in 1782*. London: Eland Publishing (2009 Edition)

Read, H. (1934) *Art and Industry*. London: Faber and Faber

Read, H. (1974) *The Contrary Experience*. London: Horizon Press

Robinson, F.N. (ed.) (1957) *The Works of Geoffrey Chaucer*. London: Oxford University Press

Shakespeare, W. (1609) *Shakespeare Sonnets: Never Before Imprinted*. London: Thomas Thorpe.

Stallworthy, J. (ed.) (1990) *The Poems of Wilfrid Owen*. London: Chatto and Windus

Summerfield, G. (ed.) (2004) *John Clare: Selected Poems and Prose*. London: Penguin Books

Tobin, J. (ed.) *Henry Vaughan: The Complete Poems*. London: Penguin Books

Trollope, A. (1857) *Barchester Towers*. Oxford: Oxford University Press (2014 Edition)

Walpole, H. (1762) *Anecdotes on Painting*. London: T. Kirgate

Wright, D. (ed.) (1981) *Edward Thomas: Selected Poems and Prose*. Harmondsworth; Penguin Books

Further Reading

Anon. *Carving Mountains* (1998) Cambridge: Kettle's Yard Publications

Backes, M. and Delling, R. (1969) *The Art of the Dark Ages*. New York: Harry Abrams

Bomford, R. and Ashok, R. (2009) *A Closer Look at Colour*. London: National Gallery Company

Bottinelli, G. (ed.) (2013) *A Vision of England: Paintings of the Norwich School*. Norwich: Norwich Castle Museums and Archaeology Service

Collins, I. (ed.) (2013) *Masterpieces: Art and East Anglia*. Norwich: Butler, Tanner and Dennis

Curtis, P. Wilson, K. (2011) *Modern British Sculpture*. London: Royal Academy of Arts

Delmare, F. and Guineau, B. (2000) Colour: *Making and Using Dyes and Pigments*. London: Thames and Hudson

Drabble M. (2009) *A Writer's Britain*. London: Thames and Hudson

Harewood, E. (2015) *Space, Hope and Brutalism: English Architecture, 1945–1975*. New Haven and London: Yale University Press

Honour, H. and Fleming, J. (1998) *A World History of Art*. London: Lawrence King Publishing

Howard, E. (2002) *Garden Cities of Tomorrow*. Cork: Attic Press

Itten, J. (1970) *The Elements of Colour*. New York: Van Nostrand Reinhold

Moore, A. and Larkin, N. (eds) (2006) op. cit. Part 1, Select Bibliography

Pastoureau, M. (2013) *Green: The History of a Colour*. Princeton and Oxford: Princeton University Press

Ibid (2008) *Black: The History of a Colour*. Princeton and Oxford: Princeton University Press

Powell, K. (2011) *21st-Century London: The New Architecture*. London: Merrell Publishers

Rice, R. (2009) *Rice's Architectural Primer*. London: Bloomsbury Publishing

Rowland, K. (1971) *The Shape of Towns*. Singapore: Times Printers

Sennett, R. (2009) *The Craftsman*. Cork: Attic Press

Soskin, R. (2009) *Standing with Stones*. London: Thames and Hudson

Toman, R. (ed.) (1997) *Romanesque: Architecture, Sculpture, Painting*. Germany: Konemann

Toman, R. (ed.) (1998) *Baroque: Architecture, Sculpture, Painting*. Germany: Konmann

Trachtenburg, M. and Hyman, I. (1986) *Architecture: From Prehistory to Post-Modernism/The Western tradition*. London: Academy Editions

Williams, G., Penz, P. and Wernhoff, M. (2014) *Vikings: Life and Legend*. London: British Museum Press

Newspapers and Periodicals

Adams, Tim (2014) Interview with Frank Auerbach discussing the 2014 Constable exhibition at the V&A *Frank's Other Half*: Observer New Review 21 September

Barrel, B. (2013) The Virtues of Topography: Review of Constable, Gainsborough, Turner and the 'Making of Landscape' exhibition at The Royal Academy, 2013. In: *London Review of Books* 3 January

Moore, R. (2015) *Are Po No's Palaces Worth Saving?* Observer New Review 20 December

Rogers, B. (2014) *Anarchy in the UK*. Guardian Review Saturday 17 January

Internet

Faithfull, S. *A melancholy journey along the A13 trunk road. Created from pixelated PalmPilot drawings* www.animateprojects.org/films/by_date/2002_2004/13

Modernist Britain. www.modernistbritain.co.uk (accessed 21 September 2016)

The History of British and Irish Towns. www.building.org/towns.shtml (accessed 21 September 2016)

Television

Goudie, L. (2015/6) *The Story of Scottish Art*. BBC4

Ramirez, J. (2015) *Saints and Sinners: Britain's Millennium of Monasteries*. BBC4

Willis, J. (2015/16) *Castles: Britain's Fortified History*. BBC4

Part 7

Select Bibliography

Auden, W.H. (1941) *New Year Letter*. London: Faber and Faber

Boast, R. (2013) *Pilgrim Flower*. London: Picador

Bunting, Basil (1966) *Briggflatts, Part 1*. London: Fulcrum Press

Causley, Charles (1988) *A Field of Vision*. London: PaperMac

Crossley-Holland, Kevin (1998) *Poems from East Anglia*. London: Enitharmon

Forman, M.B. (ed.) (1935) op. cit. Part 1

Harrison, Tony (2005) *Under the Clock*. London: Penguin Books

Hawkes, J. (1951) op. cit. Preface and Forewords

Hill, Geoffrey (1998) *The Triumph of Love*. London: Penguin Books.

Housman, A.E. (1896) *A Shropshire Lad* (1999 Edition). London: Penguin.

Lewis, David (1985) Personal Memoir in *The Art of St Ives*, catalogue to the Tate Gallery Exhibition.

Lewis, Gwyneth (2003) *Keeping Mum*. Tarset: Bloodaxe Books.

MacCaig, Norman (1990) op. cit. Part 5

MacCodrum, John (*c.*1750) transl. J.A. Love *Smeòrach Chlann Domhnaill* In: The Glasgow Naturalist (2009) Volume 25, Supplement. Machair Conservation: Successes and Challenges, 3–10.

Mackay Brown, G. *The Twentieth Stone* In: Bevan, A. and Murray, B. (eds) (2005) *The Collected Poems of George Mackay Brown*. London: James Murray.

Miller, Hugh (1828) The Old Red Sandstone, or New Walks in an Old Field. London: J.M. Dent & Co.

Tacitus: Mattingly, H. and S.A. Handford (transl.) (1970) *The Agricola and the Germania*. Harmondsworth: Penguin Books.

Nicholson, N. (1954) *The Seven Rocks* in *Pot Geranium*. London: Faber and Faber.

Oswald, Alice (2005) *Woods etc*. London: Faber and Faber.

Priestley, J.B. (1947) *The Linden Tree* (2000 Edition). London: Penguin Books.

Somhairle MacGill-Eain (Sorley MacLean) (1999) *Ceann Loch Aoineart* (*Kinloch Ainort*) (1932–1940) in *O Choille gu Bearradh* (*From Wood to Ridge*). London and Edinburgh: Carcanet/Birlinn.

Stalworthy, J. (ed.) (1985) *The Poems of Wilfred Owen*. London: Chatto and Windus.

Thomas, Dylan (1952) *Collected Poems*. London: J.M. Dent & Sons.

Watkins, Vernon (1954) New Selected Poems (2006 Edition). Manchester: Carcanet Press.

Further Reading: general; suitable for non-specialists

Andreaea, C. (2013) *Joan Eardley*. London: Lund Humphries

BGS Exploring the Landscape Series: *Assynt; Charnwood Forest and Mountsorrel; Western and Eastern Mendip; National Forest; Forest Fawr (Brecons, Black Mountain*. Keyworth: British Geological Survey

Bowden, M. et al. (2015) op. cit. Part 1

Brunsden, D. (ed.) (2003) *The Official Guide to the Jurassic Coast: A Walk Through Time*. Wareham: Coastal Press

Clarkson, E. & Upton, B. op. cit. Preface and Forewords

Clarkson, E. and Upton, B. (2006) *Edinburgh Rock: The Geology of Lothian*. Edinburgh: Dunedin Academic Press

Coulson, S. (2014*)* *Ursula Von Rydingsvard*. Italy: Yorkshire Sculpture Park

Cunliffe, B, Bartlett, R., Morrill, J., Briggs, A. and Bourke, J. (2004) *The Penguin Illustrated History of Britain and Ireland*. London: Penguin Books

Cunliffe, B. (2013) *Britain Begins*. Oxford: Oxford University Press

Dimes, F.G. and Mitchell, M. (1996) *The Building Stone Heritage of Leeds*. The Leeds Philosophical and Literary Society, Scientific Section, Special Publication

Dodd, M. (ed.) (1992) *Lakeland Rocks and Landscape: A Field Guide, The Cumberland Geological Society*. Maryport: Ellenbank Press. Comments as for Scrutton et al.

Ensom, P. (2009) *Yorkshire Geology*. Wimborne Minster: Dovecote Press

Ensom, P. and Turnbull, M. (2011) *Geology of the Jurassic Coast: The Isle of Purbeck, Weymouth to Studland*. Wareham: Coastal Publishing

Friend, P.F. (2008) *Southern England: Looking at Natural Landscapes*. London: Collins

Gannon, P. (2008) *Rock Trails: Snowdonia*. Caernarfon: Pesda Press

Ibid (2009) *Rock Trails: Lakeland*. Caernarfon: Pesda Press

Ibid (2010) *Rock Trails: Peak District*. Caernarfon: Pesda Press

Ibid (2012) *Rock Trails: Scottish Highlands*. Caernarfon: Pesda Press

Gillen, C. (2013) op. cit. Preface and Forewords

Gormley, A., Hutchinson, J., Gombrich, J. Enjatin, L. and Mitchell, W. (2000) *Anthony Gormley*. London: Phaedon Press

Goudie, A. and Gardner, R. (1985) *Discovering Landscape in England and Wales*. London: Chapman and Hall

Greaney, M. (2013) *Liverpool: A Landscape History*. Stroud: The History Press

Hill, D. (1996) *Turner in the North*. New Haven and London: Yale University Press

Hunter, A. (2001) op. cit. Preface and Forewords

Hunter, J. and Ralston, I. (eds) (2009) *The Archaeology of Britain: An Introduction from Earliest Times to the Twenty-First Century*. Abingdon: Routledge

James, R. and Russell, J. (2015) *Ravilious: The Watercolours*. London: Philip Wilson

Jenkins, D.F. and Munro, M. (2012*)* *John Piper: The Mountains of Wales*. Cardiff: National Museum of Wales

Marks, R. (1993) *Stained Glass in England During the Middle Ages*. Oxford: Routledge

Narsh, J. (2005) *William Morris and the Red House: A Collaboration Between Architect and Owner*. London: National Trust

Nield, T. (2013) *op. cit.* Part 5

Oliver, N. (2011) *A History of Ancient Britain*. London: Orion Books

Parker-Pearson, M. (2012) *Stonehenge: Exploring the Greatest*

Stone Age Mystery. London: Simon Schuster

Ibid. (ed.) (2015) *Stonehenge: Making Sense of a Prehistoric Mystery*. York: Council for British Archaeology

Pearson, F. and Stevenson, S. (2014) *Joan Eardley*. Edinburgh: National Galleries of Scotland

Postle, M. and Simon, R (eds) (2014) *Richard Wilson and the Transformation of European Landscape Painting*. Yale Centre for British Art: Yale University Press

Pryor, F. (2011) *The Making of the British Landscape: How We Have Transformed the Land, from Prehistory to Today*. London: Penguin Books

Putnam, J. (2007) *Enclosure: Andy Goldsworthy*. London: Thames and Hudson

Robertson, A. (1996) *Atkinson Grimshaw*. London: Phaedon Press

Scottish Natural Heritage in conjunction with the BGS *Scottish Landscape and Geology Guides*.

Scrutton, C. (ed.) (1994) *Yorkshire Rocks and Landscape: A Field Guide, Yorkshire Geological Society*. Maryport: Ellenbank

Shaw, G. (2011) *Graham Sutherland: An Unfinished World*. Oxford: Modern Art Oxford

Stevens, C. and Phillips, M. (2002) *The Barbara Hepworth Garden*. London: Tate Publishing

Taylor, M.A. (2007) *Hugh Miller: Stonemason, Geologist, Writer*. Edinburgh: National Museums of Scotland Enterprises

Toghill, P. (2006) *Geology of Shropshire*. Marlborough: Crowood Press

Toghill, P. (2006) op. cit. Preface and Forewords

Tomlin, R.S.O. (2016) *Roman London's First Voices: Writing Tablets from the Bloomberg Excavations, 2010–14*. London: Museum of London Archaology Monograph 72

Turnbull, R. (2009) *Granite And Grit: a Walker's Guide to the Geology of British Mountains*. London: Frances Lincoln

Uglow, J. (2002) *The Lunar Men: The Friends Who Made the Future*. London: Faber and Faber

Further Reading: advanced or specialist

BGS British Regional Geology series

Brenchley, P.J. and Rawson, P.F. (2005) op. cit. Preface and Forewords

Stone, P. (ed.) (1996) Geology in south-west Scotland. Keyworth: British Geological Survey

Trewin, N.H. (ed.) (2002) op. cit. Preface and Forewords

Woodcock, N. and Strachan, R. (eds) (2000) op. cit. Preface and Forewords

Newspapers/Periodicals

Gregory, C. (1986) *Wright of Derby: The Great Artists*, 65. London: Marshall Cavendish

Harrison, F., Edmunds, J. and Farrett, D. (1984) *Constable: The Great Artists 1*. London: Marshall Cavendish

Smith, A. (2015) *Romancing the Stone: Barbara Hepworth's work and its universe of meaning*. New Statesman 15 June

Internet

Adam Kennedy painter: www.limetreegallery.com/artists/adam-kennedy/468/www.the skinny.co.uk/art/interviews/adam-kennedy-ship-painter (accessed 21 September 2016)

Caroline Bailey painter: www.carolinebailey.co.uk (accessed 21 September 2016)

Chris Griffin painter: artinwales.250x.com/ArtistsGrif.htm

Vaughan Melzer phographer: melzer@gmail.com

Millennium Centre Cardiff: en.wikipedia.org/wiki/Wales_Millennium_CentreMillennium Centre Cardiff (accessed 21 September 2016)

Sandy Wylie painter: artuk.org/discover/artists/wylie-sandy (accessed 21 September 2016)

The Eden Benchmarks: www.edenbenchmarks.org.uk/sclptures.htm (accessed 21 September 2016)

Val Corbett photographer: val.corbett121abtinternet.com

Television

Goudie, L. (2016) *The Story of Scottish Art*. BBC4

Index

Page numbers in *italic* denote illustrations

Aberdeen 182
Acadian event 58
accretionary wedge, Southern Uplands 54
Addingham Edge Grit *36*
adits 111, *112*
aggregate 101
Agricola, Georgius (1494-1555), *De Re Metallica* (1556) 13, 114
Agricola, Gnaeus Julius (Roman general) 186, 194, 202
agriculture
    Midland Valley 185
    Neolithic 75–76
        Orkney 176
Ailsa Craig 64
air temperature *74*
alabaster 116
algae 95, 97
Alpine-Himalayan mountain belt 68
Alston massif 209, *210*
'Amesbury Archer' 25–26
ammonites 61
andesite 52
Anglesey 230, *231*, *232*
Anglian Glaciation 70, *71*
Anglo-Saxons
    building stone 97, 98
    East England 252
    monuments and churches 125, 126
    sculpture 137–138
anhydrite 107, 115, 116
anticlines 45
Antonine Wall 186
apatite analysis 64
Aquae Sulis *see* Bath
Archean Era 53
architecture 121–132
    Anglo-Saxon 125–126
    Classical 127–128
    Georgian 127–128
    Gothic 126, 139
    mediaeval 127
    modern 29, 121, 132
    Norman 125–126, 127
    Perpendicular 126
    postmodern 132

Renaissance 126
Victorian
    civic 128–129, *130*
    industrial 128, 130–131
Argyll *165*, 166–167, 179–183
Arran *190*, 191–192
Arts and Crafts Movement 21, 131, 140, 152, *251*
Askrigg massif 209, *210*
Assynt 55
    Lewisian gneiss 53
Assynt Foreland/Outer Hebrides *165*, 166, 169–173
asteroid impact 61
Atlantic Ocean 52, 64, 73
    influence on climate 2
    spreading 64, 66, 67
atmosphere, oxygenated 53
Auden, W.H (1907-1973)
    Down There in About the House (1966) 80
    In Praise of Limestone (1948) 161
    Letter to Lord Byron (1936) 161
    New Year Letter (1941) 209
Avalonia 53–54, *55*, 56
Avon Falls *181*
azurite 110, 144

Bailey, Caroline (b.1953), *Sailing out of Mallaig 12* (2013-14) *193*
Baltica 54, *56*
Bamburgh Castle *81*
Barclodiad y Gawres chambered tomb 232
barite 114
Barnack Church *255*
Barnack Stone 95, *96*, 98, *105*, 126, 255
Baroque sculpture 139
Barra 169, *170*
basalt 192, *193*
Bath (Aquae Sulis) 246, 248
    lead 114
Bath Stone 25, *98*
batholiths 7, 240
    Lake District 54, 204, 205
Beaker Folk 76
Beaker-style pottery 9
belemnites 61

Ben Nevis *179*, *180*
Ben Slioch *170*
Benbecula 169, *170*
Benfield, Eric, *Purbeck Shop* (1940) 121
Bewcastle High Cross 201, *201*
bings 119, 185
Birmingham town hall 128, *129*
Blake, William (1757-1827)
    Auguries of Innocence (1803) 156
    reaction against Industrial Revolution 32, 151
'bluestone', Stonehenge 10, 23, 97, 236
Boast, Rachel (b.1975), *Caritas* 184
Bølling-Allerød interstadial 4, *74*
Book of Kells 182
Border Reivers 199, 202
Borrowdale Volcanics 52, *205*
Botallack Mine, Cornwall *8*
Bourne, Mary (1963), *Water Cut* (1996) *208*
Brailsford, Victoria, *Red River* 28, *29*
brass 114
Brecklands 80
Brecon Beacons 236
Brett, John (1830-1902), *The Stonebreaker* (1857-58) *102*, 151
bricks 100–101
    Roman 100
    Victorian mechanically produced 129–131
Bridgewater Canal 86, 221
British Geological Survey
    'Geology of Britain Viewer' database 6, 40
    Maps and Guides 6, 41
Bronze Age society 227
    'Amesbury Archer' 25–26
    'Beaker Folk' metallurgy 76, 77, 78
    colonization 3, 76–78
    mineralogy 13, 110, 111–112, 114
    monuments 10, 123, 207
    Outer Hebrides 171, 172
    sculpture 194, 196, 207, 208
    South Wales 237
    standing stones 232
bronze manufacture 113–114
Brooke, Rupert (1887-1915)

The Chilterns (1916) 158
The Soldier (1914) 159
Browne, Sir Thomas (1605-1682)
Hydrotaphia (1658) 13
Pseudodoxia Epidemica (1646) 13
Bruce's Stone *195*
Bryn Celli Ddu Neolithic tomb *122*, 232
building stone 92–105, 242
Anglo-Saxon 97, 98
imported 97, 98
Norman 97, 98
Roman 89, 90, 97
weathering 95, 96, 97
buildings 7, *8*
Bunting, Basil (1900-1985), on landscape
160, 209
Byatt, A.S.(born 1936), *The Stone Woman*
(2005) 7

Caen Stone 97, *98*, 126, 248
Cairngorm mountains *179*, *180*, *181*
calderas 54, 189, *190*, 230
Caledonian orogeny 46, 54, 58, 232
Callanish stone circle 171–172, *172*
Cambrian Mountains 232, 235
Cambrian Period 2, *15*, 54, *55*
Camster chambered cairn *176*
Camulodunum (Colchester) 83–84
Balkerne Gate 124
canals 86, *87*, *128*
Midlands 86, 221
Capel Garmon passage tomb 234
carbonate platforms, Jurassic 64, *65*
carbonic acid 37, 38
Carboniferous Limestone 58
Goredale Scar 39
'Green Bridge of Wales' 44, 236
unconformities 47
Carboniferous Period *4*, *15*, 58, 61
Cardiff, Millennium Centre 238, *239*
Carrock Fell *42*
carseland 185
Castlerigg stone circle 206, *207*
castles 81, *82*, 126, 202, *228*
cattle droving 197
cave art 24, *25*
cement 101
cementstones 58, 101, 201
cemeteries 135–136, 140
Cenozoic Era *15*, 16, 64, *67*, 68, 69
Central Scarplands *165*, 168
Centre for Alternative Technology,
Machynlleth 132, 233
chalcopyrite *109*, 110
Chalk 64
chalk downlands 32, 68, *247*, 248, *250*
Chalky Boulder Clay 70
charcoal 113

Charnia 54
Charnwood Forest 54, 220
Charterhouse-in-Mendip, Roman silver
mine 88–89
Chaucer, Geoffrey (1342-1400),
Canterbury Tales 155
Cheshire Triassic basin *215*, 224, *225*
salt 116, 226
Chester (Deva) *225*, 226
Cheviot 201
Chilterns *247*, 248
china clay *37*, 244
Christianization 125
Argyll 181
Assynt/Outer Hebrides 172–173
Southern Uplands 194, 196
churches 23, 125–126, 139
round tower 126
cinnabar 143
Clare, John (1793-1864), on landscape
33, 158
clays 37
Mesozoic 64
claystone, Triassic, Leicester 25
Cleveland Dyke 64
climate
effect of volcanic eruptions 5
maritime 2
climate change 3–5
Bronze-Age 78
orbital theory 49
wine growing 256, 257
see also glaciation; sea-level change
Coal Measures 3, *4*, *60*, 61
Durham 210
English Midlands 220
sandstone 126, 127, 211
Somerset Coalfield 14–15, 242
South Pennines 214
South Wales 236
Welsh-English Borders 226
coal mining 14, 117
Durham 212
English Midlands 220, 221
Kent 248
Somerset 14–15, 242
South Wales 237–238
Coalbrookdale 113, 227
Cobbett, William (1763-1835), *Rural
Rides* (1830) 158
coke, use in iron smelting 113
Colchester *see* Camulodunum
Collyweston Stone 98
Columba 181, 182, 183
concrete 101
Constable, John (1776-1837) 149–150
The Stour Valley and Dedham Church
(c.1815) 149

continental drift 48–49
copper
in manufacture of bronze alloy 113–114
mining 76, 78, 89, 90, 108, 111
Anglesey 230, 232–233
smelting 110
Cornubian granite batholith 240
corries 174, 194
Cotman, John Sell (1782-1842), *Hell
Cauldron* (1803-5) 146, *148*
Cotswold Stone 98
Cotswolds 248
Council for the Preservation of Rural
England 6
countryside, access to 6
Cow and Calf, Wharfedale *36*, *42*
'Crags', East Anglia 68, *69*
Craven faults 209, *210*
Creswell Crags
Mesolithic cave 252, 254
art 24, 25
Cretaceous Period *15*, 64, *66*
Croft Quarry, Leicestershire *220*
crofting 171
Hebrides 189, 191, 192
Crome, John (1768-1821) 146
Mousehold Heath (c.1818-20) *147*
Cromer Forest Formation 68, 256
Cromer Ridge 70, *147*, 252
cromlechs 122
Cross Fell 207, 209, *210*
cross-stratification 45
crosses 11
Anglo-Saxon 137–138
Bewcastle High Cross 201
Gosforth Cross 208
Ruthwell Anglo-Saxon High Cross 25,
138, 155, 194, 196, 197
Crossley-Holland, Kevin (b.1941), *Dusk,
Burnham Overy Staithe* (1972) 252
crust, Precambrian 52–53
crystallography 13
Cuillin Hills 189, 191, *191*
Culm Basin 61, 240
Cummins, Paul (b.1977), *Blood Swept
Lands and Seas of Red* (2014) 136
cursi 122
Cuvier, Georges (1769-1832), fossils 15

dacite 52
Dalradian sedimentation 54, 55, 180
Dalriada 180, 181, *182*
Darby, Abraham (1678-1717),
Coalbrookdale iron smelting 113, 227
Dartmoor 93, 243
Bronze-Age colonization 76, 77, 78
china clay 37
Darwin, Charles (1809-1882)

as geologist 19–20
  Origin of Species by Natural Selection
    (1859) 16, 20
Death of the Lark/Marwnad yr Ehedydd
    230
deep time 2, 48, 50
Defoe, Daniel (c.1660-1731)
  on arrival in Halifax 214
  on travel in Westmorland 156
deglaciation 3
deltas
  Carboniferous 61
  Mesozoic 64
Descartes, René (1596-1650), structure of
  the Earth 13
Devensian Glaciation *71*, *74*
Devonian Period 2, *15*, 58, *59*, 240
Dewey, John, Iapetus Ocean 52
dinosaurs 61
diorite quarrying 101, *220*, 234
dip 39
dolerite 53, 101
  Edinburgh 188
  sculpture 134
  sills 81, 200
  spotted see 'bluestone'
dolmen tombs 122
dolomite 61
  building stone 95, 252
Dolwyddelan Castle *231*
Dornoch Firth 174, *175*
Doulting stone 242
downfolds 45
'drift' geological maps 39–40
Drumbeg stone circle *123*
Dryas octopetala 4
Dryden, John, (1631-1700), *Astraea
  Redux* (1660) 11
drystone walls, Bronze-Age 76, 78
'Dudley Locust' 220
Dumbarton Rock *186*
Dumfries 197
Dunadd citadel 181, *182*, 183
dunes 45
  Permian 61
Dunstable Downs, Neolithic burial *12*
Durham Cathedral 126, *127*, 139, 211,
  212
Durness carbonate shelf-platform *55*, 58
Durness Group 54, 169
Dyffryn Ardudwy burial chamber 234
dykes 7, 61, 64
  dolerite 53
  ring 189, *190*

Eakring-Duke's Wood oilfield 119
Eardley, Joan (1921-1965), *Cliffs and Sea*
  (1958) 182, *183*

East Anglia, 'Crag' deposits 68, *69*
Eastern and Central Scarplands *165*, 168,
  252–257
Ebbatson, Richard, *Landscape and
  Literature 1830-1914* (2014) 21
Eboracum (York) 84
eccentricity 49
echinoderms 61
Eden Benchmark Project 28, *29*, 208
Edinburgh 185, 188
  Lower Carboniferous sandstone 128
Eigg 192
Eliot, George (1804-86) on landscape 33
Elland Flags *94*, *130*, 214
English Heritage 6, 100
English Midlands *165*, 167–168, 219–223
Enlightenment
  geology 16–17
  mineralogy 13
Eocene Epoch 68
epiflora, building stone 95, 97
Ermine Street 255
erodability, differential 42–43, 174
erosion 37
erratics
  cultural 9–12
  glacial 70, 71, 232
estuarine drowning 3
evaporites 52, 107, 115–116
  Zechstein 61, 63
Exmoor 240
Eyjafjallajökull volcano 2010 eruption 5

farming *see* agriculture
Farrell, Terry (born 1938), *The Deep
  submarium* (2002) 29, 31, *31*, 121
faulting 42–43
  Jurassic 64
fenland 83, 256
field courses 6
Fiennes, Celia (1662-1741), on waterfalls
  156
Fingal's Cave 192, *193*
Firth of Clyde 185
Firth of Forth 185
fish, Devonian 58
Fiskerton Iron Age causeway 83
Flag Fen causeway 83, 104
flagstones 92, *94*
Flandrian Glaciation *74*
flint
  building stone 96, 101–105
    knapping 102, 104, 105
  mining 10, 102–105
    Norwich 104, 105
flint tools 9, 68, 70, 72
floodplains 83
Flow Country peatland 174, *175*

fluid inclusions 107
fluorite *109*, 114
folding 43, 45
  overthrust 54
footpaths, public 6
Forest of Dean 228
Forteviot cist grave, white quartz 10–11
Forth Sill 61
Fosse Dyke drainage system 253
Fosse Way 221
fossil fauna 52
fossils
  as carvings 11
  English Midlands 220
  as grave goods 11–12, *12*
  work of Cuvier and Smith 15–16, 18
Frankland, John, *Boulder* (2008) 23, *24*,
  142
freestone 95, 97, 134

Gaels 180, 181–182
Gainsborough, Thomas (1727-1788), *Mr
  and Mrs Andrews* (c.1750) *144*
galena 114
Galloway *165*, 167, 194–198
garden city movement 131
geochemical cycle 38
geological history 2
geological maps 38–40, *41*
Geological Society of London 6–7, 17
  Lyell Medal 19
Geological Survey of Great Britain 20
geological time-scale 2, *15*, 16
GeoRegions 165–168
Gherkin, Richard Rogers (2003) 29, 132
Gipping Glaciation 71
glacial landforms 75, 174
  Lake District 205
glaciation 4, *71*
  Anglian 70, 71, 72
  North Wales 232
  and sea-level change 68, 73
  South Wales 236
Glasgow 185, 187–188
  Permian sandstone 25, 128, 187
'Glasgow Boys' school of colourists 198
Glasgow School of Art 131
Glencoul Thrust *175*
gneiss
  banded 53
  Lewisian 5, 169, 170, 171, 175
gold mining *90*, *108*, 112, 115, 196–197,
  237
Goldsworthy, Andy (b.1956)
  Sheepfolds (1996-2003) 142
  Striding Arches 142, 194, 198
Gondwana 53–54, *55*, *59*
  ice-sheets 61

Goredale Scar *30, 39*
Gormley, Anthony (b.1950)
    Angel of the North (1998) 136, 213
    Another Place (2007) 135, 142, 227
Gosforth Cross 208
Goudie, Lachlan, *The Story of Scottish Art*
    (2015) 33
Grampian Highlands 54, *56, 165, 166–*
    167, 174, *175*, 179–183
granite 42
    building stone 8
    intrusions 180
        Cheviot 201
        Galloway 194, 195
        Lakeland 204
        North Pennines 209, 210
        South West England 240
    South West England 243–244
Great Glen Fault 58, *59*, 174, *179*
Great Langdale tuff 10
Great Orme Bronze Age copper mine
    111–112
Great Scar Limestone *30*, 47, 209
    karstic pavement 8
'Green Bridge of Wales' *44*, 236
Greensand, Lower 64, *66*
greenstone axes 10
Griffin, Chris (b.1945), *Chapel and*
    Terrace, Swansea Valley (2007) 238
Grimes Graves flint mines 102–103, 104,
    105
Grimshaw, John Atkinson (1836-1893),
    Park Row, Leeds (1882) 218
'Grotesque Chaos' style 139–140
Gruinard Bay 5
Guettard, Jean-Etienne (1715-1786),
    geological map of France (1780) 17–18
Gwna Mélange 233
gypsum 107, 116
    Permian 61
    'white burials' 11

Hadean Era 52–53
Hadid, Zaha, Glasgow Riverside transport
    museum 31
Hadrian's Wall 199–200, 208
haematite *109*, 110, 113, 204
Halifax Town Hall 128, *130*
halite 107, 115, 116
Halperin, Ilana, geological time-scale in
    art 21, 142
Hancock, Lulu (b.1966), *Dusk,*
    Mountsorrel Quarry, Leicestershire
    (1999) *27*
hand axes
    ceremonial 9–10, 206, 234
    greenstone 10
    Happisburgh 68, 70

jadeite 9–10
    Penmaen Mawr 234
    Thames river gravel deposits 72
Happisburgh hand axe 68, 70
Hardy, Thomas (1840-1928)
    During Wind and Rain (1917) 135
    on landscape 33, 158
    Tess of the d'Urbervilles (1892) 158
Harlech Dome, Cambrian turbidites 54,
    55
Harris 169, *170*
Harrison, Tony (b.1937), *Under the Clock*
    (2005) 252
Hawkes, Jacquetta (1910-1996), *A Land*
    (1951) 25, 214, 256
Hawthorne, Nathaniel (1804-1864), on
    landscape 158
headstones 11, 135-136, 140
Hebden Bridge *216*
Hebrides
    Inner 165, 167, 189–193
    Outer 165, 166, 169–173
Hedda Stone 137, *138*
Helmsdale Fault 174, *175*
henge monuments 122
    see also Stonehenge
Hepworth, Barbara (1903-1975) 140,
    142, 218, 245
    Two Forms (Divided Circle) (1969)
    245
Hercynian fold belt 236, 240, 242
Hercynian Orogeny 47, 61, *62*
High Force waterfall 209, *210*
Highland Boundary Fault *179*, 186
Highland Clearances 182, 192
hill forts 81, 84, *226, 229*
Hill, Geoffrey, *The Triumph of Love* (1998)
    219
hiraeth 2
Holman Hunt, William (1827-1910), *Our*
    English Coasts (Strayed Sheep) 151
Holmes, Arthur (1890-1965)
    plate tectonics 49
    radiometric dating 48
    The Age of the Earth (1913) 44, 48
Holmes, Katharine (b.1962) 154
    Yorkshire Dales art 28–29, 30, 213
Holocene Epoch 4, 74, 75–78
Holyhead Mountain 230, *232*
    Hut Group 233
Homo heidelbergensis 70
Homo sapiens 24, 72
Hornton ironstone 27, 134, *141*, 255
Hound Tor 93
housing, urban 131
Housman, A.E. (1859-1936), *A Shropshire*
    Lad (1896) 224, 246
Hughes, Glyn (1935-2011) on landscape

163
Hughes, Ted (1930-1998), *Remains of*
    Elmet 33, 162–163
humic acid 37
hunter-gatherers, Mesolithic 75–76
Hutton, James (1726-1797) *16*
    plutonism 16, *17*
    Siccar Point 45–46, 50
    Theory of the Earth (1788) 16–17, 45
    unconformities 45–46, 50
    weathering and soil 37
hydrocarbons 119
    Aberdeen 182
    'fracking' 89
    Kimmeridge Clay 64, 89
hydrothermal waters 107
hydroxides 38, 143

Iapetus Ocean 52, 54, *55, 56*, 171, 200
Ice Age 2, *71*, 73
ice wedges 72
Iceland mantle plume 64
Iceni, Boudiccan Revolt 250, 256
Icknield Way 83, 252, *253*
igneous rock 7
immigration 3
Impressionism 152
industrial architecture 128
Industrial Revolution 3
    reaction against 31–32
    water supply 80
industry
    Eastern England 255
    English Midlands 221–222
    Lakeland 204
    Midland Valley 185, 187
    South Pennines 216–217
    South Wales 237–238
    steel 113, 204, 222
    Tyneside 212–213
    Welsh-English Borders 224–225,
    226–228
inhabitants
    love of landscape 6
    origins 2–3
    sense of belonging 2, 32, 218
    temperament 2
inliers 53, 204
interglacial stages 3, 4, 72, 73
Iona 181, 182
iron
    meteoritic 112
    smelting 90, 113
Iron Age society 78
    Anglesey 233
    Assynt 172
    colonization 3
    Eastern England 256

gold 115
Hebrides 189
hill forts 123–124, 226, 229, 236
mineralogy 13
South West England 243
Southern England 246
iron oxide 110, 143
Ironbridge *227*
ironstone 27, 64, 113, *254*, 255
Isle of Man *165*, 167, 204–208

Jacobite rebellions, roads 86, 196
jadeite, Alpine 9–10
Jarrow-on-Tyneside priory 211
jet 9, 255
joints, building stone 92, *93*
Jura quartzite 54
Jurassic Period *15*, 64, *65*

kaolinite *37*, 244
karst landscape *8*, 242
Keats, John (1795-1821)
    on Ben Nevis 179
    on Galloway 194
    on the Lake District 157
    on Staffa 189
    To Ailsa Craig (1819) 17, 157–158
Kennedy, Adam (b.1987), *River Clyde Shipyard Study* (2011) 187
Kent, Flor (b.1961), *Kindertransport* (2003) *135*
Kimmeridge Clay 64, 89, 256
Kinderscout *215*, 218
King's College Chapel, Cambridge, fan vaulting 139
Kisdon Hill, Swaledale *38*

Ladbooke, Robert (1768-1842) 146
    Beeston Regis from the Roman Camp (1842) *147*
lake deposits 72
Lake District, volcanism 54, *56*
Lake District batholith 54
Lakeland/Isle of Man *165*, 167, 204–208
lakes, ice-bound 73, 75
Laki volcano 1783-4 eruption 5
land bridges 48–49
landforms 42–43
landscape
    art 26–31
    British love of 6–7, 32
    different perspectives 32–33
    imaginative response to 21–23
    interpretation by immigrants 32
    subjective 2
    variety of 26
    writings about 7, 33, 155-163
Landseer, Sir Edwin (1802-1873),

Monarch of the Glen (1855) 33
lapis lazuli 143
Lapworth, Charles (1842-1920), Moine Thrust 58
Last Glacial Maximum *71*, 73
Laurentia 54, *55*
Laurussia *59*
lava flows 7
Lawrence, D.H. (1885-1930), *The Rainbow* (1915) 159–160
Laxfordian Orogeny 53
Lazonby Permian sandstone 28, *29*
lead mining 89, *90*, *108*, 112, 114, *115*, 196–197
    Mendips 88, 114, 242
    North Pennines 209, 210, 211, 213
Leicester, Triassic claystone 25
Leonardo da Vinci (1452-1519), on landscape art 143
Lewis 169, *170*
Lewis, Gwyneth (b.1959), *Keeping Mum* (2003) 230
Lewisian gneiss *5*, 169, *170*, *171*, *175*
Lewisian Group 53
Lias 64
lichen 95, 97
limestone
    Carboniferous 39, 44, 47, 58, 236
    Lincolnshire 95, 96
    Magnesian 61, 84, 115, 252
    oolitic 64
        building stone 25, 95, 128, 242
    Portland Stone 29, 30
    see also dolomite
Lincoln *see* Lindum
Lincoln Cathedral 95, *96*
Lincolnshire Limestone 95, *96*
Lindum (Lincoln) 84
Lines, John (b.1938), *Distant Lutterworth* (1994) *223*
literature 155–163
    Anglo-Saxon 155
    modern 159–163
    Romantic Movement 156–158
    South Pennines 218
    The Georgians 158–159
    Victorian 158
    World War I 159
Little Ice Age 5
Liverpool 225–226
Lizard Peninsula, ophiolites 61
Loch Borralan chambered cairns 172
Loch Lomond Readvance 75
London 3, 248, 250–251
    Londinium 248, 250
London Basin 68, *247*, 248
London Clay 68
Long Meg and Her Daughters stone circle

207, *208*
Long, Richard (b.1945) 142
    A Line made by Walking (1967) 25, 133
Ludlow Castle *228*
Lunar Society, The Moonstones 222
Lyell, Charles (1797-1875) 16
    Earth's 'wobble' 49
    Principles of Geology (1830-33) 18, 19
    uniformitarianism 18–19
Lyell Medal *19*

MacCaig, Norman (1910-1996)
    A Man in Assynt (1969) 169
    Landscape and I 161
    Sandstone Mountain 161
    Two Thieves (1983) 174
    Water (1990) 92
MacCodrum, John (1693-1779), *Smeorach Chlann Domhnaill* (c.1750) 169
MacDiarmid, Hugh (1892-1978)
    On a Raised Beach (1932) 3–4, 6, 7
    Tarras Water (1934) 160
    Whita Hill (1936) 160
machair 169, 171, 172
Mackay Brown, George (1921-1996)
    Seal Island Anthology 162
    The Solstice Stone (2001) 11
    The Twentieth Stone (2005) 174
Mackintosh, Charles Rennie (1868-1928), Glasgow School of Art 131
MacNeice, Louis (1907-1963)
    Autumn Journal (1939) 95, 161
    Snow (1935) 7
    The Hebrides (1938) 161
Maeshowe passage tomb 123, *176*
Magdalenian culture 24
magma 7
magma chambers 189, *190*
magmatism
    Caledonian 54, 58
    Carboniferous/Permian 61
    Cenozoic 64
        Hebrides 189
    Ordovician 54
Magnesian Limestone 61, 84, 115, 252
malachite *109*, 144
    Bronze Age society 13, 110
Malham Cove *8*
Malvern Hills 228, *229*
mammoth, West Runton 68
mantle
    convection 53
    Precambrian 52–53
mantle plume, Iceland 64
Manx Slate Group 205
mapping 36–43

marching routes
  Roman 85, 196, 255
  Saxon 86
memorial stones 135–136
  Roman 11, 136
Mendip Hills 242
Mercia 221
Mercia Mudstone 219, *220*, 221, 224
Mesolithic society
  colonization 2–3, 171
  communities 3, 252, 254
  Hebrides 189
  homes 122, 194
  hunter-gatherers 75–76
  mineralogy 13
Mesozoic Era *15*, 16, 61, 64
metalwork, West Midlands 222
metamorphic rock 7
metamorphism, Precambrian 53
meteorites, Hadean Era 52–53
mid-ocean ridges 64
Midland Valley *57*, *60*, 61, *165*, 167, 184–188
Midlands *see* English Midlands
Milankovitch, Milutin (1879-1958), climate oscillation 49–50
military roads 86
Miller, Hugh (1802-1856) 178
mills 128
Millstone Grit *38*, *42*, 61, 214–216
  building stone 94, 128
  Roman road 86
Minches 169, *170*
mineralogy, early use 13
minerals 7, *109*
  mining 88–91, 106–116
  pigments 143–144
  precipitation 107
mining *88*
  Bronze-Age 76, 77, 78
  dangers of 110, 112
  flint 10, 102–105
  Lakeland 204
  longwall 111
  minerals 88–91, 106–116
  opencast 110–111
  pillar-and-stall 111
  Roman 88–89, 90, 112
  rural 89–90, 109
  settlements 82, 107, 108, 109, 113, 117–118
  see also coal mining
Miocene Epoch 68, *69*
Modernism, landscape painting 152
Moine Group 53, *55*, 174
Moine Schists 172, 174, *175*
Moine Thrust 45, 54, *57*, 58, 171, 174, *175*
Mold Cape 9, 227

monastic foundations
  Borderlands 202–203
  Southern Wales 236
  Tyneside 211–212
  water supply 80
  Yorkshire 255
Mons Graupius battle 179, 181, 186
Montaigne, Michel de (1533-1592), *Of the Vanity of Words* 9
monuments 7, 23–24
  see also Neolithic society, megalithic monuments
Moore, Henry (1898-1986) 27, 218
  Mother and Child, (1932) 140, *141*
moraines 70, 73, 75
Moray Firth 174, *175*
Moritz, Karl Philipp (1756-1793), *Journeys of a German in England* (1782) 156–157
Morris, William (1834-1896)
  Arts and Crafts Movement 131
  The Red House 131, 251
mortar 101
moss 95, 97
mountain building, Precambrian 53
Mountsorrel Quarry, Leicestershire *27*, 220
Mull *190*, 191, *192*
Multangular Tower, York *125*
Munro peaks 180
music, Hebrides 191
Must Farm Bronze Age hut circle 101
mylonite 58

nappes 45, 54
Nash, Paul (1889-1946) 21–23, 152
  Landscape of the Megaliths (1937) 153
  Pillar and Moon (1932-42) 21, 23, 143
national parks 6
National Trust 6
Natural England 6
natural resources 7, 8, 88–91
Natural Resources Wales 6
Neanderthal society 24, 72
Neo-Classical style 128, *129*, 140
Neo-Romanticism, landscape painting 152, *153*, 154
Neogene Period *15*
Neolithic society
  agriculture 75–76, 176
  Anglesey and North Wales 232–233, 234
  colonization 3
  Hebrides 189
  megalithic monuments 10, 122–123, 176
    Argyll 181
    Cairn Holy 194

  Callanish 171–172
  Camster chambered cairn 176
  Castlerigg stone circle 206, 207
  Loch Borralan chambered cairns 172
  Long Meg and Her Daughters 207, 208
  Maeshowe passage tomb 123, 176
  Ring of Brodgar 176
  Stenness 176
  Skara Brae village 176–177
  South Wales 236
  South West England 242
  Southern England 246
Neptunism 17
Newcastle-on-Tyne 210–211
Newton Peveril jadeite hand axe 9
Nicholson, Norman (1914-87)
  Beck (1975) 162
  Coniston Flag (1954) 204
  Wall (1975) 162
  Wednesday Early Closing (1975) 88
Norfolk Broads, peat extraction 91, 118
Normans
  architecture 125–126
  building stone 97, 98
  castles 81–82, 126
  churches 126
  sculpture 139
North Pennines *165*, 167, 209–213
North Sea 64, *65*
North Sea Basin 69
  sea-level rise 3
  subsidence 42, 64, 66, 68, 75
North Wales *165*, 168, 230–234
North West Highlands/Northern Isles *165*, 166, 174–178
North York Moors 252, *253*
Northern Isles *165*, 166, 174–178
Norwich
  Mediaeval wool trade 256
  St Michael-at-Coslany 257
Norwich Castle 81, *82*
Norwich Cathedral 97, *98*, 139
Norwich School of landscape painting 146, *147*, *148*, 256

Ochil Fault *185*
ochre, Mesolithic pigment 13, 24, 110
Offa's Dyke 224
oil production 119
oil shales 58, 119
Old Gang Smelt Mill *211*
Old Oswestry Iron Age hill fort *226*
Old Red Sandstone 46, *59*, 174, *175*, 176, 186, 194, 201, 240
  unconformities 58
oldest rocks 52–53
Oligocene Epoch 68

ophiolites 52, 61, 176
orbit, and climate change 49
Ordnance Trigonometrical Survey 20
Ordovician Period 2, *15*, 54, *55*, *56*
Orkney Islands 174, *175*, 176
    Neolithic society
        monuments 123, 176
        Skara Brae 176–177
orpiment 110, 143
Oswald, Alice (b.1966), *Woods etc* (2005)
    246
Otter Sandstone *242*
outcrops 36–37
    as settlements 81
outdoor pursuits 6
overfolding 45, 54
Owen, Wilfred (1893-1918), *The Miners*
    (1918) 106, 159, 219
Oxford Clay 64, 256
oxidation 38
oxygen isotope studies 50

painting, landscape 143–154
    contemporary 154
    Hebrides 191
    Impressionists 152
    Modernism 152
    Neo-Romantics 152, 153, 154
    Norwich School 146, 147, 148
    pigments 143–144
    Pre-Raphaelite 151–152
    Romantics 149–151
    Scotland 188
Palaeo-Tethys Ocean 59, *60*, 63
Palaeogene Period *15*
Palaeolithic society 24, *25*, 236, 252, *254*
palaeomagnetism 52
palaeontology 15–16
Palaeozoic Era *15*, 16
Palmer, Katrina, Portland Quarries 29, *30*
Pangea 49, 61, *62*, *63*, 64
    break-up 64, 66
Paris Basin 68
Parys Mountain copper mine 230,
    232–233
passage graves 122–123, 176, 234
Paviland Cave, Gower, *Homo sapiens* 24,
    72, 236
peatlands 118–119
    northern Scotland 174, 176
Pennant sandstone 61
Pennines *see* North Pennines; South
    Pennines
Penrhos Feilw standing stones 232
Pentre Ifan burial chamber *237*
Permian Period *15*, *62*, *63*
    extinction event 61
petroglyphs 136, *137*, *186*, 196, 201

Pevsner, Nikolaus (1902-1983), *The*
    Englishness of English Art (1956) 32
Picts 180–181, 186
    sculpture 138, 186
pigments 143–144
pingos 72
Piper, John (1903-1992) 152, 154
    on the colour of rocks in Snowdonia 7
    The Rise of the Dovey (1943-44) 28
plate tectonics 49, 52
Playfair, John (1748-1819) 16
    Illustrations of the Huttonian Theory of
        the Earth (1802) 17
Pleistocene Epoch 4, 68, *74*
Pliocene Epoch 68
Plutonism 16, 17
plutons 7
polyhalite 107, 115
population *4*
Porcupine rift 64
Port Askaig Group 54
Portland cement 101
Portland Stone 29, *30*, 95, 128, 250–251
potash 107, 116
Potteries 221–222, *221*
pottery, Roman 100
Pre-Raphaelite artists 21, 151–152, *251*
Precambrian 52–54
precession 49
precipitation, mineral 107
Preseli Hills, 'bluestone' 10, 23, 97, 236
Priestley, J.B. (1894-1984), *The Linden*
    Tree 224
Priestley, Joseph (1733-1804) 222
Prussian Blue 144
pyrolusite 143

quarrying
    Roman 89, 90
    rural 89–90
quartz, white 10-11, *109*
quartzite
    Cambro-Ordovician 171, 174, 175
    Dalradian 54, 180, 181
Quaternary deposits 39–40
Quaternary Period *15*, 68, 70–75
Quinag 169, *170*, *171*

radiometric dating 47–48, 53
railway buildings 128, *129*
raised beaches 3, 75
Ramblers Association 6
Ravilious, Eric William (1903-1942), *The*
    Vale of the White Horse (1938) *249*
Read, Herbert (1893-1968), on landscape
    160
reaves 7, 76, 77, 78
rebound 3

recumbent folds 45
'Redlands', Devon 241, *242*
Renaissance, mineralogy 13
Rheic Ocean 52, 58, *59*, *60*, 61, *62*
ridge routes 83, 84
Ridgeway 83
rift basins, Carboniferous 58, 61
rift tectonics, Mesozoic 64, *65*
Ring of Brodgar 176
ripples 45
river terrace deposits 72, 101, 171, 246
River Tyne 210–211
River Wear 211
riverine sedimentation 59, 61, 64
Rochdale Canal 87, *128*
Roche Rock *244*
rock identification 7, 9
Rockall rift 64
Rococo style 139, 140
Rogers, Richard, *The Gherkin* (2003) 29,
    132
Romans
    bricks 100
    imported building stone 97
    mineral smelting 114, 115
    mining 88–89, 90, 112, 237
    quarrying 89, 90, 97
    roads 83–86, 196, 201, 221, 224, 255
    settlements 81
        architecture 124–125
    South Wales 236–237
    South West England 243
    Southern England 246, 248, 250
    Welsh-English Borderlands 224
roofing stone 92, *94*, 98, 99, 100
Roseberry Topping 252, *254*
round tower churches *126*
roundhouses 3, 75, 76, 172, 194
    Skara Brae 176–177
Rùm 192
rural life 31–32
Ruskin, John (1819-1900) 21
    In the Pass of the Killiecrankie (1857)
        22, 151
Ruthwell Anglo-Saxon High Cross 25,
    138, 155, 194, 196, *197*
Rydingsyard, Ursula von (b. 1942), *Bronze*
    Bowl with Lace (2013-14) *133*

St David's Cathedral 238, *239*
St Ives School of Artists 244–245
St Just tin mining *243*
St Mary's Church, Wreay, sandstone
    carving *11*
Salisbury Cathedral 139
Salmen, Keith (b.1959), *Evening Light on*
    Cul Mor, Assynt (2014) *173*
salt 107, 115, 116

Oligocene 68
Permian 61, 63
Triassic, Cheshire 89, 116, 226
sandstone
   building stone 92, 94, 98, 100, 126–
     129, 130, 131, 226
   Carboniferous 61
   carving 11
   Permian
     Glasgow 25
     Lazonby 28, 29
sarsen stones 97
   Stonehenge 23, 123
Saxons
   building stone 97, 98
   church-building 125–126
   marching routes 86
Scapa Flow
   World War I 177
   World War II 177, 178
scarplands 43
Scott, Sir Walter (1771-1832)
   Heart of Midlothian (1818) 184
   Waverley novels 198
Scottish Heritage 6
Scottish Natural Heritage 6
Scottish-English Borderlands 165, 167,
   199–203
Scourian event 53
scree 75
sculpture 133–142
   Anglo-Saxon 137–138
   Baroque to Neo-Classical 139–140
   direct carving 140, 141, 142
   in situ stone 142
   materials 134
   modern 140, 141, 142
   Neolithic-Bronze Age 136, 137, 201
   Norman 139
   post-modern outdoors 142
   Roman 136–137
   siting 134–136
   South Pennines 218
   Victorian 140
sculpture parks 28
sea-level change 3
   Quaternary 68, 72, 73, 74, 75
sedimentary rock 7
sedimentary structures 45
Sekhmet, Egyptian goddess (c.1350 BC)
   134
Serpent Stone 186
settlement 80–87
   abandonment of 80–81
Sharpitor 93
Shaw, George (b.1966), Tile Hill paintings
   26–27, 154
Shelter Stone Crags 181

Sherwood Sandstone 219, 220, 221, 224,
   226, 252
Shetland Islands 174, 175, 176
Siccar Point 45–46, 46, 50, 194
siderite 89, 113
Sidlaw anticline 185
silcrete 123
sills 7, 61, 62, 81
Silurian Period 2, 15, 57
   Iapetus Ocean 54
   Southern Uplands 194, 195
silver mining, Charterhouse-in-Mendip
   88–89
Sinclair, George (d.1696), A Short History
   of Coal 14
Sites of Special Scientific Interest (SSSIs)
   6, 174
Skara Brae village 176–177, 177
Skiddaw Group 54, 205
Skye 189, 190, 191
slate
   North Wales 232, 233
   roofing 92, 94, 98, 99, 100
slides, submarine 54
Smailholm peel tower 200
Smith, Sydney (1771-1845), on the
   Cotswolds 157
Smith, William (1769-1839)
   geological map 17–18
   Rotunda Museum, Scarborough 18
   strata 15–16, 18
'Snowball Earth' 54
Snowdonia 234
Snowdonia Volcanic Group 52, 56, 230
soil formation 37–38, 75
'solid' geological maps 39, 40
Somerset Coalfield 14–15, 242
Somerset Levels 241–242
   Sweet Track 10, 83
South Pennines 165, 167, 214–218
South Wales 165, 168, 235–239
South West England 165, 168, 240–245
Southern England 165, 168, 246–251
Southern Uplands 54, 56, 57, 165, 167,
   186–187, 194–198
Southern Uplands Fault 184, 187
Sowerby Bridge, West Yorkshire 34
Speedon Clay 256
springs 107
Staffa 189, 192, 193
Stainmore Basin 209, 210
Stainmore Group 209
Starr Carr Mesolithic village 75
Start Point, coastal bench 72
steam engines 90
steel industry 113, 204, 222
Stenness standing stones 176
Steno, Nicolaus (1638-1686), De Solido

(1669) 14
stone, as building material 7, 8
stone circles 122, 123, 171–172, 176, 192,
   206, 207
   see also Neolithic society, monolithic
     monuments; Stonehenge
Stonehenge 23, 97, 122, 123, 136
   'bluestone' 10, 23, 123, 236
Stonesfield Slate 98
Storegga tsunami 75, 76
Strachey, John (1671-1743), Somerset
   Coalfield 14–15, 14
Stranraer 196, 197
strata 14–16
   correlation 45
   'younging' 45
Strathmore syncline 185
stratigraphy, William Smith 15–16, 18
strontium isotopes, in teeth and bones 25
subduction zones 52
   Iapetus Ocean 54, 56
subsidence 64
   North Sea Basin 42, 64, 68
sustainability 132
Sutherland, Graham Vivian (1902-1980)
   152
   Western Hills (1938) 153
Sutton Hoo ship-burial 137
Swaledale 38, 209, 210, 211
Sweet Track , Somerset Levels 10, 83
Sweetheart Abbey 202
synclines 45

Tambora volcano 1816 eruption 5
Tate Britain, 2014 exhibition Ruin Lust 21
Tay Nappe 45
Taynton Stone 90, 98
Tayvallich Volcanics 54, 180, 181
tectonics 2
temples, Roman 23
Tethys Ocean 52, 64
   see also Palaeo-Tethys Ocean
Thames river gravel deposits 72, 246
Thomas, Dylan (1914-1953), The force
   that through the green fuse drives the
   flower (1934) 235
Thomas, Edward (1878-1917)
   The Chalk-Pit (1915) 91
   The Shieling (1916) 159
   Wind and Mist (1915) 159
tiles, Roman 100
till 70, 71, 73, 73, 256
tilting
   crustal 40, 42
   regional 43
tin 110
   in manufacture of bronze alloy 114
   mining 76, 78, 89, 90, 108

Cornwall *8, 112*, 243
Todmorden Turnpike *87*
toll roads 86, *87*
tombs
    cist, Forteviot 10–11, *10*
    megalithic 122, 232, 237
    passage 122–123, 176, 234
tombstones 11, 135–136, 140
tonalite 53
topography 40, 42–43
Torridonian Group 53, 169, *170*
Torridonian Mountains *171*, 172, *173*
tors *93*, 241, *244*
tourism
    Hebrides 192
    Lakeland 204
    North Wales 234
Towie Stone 136
town halls 128, *129*, *130*
town planning 131
Traprain Law 81, 186
Triassic Period *15*, 64
trilobites 54, 252
Trollope, Anthony (1815-1882),
    Barchester Towers (1857) 158
Trueman, Professor A.E. (1894-1956),
    Geology and Scenery in England and
    Wales (1949) 6
tsunami, Storegga slide 75, 76
tsunamite 76
tuff, Great Langdale 10
turbidites 52, *57*, 61
    Harlech Dome 54, 55
Turner, J.M.W. (1775-1851)
    Keelmen Heaving in the Coals by
      Moonlight (1835) 150
    landscapes 27, 150–151, 199, 203
    Norham Castle, Sunrise (c.1787) 203
turnpikes 86, *87*
Tyne Bridge 211, *212*

Uist 169, *170*, 171
Ulster 196
unconformities 17, 45–47, 50, 58
uniformitarianism 18–19
upfolds 45
uplands 42
uplift 42, 64, *67*, 205
urban housing 131–132

Variscan mountain belt 61, *62*
Vaughan, Henry (1621-1695), *Silex*
    Scintillans (1650) 155–156
Vespasian , Roman invasion 84, 236
Viking rift 64, *65*, *66*
Vikings
    East England 252
    Outer Hebrides 173

sculpture 138
Virconium (Wroxeter) Roman walls *100*
volcanic crags
    Midland Valley 185–186, *187*
    Scottish-English Borderlands 200
    as settlements 81, 186
volcanic eruptions, effect on climate 5
volcanic islands, Inner Hebrides 189–193
volcanism
    Cambrian 230
    Cenozoic 64
    Hebrides 189–193
    Jurassic 64, 65
    Ordovician 54, 56
    Palaeocene 42

Wade's Roads 86
walls 7, *8*, 76, 78, *100*
    see also building stone
Wastwater *206*
water supply 80
Watkins, Vernon (1906-1967), *Taliesin in*
    Gower (1954) 235
Watling Street 86
Weald *247*, 248
    iron smelting 89, 113
Weald-Artois anticline 68
weathering 37–38
    building stone 95, 96, 97
    and landform 42
Wedgwood, Josiah (1730-1795) 140,
    221–222, *244*
Wegener, Alfred (1880-1930)
    continental drift 48–49
    The Origins of Oceans and Continents
      (1915) 49
Welsh Heritage 6
Welsh-English Borderland *165*, 168,
    224–229
Wenlock Edge 57
Wensleydale hay barn *8*
Werner, Abraham Gottlob (1749-1817),
    Neptunism 17
West Runton mammoth 68
wetlands 75
    communications 82–83
Wharfedale, Cow and Calf rocks *36*, *42*
Whin Sill 61, *81*, 200, 209
White Horses 248, *249*
White Scar Cave, Ingleton *47*
Whiteread, Rachel, *Ghost* (1990) 29
Wilson, Richard (1714-1782)
    Llyn-y-Cau, Cader Idris 145
    Snowdon from Llyn Nantlle (1765-7)
      234
Windermere Group 206
Wirral 224–225
'wobble' 49

Wolds 253, 256
woodland 75, 76, 248
Wordsworth, William (1770-1850)
    Prelude (1850) 33, 156
    The Excursion (1814) 33, 157
Wright, Joseph (1734-1797)
    An Iron Forge (1772) 217
    Matlock Tor by Moonlight (1777-
      1780) 145, 146
Wroxeter, Roman walls *100*
Wylie, Sandy, *Air Raid in Scapa Flow 178*
Wytch Farm oilfield 119

Yoredale Series *38*, 209
York *see* Eboracum
York stone 92, *94*
Yorkshire Sculpture Park *133*, 142
Younger Dryas stadial 4, 73, *74*, 75

Zechstein evaporites 61, *63*, 107
zinc mining *108*, 114, 196, 242